2004

THE CLONING SOURCEBOOK

THE CLONING
SOURCEBOOK

Edited by

Arlene Judith Klotzko

OXFORD
UNIVERSITY PRESS

OXFORD
UNIVERSITY PRESS

Oxford   New York
Auckland   Bangkok   Buenos Aires   Cape Town   Chennai
Dar es Salaam   Delhi   Hong Kong   Istanbul   Karachi   Kolkata
Kuala Lumpur   Madrid   Melbourne   Mexico City   Mumbai   Nairobi
São Paulo   Shanghai   Taipei   Tokyo   Toronto

Published by Oxford University Press, Inc.
198 Madison Avenue, New York, New York 10016

First issued as an Oxford University Press paperback, 2003

www.oup.com

Oxford is a registered trademark of Oxford University Press

Library of Congress Cataloging-in-Publication Data
The cloning sourcebook / edited by Arlene Judith Klotzko.
    p.   cm.
Includes bibliographical references and index.
ISBN 0-19-512882-6; 0-19-512883-4 (pbk.)
1. Human cloning.   2. Human cloning—Moral and ethical aspects.
3. Human reproductive technology—Moral and ethical aspects.
I. Klotzko, Arlene Judith.   II. Title.
QH442.2 C568   2000
176—dc21   00-025521

9 8 7 6 5 4 3 2 1

Printed in the United States of America
on acid-free paper

In loving memory of
my dear parents,

Judith & Charles Klotzko,

for gifts beyond measure and
love without end.

Dolly (1996–2003)

ACKNOWLEDGMENTS

Earlier versions of chapters 5, 7, 12, 18, and 23 appeared in the *Cambridge Quarterly of Healthcare Ethics*, vol. 7, no. 2, spring 1998, special section: "Cloning: Technology, Policy and Ethics," and are reprinted with the permission of Cambridge University Press.

Material from chapters 1, 9, 21, 25, and 26 appeared in the same special section, and it too is reprinted with the permission of Cambridge University Press.

Chapter 6 appeared first in the *New England Journal of Medicine*, vol. 340, no. 6, February 11, 1999, and is reprinted here with permission of the New England Journal of Medicine.

Chapter 10 includes material that appeared previously in *Commonweal Magazine*, 88:257–261, 1968, and subsequently in the *Annals of the New York Academy of Sciences*: (1971)184: 103–112. The material is reprinted with permission.

An earlier version of chapter 11 appeared in *Bioethics*, vol. 11, no. 5, October 1997, and is reprinted with the permission of Blackwell Publishers.

Chapter 13 appeared first in *The Human Cloning Debate*, edited by Glenn E. McGee, and is reprinted here with the permission of Berkeley Hills Books.

Some of the material in chapter 14 appeared as the editorial in *Bioethics*, vol. 11, no. 5, October 1997, and is reprinted with the permission of Blackwell Publishers.

An earlier version of chapter 15 appeared in the *Chicago Tribune*, Sunday Perspective section, September 6, 1998, and is reprinted with the permission of the *Chicago Tribune*.

Chapter 17 appeared first in the *Journal of the Royal Society of Medi-*

cine, vol. 92, January 1999, and is reprinted with permission of the Royal Society of Medicine.

An earlier version of chapter 19 appeared first in the *Cambridge Quarterly of Healthcare Ethics*; vol. 4, no. 3, summer 1995, and is reprinted with the permission of Cambridge University Press.

An earlier version of chapter 20 appeared first in the *Kennedy Institute of Ethics Journal*, vol. 9, no. 3, September 1999, and is reprinted here with permission of Johns Hopkins University Press.

Chapter 27 appeared first as *Cloning Issues in Reproduction, Science and Medicine: A Consultation Document*, and is reproduced with the permission of the controller of Her Majesty's Stationery Office.

The editor also wishes to acknowledge all those who aided immeasurably in the realization of this project: Alan Fleischman, M.D., and his assistant, Gary Reed of the New York Academy of Medicine, Arthur Caplan, Ph.D., David Magnus, Ph.D., Glenn McGee, Ph.D., and Joseph Savage of the Center for Bioethics, University of Pennsylvania; Dr. Anne McLaren of the Wellcome CRC Institute; Tony Rawsthorne, Clerk, House of Lords Select Committee on Stem Cell Research; Christopher Higgins, Director of the MRC Clinical Sciences Centre, London; Prof. Robin Weiss and his assistants Nicola Gilbert and Liz Thomson of the Windeyer Institute, University College, London; Elizabeth Graham, Information Service Manager, Wellcome Trust; Susan Joffe, Margaret Davino, Kathleen Jensen, Laurie Petrick, Linda Pullan, Joanne Buzaglo and Michael Schultz. And last, but certainly not least, Kirk Jensen, Lisa Stallings, and Jael Wagener of Oxford University Press.

CONTENTS

CONTRIBUTORS

LORI B. ANDREWS, J.D., is Professor of Law at Chicago-Kent College of Law and Senior Scholar at the Center for Clinical Medical Ethics at the University of Chicago, Illinois.

ANDREA BONNICKSEN, Ph.D., is Professor and Chair of the Department of Political Science at Northern Illinois University, where she teaches biomedical and biotechnology policy. She is a member of the Ethics Committee of the American Society for Reproductive Medicine.

KENNETH M. BOYD, M.A., B.D., Ph.D., is Senior Lecturer in Medical Ethics, Edinburgh University Medical School, Research Director of the Institute of Medical Ethics, and Associate Minister of the Church of St. John the Evangelist, Princes Street, Edinburgh, Scotland.

GRAHAME BULFIELD, Ph.D, is Director and Chief Executive of the Roslin Institute, and Honorary Professor of Genetics, University of Edinburgh.

DANIEL CALLAHAN, Ph.D., is Director of International Programs for the Hastings Center.

KEITH CAMPBELL, D. Phil., formerly Senior Research Scientist at the Roslin Institute, and Head of Embryology at PPL Therapeutics, is Professor at the School of Biological Sciences, Division of Animal Physiology, University of Nottingham, UK

ARTHUR L. CAPLAN, Ph.D., is Trustee Professor and Director of the Center for Bioethics, University of Pennsylvania Health System.

RUTH DEECH, M.A., is Chair of the United Kingdom's Human Fertilisation and Embryology Authority. She is the Principal of St. Anne's College, Oxford.

LEON EISENBERG, Ph.D., is Professor Emeritus in the Department of Social Medicine, Harvard School of Medicine.

RAANAN GILLON, M.D., is Professor at Imperial College, Medical Ethics Unit, School of Medicine, University of London. He is also a part-time general practitioner and the editor of the *Journal of Medical Ethics.*

ELIZABETH GRAHAM is Manager, the Wellcome Trust, specializing in research ethics and science policy.

RICHARD HARRIES chaired the House of Lords Select Committee on Stem Cell Research. He has been Bishop of Oxford since 1987, before which he was Dean of King's, College, London. He is the author of a number of books including, most recently, *God Outside the Box: Why Spiritual People Object to Christianity.*

KURT HIRSCHHORN M.D., is past president of the American Society of Human Genetics. He is the former chair of Pediatrics at Mount Sinai School of Medicine and chairs its Ethics Committee.

SOREN HOLM, M.D., Ph.D., is a Reader in Medical Ethics at the Institute for Medicine, Law, and Bioethics, University of Manchester. He is a member of the Danish Council of Ethics.

RON JAMES, Ph.D., is Managing Director, PPL Therapeutics.

ARLENE JUDITH KLOTZKO, M.Phil., J.D., is Writer in Residence at the Science Museum, London, Advisor on Science and Society to the MRC Clinical Sciences Centre, and Visiting Scholar in Bioethics at the Windeyer Institute, University College, London, England.

M. SUSAN LINDEE is Associate Professor of History and Sociology of Science at the University of Pennsylvania, Philadelphia.

DAVID MAGNUS, Ph.D., is Assistant Professor and Director of the Graduate Program at the University of Pennsylvania Center for Bioethics.

GLENN MCGEE, Ph.D., is Assistant Professor and Associate Director for Education at the University of Pennsylvania Center for Bioethics.

ANNE MCLAREN, D.Phil., a biologist, is a member of the Nuffield Council on Bioethics, London, the Human Fertilisation and Embryology Authority, and the Group of Advisers on the Ethical Implications of Biotechnology of the European Union.

DOROTHY NELKIN holds a University Professorship at New York University, where she teaches in the Department of Sociology and the School of Law.

BARONESS ONORA O'NEILL, Ph.D., has been a member of the UK Human Genetics Advisory Commission since 1996, and Acting Chairman from February 1999. She has been the Principal of Newnham College, Cambridge, since 1992.

ANTHONY PERRY is a molecular embryologist at the Rockefeller University, New York. He completed his Ph.D. in molecular microbiology at the University of Liverpool. In 1996, he was awarded a European Molecular Biology Organization Long Term Travel Fellowship. His research uses the mouse as a model system to study the molecular events that initiate embryonic life in mammals.

ROSAMOND RHODES, Ph.D., teaches bioethics at the Mount Sinai School of Medicine in New York.

EGBERT SCHROTEN, a theologian and philosopher, is Professor of Christian Ethics and Director of the University Center for Bioethics and Health Law, Utrecht University, The Netherlands. He is also Chairman of the Dutch National Committee on Animal Biotechnology, and a member of the Group of Advisers on the Ethical Implications of Biotechnology of the European Commission.

HAROLD T. SHAPIRO is President of Princeton University and Chair, National Bioethics Advisory Commission.

LEE SILVER, Ph.D., is a Professor at Princeton University in the Departments of Molecular Biology and Ecology and Evolutionary Biology, and in the Program in Neuroscience.

PETER SINGER is the Ira W. DeCamp Professor of Bioethics, Princeton University.

BARONESS MARY WARNOCK, former Mistress of Girton College, Cambridge, England, has chaired the British Committee on Human Fertilisation and Embryology, and a European Parliament Ethics Committee.

STEEN WILLADSEN, D.V.M., Ph.D., D.V.Sc., is an experimental embryologist. During the 1970s and 1980s, he developed methods of freezing, culture, and radical micromanipulation of sheep embryos. Using these methods, he successfully cloned first sheep embryos (in 1984), and then cattle embryos (in 1985), by nuclear transplantation. His experiments established a firm basis for the cloning of mammals.

TOM WILKIE, Ph.D., is Director of Bioethics and Senior Policy Analyst in the Wellcome Trust's Unit for Policy Research in Science and Medicine.

IAN WILMUT, Ph.D., is Head of Department, Gene Expression and Development, Roslin Institute. He was the team leader of the group that produced Dolly the sheep.

INTRODUCTION TO THE REVISED EDITION

Arlene Judith Klotzko

The *Cloning Sourcebook* was originally published in the summer of 2001. There was plenty to say about cloning then. There is quite a bit more now. The cloning landscape has changed in two central respects. The focus of discussion and debate has broadened. Therapeutic cloning (also termed cell nuclear replacement) to derive human embryonic stem cells, not to produce a new person, has joined reproductive cloning as a subject of public and policy interest. Stem cell research has become one of the most exciting and morally contentious areas of biology. The new final chapter of this revised edition gives due recognition to these crucial developments and addresses both the science and associated ethical and policy issues.

The second major change in the cloning landscape is the absence from now on of the creature that set us all talking in the first place. Dolly the sheep was the first mammal to be reconstructed from a single cell of a being that had already lived—and in her case, already died. Sadly, Dolly has also died. In February 2003, just five months short of her seventh birthday, she was euthanized in order to end the suffering she was certain to endure as a result of a lung infection. A figure of public interest and consternation to the very end, Dolly was erroneously assumed by many to have succumbed due to some cloning-related defect. She did not.

Suspicions of a causal link with cloning were actually quite understandable. Animals of every species cloned thus far have exhibited severe abnormalities. There have been miscarriages, still births, and early deaths. While some cloned animals appear to be normal, the procedure has produced many that are defective and deformed. We have seen lambs that could not catch their breath—unable to propel their blood through enormous blood vessels that were twenty times larger than normal. Autopsies have revealed shrunken kidneys and undifferentiated liver cells unable to do their proper job. The incidence of abnormalities and their severity argue very strongly against any attempts to clone human beings; we neither fully understand the causes of the problems nor have the capacity to prevent them.

Even though Dolly is no more, the questions raised by her existence engage us still. What does reproductive cloning by nuclear transfer mean for science? For society? What are the ethical issues raised by cloning animals—and in the future, humans? How should the makers of public policy cope with the stunning fact that an entire organism had been reconstructed from a single adult cell and that humans might well be next? Why are so many of us so disturbed by cloning and the idea of having clones in our midst?

The Cloning Sourcebook addresses all these questions and does so in a way that is unique in the cloning literature—by grounding what is effectively an interdisciplinary conversation upon a solid scientific foundation. In part I, the key scientists responsible for the early and crucial developments in cloning speak to us directly. Here, too, are other scientists who evaluate and comment upon these developments.

Part II explores the context of cloning and includes sociological, mythological, and historical perspectives on science, ethics, and policy. The authors also examine the media's treatment of the Dolly story and its aftermath—both in the United States and the United Kingdom.

Part III—on ethics—contains a broad range of chapters, written by some of the major commentators. Up to this point, cloning animals has been seen largely as a precursor for human applications, not as a subject deserving of its own moral analysis. Here, however, the foremost authority on animal ethics presents his views.

Part IV addresses legal and policy issues. It features individual and collective contributions by those who have actually shaped public policy on reproductive cloning, therapeutic cloning (involving the derivation of human embryonic stem cells),[1] and related contentious bioethical issues— in the United States, Britain, and the European Union. The legal contexts in the United States and United Kingdom are explored and a range of regulatory schemes proposed. Again, animal issues are not neglected. Here are regulatory frameworks that address both cloning and transgenics.[2] Therapeutic cloning and stem cell research are addressed in the new final chapter.

This book is directed toward a variety of readers. The chapters will be of interest to scholars in all the subject areas represented. But they have also been written to be accessible to the intelligent nonspecialist reader. This second audience is of great importance. Cloning and genetic modification must not remain the exclusive concern of scholars, committees, and the politically powerful—the sort of people who are often described by the British, with a delightful sense of irony, as "the great and the good." Biomedical advances and the moral and policy difficulties that they engender are everybody's business. For these reasons, the curious lay reader is most welcome here.

Several widely publicized efforts are underway to clone a child. There may well be covert attempts going on as well. Sheep, mice, cows, goats, pigs, rabbits, and a cat have all been cloned by nuclear transfer. There is no scientific reason to suppose that a human might not be next. Indeed, human reproductive cloning has become a matter of "when," not a matter of "if."

Animal cloning continues to be used for agricultural and medicinal purposes—the latter in combination with transgenics. The hard-won ability to clone pigs—demonstrated by several teams of researchers in 2000—was certainly very good news, raising hopes that one day xenotransplantation[3] might play a significant role in ameliorating the shortage of transplantable human organs. In early 2002, a team of scientists in the United States published a paper describing the creation of a cloned single knockout pig ("knockout" is the removal of a functional gene). One of two genes that cause a sugar called alpha-gal to be added to the surface of pig cells had been eliminated. This was an important and eagerly awaited step. The immune systems of humans and Old World monkeys, having lost this enzyme during the process of evolution, would recognize the sugar-coated pig organ as foreign and reject it.

Approximately one year later, another team of scientists announced the creation of double knockout pigs. Such precise genetic modifications may be successful in moderating or even defeating acute vascular rejection of the pig organ by its human recipient. Or it may not—it is far too early to say. But now there is great hope. It may even be possible to eliminate porcine endogenous retroviruses (the viruses in pigs that cannot be bred out) from their genome and thereby remove a potential threat.

In 2001, the United Kingdom positioned itself in the vanguard of therapeutic cloning research and the regulation of that research. In response, a pro-life group mounted a legal challenge that made its way through three layers of legal proceedings. In March 2003, the matter was finally resolved by a 5–0 decision by the Law Lords. Licenses have been issued for research projects involving human embryonic stem cells but, as this volume goes to press, not for therapeutic cloning research. The United Kingdom has established a stem cell bank enabling both publicly and privately funded researchers to obtain the best cell lines available. Many countries have embraced embryonic stem cell research, but not therapeutic cloning. In contrast, limits on stem cell research policy in the United States have discouraged and even demoralized researchers. Government-funded scientists are severely restricted and intellectual property issues problematic. As this volume goes to press, there is no legislation that forbids either reproductive or therapeutic cloning. Both remain a subject of public and political controversy.

This volume will serve as a resource for the study of one of the most extraordinary developments of the late twentieth century. But it can also be useful in a more proactive sense when the next stunning, or even revolutionary, biotechnological feat is announced to an astonished world. For, beyond cloning—and to a great degree behind the cloning fears—are the twin specters of eugenics and genetic determinism.

The dawn of the millennium put before us is the full sequence of the human genome—what many have called the operating system for the human body. The computer analogy seems irresistible. But it is also flawed, if it creates a vision of programmed creatures whose destiny is irrevocably written in the book of life. Given persistent misunderstandings about the nature of clones—who are, after all most akin to later-born identical twins, with even a higher degree of variability than we know twins to have—a major challenge remains.

And, again, it is education. The challenge is to convince the wider public that we are far more than the sum of our genes. For clones—and for all the rest of us who have come into being in a more conventional manner—environment plays a crucial role in determining who we are, and how and when we fall ill.

When asked whether completion of the entire human genome sequence would elicit fears and misunderstandings similar to those evoked by the appearance of Dolly, Dr. Francis Collins, director of the Human Genome Project, had a simple yet eloquent response: "The best antidote to the incorrect reductionist view that is fed by the media is to get the data."[4]

In the spirit of curiosity, and an inquiry that begins with the facts, I invite you to read on.

NOTES

1. Medical and scientific applications of cloning technology which do not result in the production of genetically identical babies.

2. Alteration to the genome of an animal by the addition of a gene or genes from another species.

3. The transplantation of cells, tissues, and organs from one species into another.

4. Personal communication.

THE SCIENCE OF CLONING

I

VOICES FROM ROSLIN

The Creators of Dolly Discuss Cloning Science, Ethics, and Social Responsibility

Arlene Judith Klotzko

olly is a very remarkable sheep. Not because of what she is, but because of the mode by which she appeared in our midst. Dolly was cloned[1] in a laboratory by a technique called nuclear transfer; she is virtually genetically identical to a sheep born six years before she was. And we—and the ways in which we view the world and each other—will never be the same again.

Media coverage of Dolly has been extensive, and her image is instantly recognizable throughout the world. In the words of biologist Stephen Jay Gould, she is "the most famous member of her species since John the Baptist designated Jesus as the Lamb of God."[2] Unfortunately, however, much of the reporting can be characterized as sensational, even irresponsible and alarmist.[3] And from the beginning the primary focus of attention has been on the applications and implications of the Roslin work for the cloning of human beings.

Inappropriately so, for purely factual as well as moral reasons. The cloning of existing human beings by nuclear transfer, although now believed to be theoretically possible, is still far from being empirically feasible. As Professor Grahame Bulfield, the director of the Roslin Institute, will tell us shortly, given existing technology and levels of efficiency, the cloning of one human being could well require the use of 1,000 eggs (oocytes) and 20 to 50 surrogate mothers.

3

Moreover, many of the items in the catalog of human cloning horrors—or of wonders, depending upon who is doing the cataloging—remain in the realm of science fiction. So, as one observer with enviable literary flair has noted (specifically in relation to President Clinton's "emergency mode" reaction to the cloning of Dolly), there has been "much ado about mutton."[4] Or at least much ado about a highly speculative subject. And far too little ado about two areas of legitimate moral concern. First, the morality of the genetic manipulation of animals, not just through cloning, but through animal transgenics (the creation of animals with human genes or those of other species)—the context and the focus of the research that gave rise to Dolly. Second, the morality of human therapeutic which would necessitate the deliberate creation and destruction of embryos. Dr. Ian Wilmut, the leader of the group that produced Dolly, will describe the extraordinary promise of such work. [Editor's note: Subsequent discussions of the potential of therapeutic cloning involve the creation of embryonic stem cells. See chapter 27.]

Even the scientific significance of Dolly has been widely misunderstood. As Prof. Bulfield will tell us, the true breakthrough achieved by the Roslin Institute team did not come with Dolly. It came almost exactly one year before, with the creation of Megan and Morag, sheep cloned from differentiated embryo cells. According to Dr. Keith Campbell, Dolly was, scientifically speaking, merely "the icing on the cake."

Many myths and misunderstandings persist about what exactly has been done at the Roslin Institute. And why it was done. And what this work truly implies, scientifically and medically, for the future of animals and more troubling, of course—for the future of humans. It is our hope that "Voices from Roslin," offering as it does a unique opportunity to meet the participants, will help to dispel those myths and misunderstandings, and allow both moral analysis and public policy discussion and formation to proceed on a more secure factual foundation.

THE PARTICIPANTS

Grahame Bulfield, director and chief executive of the Roslin Institute, holds a Ph.D. in genetics from the University of Edinburgh, where he is an honorary professor of genetics. His early training was in agriculture. In 1993, Prof. Bulfield was appointed to the Ad Hoc Committee to consider the Ethical Implications of Emerging Technologies in the Breeding of Farm Animals. The Committee presented its final report in 1995.

Keith Campbell, an embryologist and cell biologist, was formerly a senior research scientist at the Roslin Institute, and then director of embryology at PPL Therapeutics, where he served as the coordinator for all projects involving nuclear transfer in cattle, sheep, and pigs. Dr. Campbell, now a professor at the School of Biological Sciences, Division of Animal

Physiology, University of Nottingham, UK, holds a D.Phil. from the University of Sussex.

Ron James, Managing director, PPL Therapeutics, holds a Ph.D. in organic chemistry from Imperial College, London. He worked as an organic chemist before moving into a technology transfer role. Dr. James was involved with the formation of PPL in 1986–1987, at which time he was part of Prudential's venture capital team, and responsible for health care and biotech investments.

Ian Wilmut, Ph.D., is a mammalian embryologist who was the team leader of the group that produced Dolly. He was coleader of the successful project to produce human therapeutic proteins in the milk of transgenic sheep. His subsequent research has involved the development of new methods for the introduction of genetic change into livestock using nuclear transfer.

THE ROSLIN–PPL COLLABORATION: THE WORK OF ANIMAL TRANSGENICS

A J K: What is the Roslin Institute's remit?

G B: To conduct basic and strategic research. To create new opportunities for livestock products and animal biotechnology industries. In addition, the institute carries out research on animal welfare, both to inform government policy and address public concern. As things are moving on, we are very much centering ourselves on all aspects of genetics. We have as our objective to be the leading center for animal biotechnology. That's what our objective is to become in the longer term.

A J K: Could you tell us about the collaborative relationship between the Roslin Institute and PPL Therapeutics?

R J: PPL was founded in 1987 to commercialize an earlier Roslin Institute invention—the idea that animals could be genetically modified so that they produced therapeutically useful proteins in their milk. That project, led by Dr. John Clark, OBE, was centered on the genetic engineering necessary to achieve this. PPL has had an ongoing relationship with the institute ever since, and we have funded a variety of projects there, in exchange for rights to use the results in PPL's field of interest.

A J K: Was Ian Wilmut involved in that first project?

R J: Yes. He was involved in the development of new embryo handling techniques, without which the concept would not have been realizable.

A J K: How was PPL financed?

R J: In common with most other high technology startup companies, PPL was financed by venture capital funds, which take shares in the company in the hope of seeing shares increase in value and making a capital gain by selling them years later once the company is successful. The majority of the cash that PPL raised in that first round was passed to the Roslin Institute to pay for ongoing research to prove the concept worked in sheep following a feasibility demonstration using mice.

AJK: Aside from the rights to use the techniques developed at the Roslin Institute, has your collaborative relationship provided any other benefits to PPL?

RJ: Yes. The relationship has been seen as beneficial by the venture capitalists upon whom we depended for the successive rounds of funding that biotech companies typically need to survive the ten or so years that it takes to develop a new pharmaceutical product and get it to the market, and so finally to make profits.

AJK: How do you produce proteins for therapeutic use in humans in animal milk?

RJ: The key lies in the introduction of a hybrid gene into the animal. All genes are composed of two parts. The main part carries the DNA code for whichever protein that gene controls the production of. The second part is a DNA sequence that acts as a switch, controlling when and in what cells that protein will be made.

   The hybrid gene, invented by John Clark and his team at the Roslin Institute and used by PPL, comprises the switch DNA taken from a gene for a sheep milk protein joined to the coding DNA from the human gene for whatever human protein we wish to produce. Although the hybrid gene exists (as do all genes) in every cell in the transgenic animal, the presence of the sheep milk switch means that production is directed exclusively to its mammary gland when the animal is lactating to produce milk to feed its young. Milk is collected from these animals in the normal way, and the additional human protein it contains is purified from the milk for use in human medicine.

AJK: What, if anything, has all of this got to do with Dolly?

RJ: Dolly was the end result of a series of experiments designed to develop a better way of producing transgenic large animals central to PPL's business. Earlier experiments, reported in the same *Nature* paper,[5] had produced four lambs from differentiated embryo cells and two lambs from fetal fibroblast cells [connective tissue].

   The fact that Dolly is a clone—a genetically identical copy of the ewe that provided the mammary cell from which she was derived—was less important to us than the demonstration that it was possible to reconstruct a whole animal from a differentiated cell.

AJK: So, the aspect of Dolly that gripped the world's imagination—the fact that she is a genetic copy of a being that had already existed—is actually peripheral to your scientific and commercial agenda?

RJ: Yes, peripheral in the sense that being able to alter the genes in the cell and then to produce a transgenic animal from the cell was always our key objective. But not unimportant, in that being able to produce clones—

a small number of identical animals produced from identical genetically modified cells—will allow PPL to obtain product for clinical trials more quickly than would be the case if conventional breeding had to be used.

AJK: What do you see as the ethical issues arising from this new ability to make germline (inheritable) genetic changes in animal cells in culture?

GB: The major ethical issues in regard to farm animals are really about playing God and about animal welfare. My view is that human beings, for the last four thousand years, have been moving genes around in farm animals and selecting farm animals. The modern pig looks nothing like the wild boar, and the Holstein cow looks nothing like the European bison. With genetic modification, we are not moving around hundreds of genes, we are only changing one gene and keeping everything else constant.

The view I take on animal welfare is that the technology itself is a red herring. If an animal is lame because of genetic modification or selective breeding or poor nutrition, or because I kick it, it is wrong that it's lame. So you have to pay attention to the phenotype—that is, to the animal itself—rather than the technique that produces the problem.

AJK: How do you respond to the concerns of animal rights advocates, as distinct from their tactics?

GB: There are some things that one should never do to animals. Amputate legs, transplant heads from one animal to another, for example. There are other areas where you have to have a cost-benefit analysis, a consequentialist analysis, where the possible suffering is outweighed by the good. No matter what, you should treat animals in the most humane way possible. In my experience, most people who work with animals in laboratories care quite a lot about animals. I'm sure you've seen that for yourself during your time here.

AJK: Are nuclear transfer using a genetically modified donor nucleus [the technique developed by the Roslin Institute and PPL Therapeutics] and microinjection (injecting DNA into a fertilized egg) the only mechanisms for making changes in an organism's germline?

GB: Generally speaking, in mammals, microinjection or nuclear transfer are the only ways. There have been attempts in the past to use viral vectors to take in bits of DNA, retroviral vectors in particular. This has happened in chickens. But they generally have not been successful. They have been used in *Drosophila* as well, with some success. But generally speaking, they are not very good, not very stable; people do not like using viruses. I think you would only use these techniques if everything else failed [Editor's note: It is now possible to generate transgenic mammals by microinjecting unfertilized eggs.]

AJK: In regard to the introduction of human genes into animals and the production of abnormal animals, in your lecture at Utrecht University, you said that with current techniques [not including the new Roslin technique of gene targeting] the chances of producing an abnormal animal are about 10% because only 10% of the genome has genes in it. If you hit a gene, you cause a problem. Could you expand upon this?

G B: You are only really going to cause abnormalities if you disrupt an existing gene. And there are about 70,000 genes in an animal. And that's only about 10 percent of the DNA. In other words, 90 percent of DNA is junk. And, obviously, if you are doing microinjection—injecting DNA—and the gene that you are injecting goes into the junk area, it will not cause any problem. It can cause problems if it hits a gene. And if it is an important gene, it is going to kill the embryo.

A J K: Would the chances of producing an abnormal transgenic animal lessen if your technique were used?

G B: Yes. What we are trying to do with gene targeting is to line the gene we are putting in alongside the existing gene and we are only putting one copy in, so it is directly replacing the existing gene. So, in this case, we wouldn't be causing any problems. All we would be doing is replacing the natural gene with the gene we've modified. This is a very specific alteration. So the problem with microinjection, where the gene goes in randomly, and maybe in 10 percent of the cases knocks out a gene, and causes lethality, is not likely to be an issue with the new technique.

A J K: Dr. Wilmut has told me that you intend to use your gene targeting techniques to produce sheep to serve as animal models for cystic fibrosis. Could you describe the rationale for this work?

G B: The problem with cystic fibrosis research is that mouse models are not good. Mice do not have quite the same physiological changes. Their lungs seem to be different. The view is that the lungs of sheep might be very similar to human lungs, so that we would create an animal model of cystic fibrosis in sheep, and then attempt different methods of therapy.

This could be to develop somatic gene therapy that is, to target genes to the lungs. But it also could be to develop different types of sprays to alleviate the problems in the lungs. There might be a way to develop prophylactic methods.

A J K: Would gene knock-out in culture and then nuclear transfer achieve genetic changes that could be transmitted in the germline?

G B: Yes, that is exactly what Polly is leading up to. Being able to add genes, modify genes in culture, or do knock-out and then using nuclear transfer. [Editor's note: Polly, who was born in July 1997, is a cloned transgenic lamb. She has a human gene in every cell of her body.]

A J K: Why is Polly important?

G B: The important point about Polly is that she was derived by nuclear transfer from cells that have been grown in culture and had been genetically modified.

AJK: What are you going to do next?

GB: We want to try and adapt the technology to pigs and cattle, and that's ongoing. After Dolly, we wanted to see whether we could achieve genetic modification with cells in culture. We've done that, and that's Polly. So in many ways we have probably accomplished most of the scientific breakthroughs with the exception of getting it working on pigs. Also, we want to make it work on mice so we could study, for example, aging in mice. In a way, the future will not be as dramatic as it has been up to now. It will be more about improvement and consolidation.

AJK: Will you be working with sheep or will you transfer your focus to cattle and pigs? Or even rabbits?

GB: Probably we will carry on working on sheep for some time. But the commercial applications, both in biomedical research and in agricultural research, are going to be on cattle and pigs, probably, and on sheep maybe for pharmaceutical proteins. Because cattle produce so much more milk, in the long term it will probably be cattle. And for animal breeding, it will be cattle and pigs.

Ian believes that rabbits have some use as an experimental tool to look at some of the questions of aging and differentiation. And mice, from my point of view, are the animal model par excellence. And even rats. I suspect that for a lot of the more basic research, it will be mice and rats and rabbits that will be used.

AJK: Could you describe some of the uses that you foresee for the new technology of genetic modification of cells in culture, the technology that resulted in the birth of Polly?

GB: First of all, PPL would like to use it because they would like to be able to knock genes out as well as add genes—which they can do with existing technology. It would be very much more efficient if you make genetic modifications in culture because, presumably, all the cells whose nuclei will be transferred are genetically modified. When you do the direct DNA microinjection, only about 1 percent of the oocytes [eggs] injected turn out to be genetically modified, and only about 10 percent of the animals born. So it would improve the efficiency.

Also, of course, if they had a transgenic line, they could clone animals from that line—so they could reproduce the transgenic animals far more effectively and quickly. They could also produce "nutraceuticals" by modifying milk to add proteins or chemicals of importance for milk. For example, cow's milk is not very good for premature babies. So you might want to add particular proteins or something else to it.

The technique could be used to produce transgenic animals by gene knock-out [to avoid rejection] in xenotransplantation [the transplantation of animal organs into humans]. The technique would provide the additional ability to remove genes, not just add them. We have already discussed the creation of animal models for genetic diseases and the breeding of dairy cattle and probably pigs. There will also be uses for the technique in academic science for studying aging and the mechanisms of development and differentiation.

A J K: In the United Kingdom, so-called "mad cow disease," BSE, has been a serious health and economic problem. Could you describe the work that you intend to do in relation to the prion protein implicated in BSE? Does it involve disrupting the gene to see whether that would prevent the disease?

G B: As I understand it, there is a gene that produces the prion protein that seems to be involved in BSE. And the prion protein sort of goes wrong. This is not an area that I am an expert in. One of the hypotheses could be that if you knocked out the gene that produces the prion protein, maybe this would prevent the disease. If the protein is not there, it cannot start to go wrong and form crystals and damage cells.

A J K: Are any other scientists at other institutes doing research using the techniques that you developed? If so, under the patent laws, would they have to obtain a license from the Roslin Institute?

G B: There are about half a dozen labs working in this area. With respect to patenting, scientists at other labs do not have to have a license from us to do work for experimental purposes. Anyone can do any experiments in a lab; we cannot influence that at all. The only thing a patent allows you to do is prevent people from commercially exploiting what you've done. So there is no way we would want to or could stop other laboratories from repeating our experiments.

In fact, in many ways, it is in our interest that people do it. You should never believe an experiment from one lab. So we would encourage people to try and repeat the experiments. However, if scientists wanted to use our technique for commercial purposes—to do xenotransplantation, for example—they would have to have a license from us.

THE TWO SETS OF EXPERIMENTS: MEGAN AND MORAG (1995); DOLLY (1996)

A K J: Could you define cloning and nuclear transfer?

K C: Cloning is a term applied to producing an exact copy. Nuclear transfer is a technique whereby an animal may be produced from a nucleus, originating from a single cell, that is transferred to an enucleated egg, an egg from which the genetic material had been removed. The cells providing the nucleus can be from embryos, fetuses, or adults.

A J K: So Dolly and Megan and Morag are all clones produced via nuclear transfer?

K C: In the strict sense of the meaning, the animals produced by nuclear transfer are not true clones. Account must be taken of possible changes that occur in the genome during embryo and fetal development or while the cells are in culture.

In the case of Megan and Morag—sheep that we cloned by inserting identical nuclei from an embryo-derived differentiated cell line into separate enucleated eggs—differences in the components of the egg cytoplasm would result in differences in the offspring. For example, differences in the mitochondrial genome.

A J K: Is nuclear transfer a new concept?

K C: No. The technique of nuclear transfer was described as far as back as 1938 by Hans Spemann. He described it as a method of studying cell differentiation. In the early 1980s, similar techniques were described in the mouse. Then in the mid-1980s Steen Willadsen demonstrated the technique in sheep and produced lambs. [Editor's note: See Dr. Willadsen's paper, which appears as chapter three in this volume.]

A J K: What is differentiation?

K C: Differentiation is the specialization of cells to perform particular functions. Although all cells contain a complete copy of the genome [all the genes and other DNA], the specific genes that are turned on in a cell are those required for it to fulfill its particular function.

A J K: What were the goals of the Megan and Morag and the Dolly series of experiments?

K C: The main goal of the first experiment was to produce animals from an embryo-derived cell line that had differentiated in culture. The extension of our work involved looking at other differentiated cell types. To this end, in the second experiment, fetal and adult cell types were tested. Dolly is the first demonstration that the genome of an adult somatic cell can recontrol development following nuclear transfer. Her existence demonstrates that there is no loss or permanent inactivation of the genome during development.

A J K: What was the goal behind this goal?

K C: The main interest at Roslin was to find a method for the production of offspring by nuclear transfer from cells that could be maintained in culture and used as a route to precise genetic modification in animals for transmission through the germ line.

A J K: How would you characterize the difference between the first set of experiments (which produced Megan and Morag) and the second (which produced Dolly)? Did it relate to technique or just to cell type?

I W: The actual method of nuclear transfer that produced all of the lambs is the same. The difference is the nature of the cells that we used as nuclear donors. In the first series of experiments, we used embryo cells. In the second, we used embryo, fetal, and adult cells. Clearly, using adult cells is different in two respects: the number of cells available and the fact that when you have an adult animal you already know what it is going to be like. So you can choose which ones you want to copy. But as far as the actual technique is concerned, it is exactly the same.

AJK: Was the breakthrough achieved in the first series of experiments rather than in the second?

GB: Yes. The breakthrough was the Megan and Morag experiment. The reversal of differentiation was demonstrated in that experiment. All that Dolly and the sheep cloned from the fetal fibroblast cell lines did was show that you could do it with even more differentiated cells, whether fetal cells or adult cells.

AJK: Have you or anyone else repeated the Megan and Morag experiment?

KC: Yes. We have repeated the experiment, as has a team in New Zealand.

AJK: Can embryo cell lines, fetal cell lines, and adult cell lines allow for genetic modifications equally well?

IW: We really do not have accurate information yet about genetic modification in cultured cells. The next range of experiments that we and no doubt others will be carrying out will be done to find out the answer to exactly this question. There is very little information available yet about the efficiency of genetic modification in various types of cells.

AJK: You told me once that you believed it would be easier to do genetic modification in fetal cells.

IW: Yes, probably because of the number of cells. Cells in culture only have the ability to go through so many cell divisions before they start to accumulate errors. The big drawback to embryo cells is that you start with a small number—by definition—and you would need quite a large number in order to be able to do gene targeting. You also need the ability for them to keep on growing in culture for several more divisions to select those with the change.

At the present time, we think it will be more likely that we could successfully use the technique with fetal cells because you already have enough when you start. [Editor's note: Polly—the cloned transgenic lamb with a human gene in every cell of her body—was produced using fetal fibroblast cells.]

If you start with an embryo, you have so few [cells] that you lose a large part of that opportunity in just growing up to a minimum number of cells necessary for gene targeting. So that by the time you have gone from the few thousand cells that you started with up to the number that you need for gene targeting, you no longer have enough time in culture during which the cells will remain normal to allow you to carry on.

AJK: These errors in cells will accumulate if they divide enough times?

IW: Yes. There is a limit that varies in different species and actually relates to how long the species live. There is a very short [period] for mice—much shorter than it is for humans—before things begin to go wrong.

A J K: Is this why it might be problematic to clone from adult cells that have divided so many times?

I W: That's right, it is one reason.

A J K: Do you think that whatever type of cell line (embryo, fetal, adult) turns out to be best for genetic manipulation in animals, the same would hold true for their human counterparts?

G B: When one looks at cell lines, fibroblasts, for example, which are the wound healing cells that you can grow from adults or from embryos, grow very well in culture because their natural function in the body is to grow and cover up cuts and things. You can have a fibroblast cell line from chickens or humans or from pigs or cattle—they behave more or less the same in culture. So many of these cell lines do behave similarly across species.

NUCLEAR TRANSFER, THE CELL CYCLE, AND QUIESCENCE: THE BIRTH OF AN IDEA

A J K: You said that, in the Dolly series of experiments, you used adult, fetal, and embryo cells?

I W: That's right.

A J K: Did you find that your success—with all the cell types—was, in fact, due to the method of achieving quiescence?

I W: Yes. [Editor's note: The significance of quiescence for successful cloning is now in dispute. See, for example: A Wakayama, T., Rodriguez, I., Perry, A C F, Yanagimach'i R., and Mombaerts, P. (1999) "Mice cloned from embryonic stem cells." Proc. Natl. Acad. Sci. USA. 96, 14984–14989]

A J K: What was the scientific consensus with respect to the ability to clone a mammal from a differentiated cell before the Megan and Morag experiments?

K C: In my opinion, the general scientific consensus was that this could not be done. In general, it was thought that specific undifferentiated cell types, such as germ cells or embryonic stem cells, would be required for successful nuclear transfer.

A J K: Did you believe that the consensus was wrong? If so, why? Was your conviction intuitive or strictly rational?

K C: I personally did believe that the consensus was incorrect. My conviction was both intuitive and rationally based upon my knowledge gained in a range of scientific environments, including cell cycle work in amphibians and yeast, prior to moving into mammalian embryology. In certain aspects, there are great similarities among amphibia, yeasts, and mammals.

A J K: Could you give us an example of any prior research experience that contributed to your rationally based conviction?

KC: In an earlier life, I worked for the Marie Curie Memorial Foundation, a cancer research charity that also carries out research. I came to believe that we could use differentiated cells as donors for nuclear transfer because, in certain tumors, there are many different cell types that most probably originated from a single [adult or differentiated] cell type. This suggested to me that the differentiated fate of cells is not fixed.

AJK: Please describe the cell cycle and how it is relevant to nuclear transfer.

KC: During the lifetime of a single cell, it is subject to two interdependent sets of events. First, there are the events associated with growth. This is dependent on environmental and nutritional conditions. Second, there are the events occurring within the nucleus. (The nucleus contains the genetic material, or DNA, that is associated with proteins to give it a structure, termed chromatin.) These events occur in two phases: the S phase, during which DNA is replicated, and the M phase, during which the DNA is condensed into chromosomes and equally segregated to the two new daughter cells. Following this, cell division occurs.

AJK: Dr. Wilmut has said that your contribution to the team's efforts concerned your realization that the effectiveness of nuclear transfer depended upon when it was done in relation to the cell cycle of the donor nucleus. Could you explain this?

KC: This is true; however, we must also consider the cell cycle phase of the recipient cytoplasm [the egg] and interactions between the nucleus and the cytoplasm. If nuclear transfer is attempted in the S phase, while the DNA is replicating, it will not work. But if a cell can be made to enter an inactive state called quiescence, in which there is no growth occurring and DNA replication has not occurred, we can perform nuclear transfer.

We achieved quiescence by depriving cells in culture of needed nutrients. Few if any genes remained switched on. When their nuclei were removed, placed next to the enucleated egg cells, and fused by an electric current, the eggs were able to reprogram the donor nuclei so that they behaved as if they had come from undifferentiated cells.

AJK: I've read that the sheep embryo activates its genome relatively late compared to humans. It's been suggested that this late activation of the genome may have been crucial to your being able to reprogram the transplanted nucleus, providing time for chromatin remodeling of that nucleus. Please define chromatin remodeling and comment upon this theory.

KC: Chromatin remodeling refers to the reversal of the chromatin structure [in the nucleus] to an embryonic state so that it may recontrol development in the spatial and temporal manner required. The processes involved and the changes, that occur are poorly understood. [Editor's note: A recent paper suggests that a protein (called ISWI) already known to play a role in remodelling chromatin could be an oocyte factor involved in reprogramming a donor nucleus. The work was done in frogs, so its relevance in mammals is not known. Kikyo, N., Wade, P, Guschin, D., Ge, H. and Wolffe, A. "Active Remodeling of Somatic Nuclei in Egg Cytoplasm by the Nucleosomal ATPase ISW.I. *Science* 2000 289: 2360–2362.] Sheep activate their genome between the 8- and 16-cell stage—between the third and fourth cell divisions. Humans apparently activate their genome at the 4- to 8-cell stage—between the second and third division. One theory that I have expressed is that in species that activate the genome at a later stage, the transferred nuclei have longer to undergo chromatin remodeling. The differences in humans and sheep are fairly small, but they may be highly significant.

AJK: How does differentiation relate to chromatin remodeling?

KC: Differentiation is the specialization of cells to particular functions. As all cells contain a complete copy of the genome, alterations in chromatin structure are thought to be involved in the control of the use of specific parts of the genome to fulfill the functions of the particular differentiated cell type.

AJK: Would the difference in the amount of time available for chromatin remodeling with respect to sheep and humans make human cloning more difficult than sheep cloning?

KC: A difficult question to answer. In theory, it may make the cloning of humans more difficult or less successful.

WHY AND WHITHER DOLLY?

AJK: When did you and Dr. Campbell decide to try to clone an adult animal?

IW: One of the misunderstandings in the media has been the belief that we originally set out to do that. We didn't. We set out to be able to make genetic modifications. I guess that if I'd been asked, I would have said that at some time we would have been able to clone an adult animal. But I would have expected it to take longer. It only became a stated objective in 1995—when the Megan and Morag generation of lambs were in utero. It dawned on us then that the new technique was indeed very powerful, and we might be able to clone an adult mammal.

AJK: Do you remember when you first came to believe that your technique might enable you to clone an adult sheep?

KC: I always believed that this would be possible. The birth of Megan and Morag and the others to me was the proof. Producing Dolly was just like the icing on the cake. [Editor's note: The "others" to which Dr. Campbell refers were produced in the same set of experiments that gave rise to Dolly. There are four cloned sheep produced from an embryo cell line, and two cloned sheep produced from a fetal cell line. All are alive and well.]

A J K: Dolly was born in July 1996, but the world did not learn of her birth until the following February. Was it difficult to keep such a secret? Were there practical considerations (such as securing a patent) that caused you to keep that secret, or were the reasons tied to publication considerations?

G B: The patents were submitted at the beginning of the Megan and Morag experiment in 1995, well before Dolly was born. When you do an experiment, it takes you about a year to get the work into a publication. And that is due to two things. First, you have to repeat the experiments and be absolutely certain, and second, it takes up to three to six months to get a paper even in the best journal. There is a publication delay while it is refereed. The timing between July and February was actually quite quick. Bearing in mind that this was a very important discovery, we had to be absolutely certain that Dolly is what she is, do all the tests. So, in my view, we rushed to publication, we did not delay.

A J K: Was the cloning of Dolly a kind of midlevel goal in your quest to clone transgenic sheep? Please explain how she fits into the larger intellectual and research scheme.

G B: Creating Polly was our true goal. And it was Megan and Morag who were our halfway house to doing genetic modification. We are likely to be using fetal fibroblasts to do Polly-type experiments in the future. In a way, Dolly was a sort of detour—using adult cells. But using adult cells is more likely to be useful in animal breeding. I suspect that in trying to get genetic modification for biomedical uses, you can really see the line going from Megan and Morag to Polly.

A J K: What is so important about Dolly from an animal husbandry point of view?

G B: As far as Dolly is concerned, the important thing is that we have shown that we could clone from a live animal. And of course in dairy cattle breeding that is exactly what you want to do. You want to be able to have a dairy cow, know that she is a good producer, and after seven or eight years, clone from her. So the two things are for rather different purposes. The genetic modification [using the fetal fibroblast cell line, as was done to produce Polly] is for biomedical applications, mainly. The second part, cloning from an adult animal, is mostly for animal breeding, in pigs and cattle.

A J K: You were able to clone Dolly with great difficulty, or, at least, some inefficiency. Could you discuss this?

G B: Dolly was derived from one out of 277 oocytes that we used. We don't know if that was unlucky and we could get that rate higher, as we have done with the embryo cells. We don't know that it wasn't a lucky one out of 277. Maybe the true rate is one out of 10,000. Now that we have produced Dolly and Polly, we have to go back and work on improving the efficiency.

AJK: Do you know the precise state of Dolly's health—whether she has genetic flaws, for example?

GB: As far as I can tell, she looks perfectly normal. As you've seen for yourself, she is spoiled rotten. The only thing wrong with her is that she's overweight because too many people have been feeding her. We do want her to breed. And we do hope that she can have a quieter life.

KC: Dolly has not been subjected to invasive examination. She is apparently healthy and normal. She will remain at the institute, living as normal a life as possible, given the media interest.

AJK: What have you done to ascertain Dolly's true age? Having just passed her first birthday, is she one year old or (because she was derived from the DNA of a six-year-old sheep) seven years old? What signs would you look for in order to answer this question?

GB: I think Ian has got somebody starting to look at Dolly's chromosomes. One theory of aging holds that as cells get old, the telomeres—the repeated units of DNA at the ends of chromosomes in somatic cell nuclei—start to break down. It should be possible to see if this is happening in Dolly's cells. But, of course, we don't know what age sheep live to. [Editor's note: Subsequent to this interview, a letter was published in *Nature* describing research results showing that Dolly's telomeres (as well as those of two cloned sheep, one derived from a fetal cell and one derived from an embryo cell) were shorter than those of age-matched control animals (Shiels, P G, Kind, A. J., Campbell, K H, Waddington, D, Wilmut, I, Colman, A., and Schnicke A. E. "Analysis of telomere lengths in cloned sheep." *Nature* 1999 May 27; 399 (6734): 316–7).

Inferences were then drawn by many that cloning would shorten telomeres in all species, and perhaps increase the rate of aging. However, research in cows and mice has shown that cloned animals (at least of the two species) have longer telomeres.

One article reported the birth of six healthy cloned calves derived from senescent donor somatic cells. Their telomeres were found to be extended beyond those of newborn and age-matched control animals (Lanza, R P, Cibelli, J B, Blackwell C, Cristofalo, V J, Francis M K, Baerlocher G M, Mak J, Schertzer M. Chavez E A, Sawyer N, Lansdorp P M West M D. "Extension of cell life-span and telomere length in animals cloned from senescent somatic cells." *Science* 2000 Apr. 28: 288 (5466): 665–9).

A brief communication, published in *Nature* by Teruhiko Wakayama et. al ("Cloning of mice to six generations." *Nature* 2000 September 21: 407: 318–319) reported that two independent lines of mice were cloned to four and six generations. Successive generations showed no signs of premature aging and there was no evidence of telomere shortening. In fact, the researchers said, the telomeres appeared to increase slightly in length.]

Maybe a six-year-old is not a very old sheep. People have told us that sheep can live fifteen or twenty years. So it may not be possible to answer this question with Dolly. We might have to wait until experiments are done on mice or rats. A three-year-old mouse is an ancient mouse. So if you took nuclei from such a mouse and did this experiment, that might be the way to go.

KC: PPL is currently investigating Dolly's DNA. There are conflicting theories on telomeres and aging. It is thought that their function is to protect the DNA from damage. We are currently examining and comparing the length of Dolly's telomeres and also those of all the other animals produced by nuclear transfer. Telomeres can be measured by extracting DNA from blood and using an electrophoresis assay.

AJK: Is it true that when Dolly dies, her presence has been requested—stuffed—at the Science Museum?

GB: I have heard people say this, but no one has written to me with such a request.

AJK: When do you intend to attempt to mate Dolly?

GB: She will be mated at some point. But we have to slim her down a bit first. [Editor's note: As this volume goes to press, Dolly has given birth to six lambs.]

AJK: Do you intend to repeat the Dolly experiment, and try to clone another sheep from an adult cell?

KC: No. There are no plans to repeat the Dolly experiment at present.

THE NATURE OF SCIENCE

AJK: Does science develop to some degree by serendipity?

IW: Yes, I'm sure of that. But what I'm not sure about is what proportion of the scientists who have achieved breakthroughs actually set out with the particular objective in mind. And what proportion made a relatively small step, building on the work of other people. I suspect that a rather large proportion is of our type, in which we built on the earlier work and were fortunate. That's encouraging. But to let Lady Luck come your way, you have to be systematic and precise, because if you are not, you won't understand the significance of something unexpected.

GB: I think that serendipity is overrated. There are one or two famous cases of serendipitous discoveries like penicillin, but generally speaking, science appears and moves by prediction. I don't mean to say that you can predict what is going to happen in the next year or two or even in the next week because you can't.

In Ian's case he had a clever idea—to turn the problem around. Instead of trying to produce the perfect cell type, he decided to take a rather imperfect cell type and find out how he could alter the conditions under which it was being grown in culture so that it became totipotent; that is, capable of being transferred [to produce an entire animal]. And he discovered that it was quiescence or resting that did that.

AJK: What are the absolutes of science?

IW: In terms of the actual knowledge itself, the absolutes of science have to do with the accuracy and carefulness with which you have done an experiment. We must know when we have actually done the experiment and produced Dolly, that the success either is or is not the result of having used the method of achieving quiescence. That it either is or is not related to the particular cell type used.

AJK: What role do pragmatic considerations play?

IW: I would guess that there are relatively few people who actually approach their science without a motivation other than merely seeking pure knowledge. In my case, when I started in agricultural science, I would have had the mixed motivation of wanting to understand animals and wanting to improve agricultural productivity. The Agricultural Research Council was set up after the end of the Second World War, when Britain was seriously short of food, and that is the climate in which this sort of research would have started.

Now, it is fascinating to understand the biology. But the dominant idea is that the technique may be useful. So, when I first froze boar semen, I tried to understand what happens to a cell when you freeze it. But I also went to a pig farmer and told him that this is how you freeze semen and you can use it for your pig breeding company. For me that has always been an important mix.

AJK: Prof. Lewis Wolpert has described the nature of experimental work as "far from being either the glamorous or dangerously uncontrolled activity many people imagine. It is also a great deal more laborious and time-consuming. The ratio of results to effort is frighteningly small. It usually takes hundreds or thousands of tedious hours of work to obtain a result that can be described in a few minutes. . . . The thrill of success is rare. It requires a certain spiritual fortitude, quite often simply to keep going. Many—most—of your ideas turn out to be wrong, and months of experimental work can be fruitless."[6] In the experiments leading up to the creation of Dolly, what gave all of you the "spiritual fortitude" to persevere in the face of frustration, discouragement, and even boredom?

IW: Optimism. It has been ten years since we started this project. And even in this present experiment, Dolly was derived from 277 embryos, so the other 276 didn't make it. The previous year's work, which led to the birth and survival of Megan and Morag, used more than 200 embryos. We have success rates of roughly one in a hundred or less.

AJK: James Watson [whose work more than forty years ago began the science of molecular biology], responding to the criticism of scientists involved in genetics playing God, said, "If we don't play god, who will?"[7] He went on to say that if gene therapy achieves its promise, it will enable us to reverse fate. And, if we do that, we are God. Would you care to comment on his statement?

G B: It sounds like one of those comments that are made on late-night TV shows. I don't see scientists as playing God. Of course, the public only sees scientists when they have made major and significant discoveries, and thinks "Good God, look what's happened!" I see science as very, very slow. You know that Ian's work has taken the best part of ten years. It probably seems to him a long time. He probably looks back and sees it as having taken him almost the whole of his career.

A J K: In your lecture at Utrecht University, you said that it was Pasteur who actually let the genie out of the bottle, through application of science to solve medical problems. What exactly did you mean by that?

G B: As soon as you start applying science to life, which Pasteur could be seen as being the first to do in the nineteenth century, then you started interfering with life in a Godlike way. I mean a blood transfusion is playing God, isn't it? Inoculations, vaccinations. All those are interfering. We have been playing God with animals for 4,000 years. And really since Pasteur, we have been playing God with humans. The Victorians who built drains that stopped cholera and typhoid were playing God because they were eradicating diseases that had been endemic. All of us, every day, are doing things and using things that have greatly improved our lives and have greatly interfered with the so-called natural order.

A J K: Is science able to critique itself or must there be an outside critique—a social or ethical one, for example? Can the full implications of a scientific experiment even be known at the outset?

I W: The validity of a scientific idea must be assessed within the scientific community, because other people lack the knowledge to do it. The priorities as to what science should be carried out, that clearly is a social judgment for everybody is that it is not possible to predict the outcome of experiments. Using Dolly as an example, in our justification we did not specifically say that we were going to set out to clone an adult animal. Obviously, ethical judgments were made that it was an appropriate experiment to do because of the benefits of being able to make precise genetic modifications in livestock.

In hindsight, people have commented, "Shouldn't you have been considering the human implications?" My view, shared with my colleagues, is that this would have been impracticable because you simply cannot predict the outcome of experiments. The very best experiments come up with far more than you could possibly have imagined.

Therefore, the best approach is to take an informed view about science and to go out in general directions—you wish to develop treatments for disease, et cetera, et cetera. In terms of the regulation of the use of knowledge, of its impact on society, you have to regulate after the event.

You say, we now have an opportunity to shoot one another with guns. We don't like that idea and, therefore, we will prohibit the availability of guns. Or cloning, or whatever.

CLONING: THE SOCIAL AND REGULATORY ISSUES

AJK: Can science be regulated? If so, how? Must we wait until the discoveries are made before we decide how—or, indeed, whether—to allow them to be implemented?

GB: Yes, of course, we can regulate science. But it is quite difficult to do it before the fact. The Asilomar Conference on genetic modification was an attempt to do that. We ourselves have had similar efforts in this country on this subject, as you have in the United States. There can be a general, very broad law and, within that, an organization that judges the experiments. So individual experiments have to go to the committee that has a broad structure, and that committee can decide whether to approve them. We do that in this country with health and safety, we do that with genetic modification, and we do it for human fertilization and embryology.

Science is moving inexorably on a broad international front. It is going on all over the world. We, ourselves, are only one or two years ahead of everybody else in any area, even in the best areas—sometimes only one or two months ahead. So if we stop doing any work, it will carry on somewhere else. So it is difficult to regulate the science. It is much easier to regulate the use of that science.

KC: We could try to guess what may be the next discovery and then plan for it. However, scientific discoveries are like many other things in life: unforeseen benefits and negative consequences may arise and have to be met at the time.

AJK: Society's views of technology change, usually, but not always, in the direction of increased tolerance. What should be the relative weights of the moral views of scientists, on the one hand, and the prevailing societal view of the morality of any particular scientific application, on the other?

GB: Scientists do have personal ethical positions, but I don't think they are any more valid than anybody else's. I think the important thing is society's views. And different societies will take different views; even the same society will take different views at different points in time. I heard someone say on the radio this morning that the chief engineer of the BBC was sacked in 1933 because he got divorced, and they couldn't have someone with such immorality working the controls of the BBC. So you can see how things have moved on.

AJK: James Watson has said that the risks of the ability to do genetic interventions is not its misuse; it is not using it to achieve benefit. Do you agree?

GB: It certainly would be a tragedy if the public response to this technology meant that it could not be used for good. Technology can be used for good, but it could also be misused. The important thing is to separate the uses from the misuses. This applies to information technology. And the motor car as well, for heaven's sake. That can be pretty well misused. It would be terrible to take a Luddite view—a back-woodsman view—and say that we are not going to have any uses of modern technology because on the odd occasion it could be used for something dangerous.

KC: The positive aspects of any technological invention must be measured against the negative aspects. People will inevitably find a negative use for most inventions. Should this restrict our quest for knowledge and understanding?

AJK: Dr. Tom Wilkie, the former science editor of the *Independent* newspaper in Britain, wrote a book about the human genome project, entitled *Perilous Knowledge*.[8] What does the expression "perilous knowledge" mean to you?

IW: I suppose that perilous knowledge would be that knowledge which is so dangerous that the potential benefit could well be outweighed by disadvantages. Atomic energy and nuclear weapons must come quite close to perilous knowledge. Certainly, "perilous" implies that you have to be extremely cautious, so having this knowledge is not necessarily going to produce a calamity. But we would be at risk if that knowledge is misused.

AJK: As the director of the Roslin Institute and a scientist yourself, how do you view your obligation to society with respect to the techniques invented at Roslin?

GB: I believe the obligation of scientists is to make sure that any research that is being done finds its way into the public domain as soon as possible so that a regulatory view can be taken on it, both nationally and internationally. The obligation of scientists is to pass that information on. It is not to make those ethical judgments themselves.

AJK: Professor Wolpert believes that Mary Shelley's Dr. Frankenstein is the powerful image that best epitomizes the scientist unleashing forces that he cannot control. How responsible should scientists be for the use made by others of their discoveries? And how concerned were you—are you—with possible use by others of your technique to clone humans?

IW: This is the anxiety, isn't it? There is an assumption that is useful to highlight—almost an assumption that a new development will be dangerous—and yet, sitting in this room, where there is a telephone, a tape recorder, electric light, central heating, we are almost totally dependent on previous inventions. It would be true for most inventions that there would be a mix of beneficial opportunities and potentially dangerous opportunities. You can go back to something as simple as an axe. An axe must have been seen as incredibly useful for chopping up fire wood, but you can also kill people with it.

On the question about how responsible scientists should be for the uses made by other people, we clearly have something of a responsibility if this technique is used in a way that we are not comfortable with. Because we have made it available. But in the end, the greater responsibility would lie with the person who has done the actual work—made the new application.

AJK: How can scientists behave in a socially responsible manner?

IW: Our primary responsibility is to inform people and make them think. There is a danger of scientists being so wrapped up in their own research that they become arrogant and unconcerned about the views of others and about society generally. That would be a serious mistake. We recognized that we were developing a technique with enormous implications and set out to inform society—not just people in the media, but also regulatory authorities, such as the Human Fertilisation and Embryology Authority. Unfortunately, many of those communications didn't get out because of the premature publication of the story.

AJK: You were in Washington, D.C., having just given televised testimony to a U.S. Senate subcommittee, when the U.S. National Bioethics Advisory Commission—a group asked by President Clinton to write a report on the ethical and legal issues raised by the cloning of Dolly—held its first public meeting on cloning. Why didn't you take the opportunity to explain your work and its scientific implications to the commission?

IW: I was not invited to attend. I am surprised that my colleagues and I have not been asked to present any information to the National Bioethics Advisory Commission either in person or in writing.

AJK: When you testified before the Senate subcommittee, did you discern the senators' legislative approach with respect to human cloning?

IW: Before the hearing I met with Senator Frist [Bill Frist, R-Tenn., the chair] and Senator Kennedy [Edward Kennedy, D-Mass., the ranking Democrat]. It was clear that their aim was to allow careful thought before legislation. They were concerned that there was a knee-jerk response that said "we must stop that," which could inadvertently prohibit uses that society would accept. One difficulty with this subject is that people use words carelessly. There may be uses of nuclear transfer with human cells that do not involve producing a new person. It is very important to make sure that in considering the prohibition of the production of new people you don't inadvertently prohibit acceptable uses.

HUMAN CLONING: ETHICS, LOGISTICS, AND MEDICAL IMPLICATIONS

AJK: What is your view of the moral acceptability of the cloning of human beings?

GB: The general view that we have taken on the cloning of human beings is a fivefold stance. First, we don't know whether we can do it on humans. Second, we have no intention of doing it on humans. Third, we don't believe that there is any justified clinical reason for doing it on humans. Fourth, we have no intention of licensing our technology for anybody to do it on humans. Fifth, it is illegal, at least in Britain.

IW: I am uncomfortable with copying people, because that would involve not treating them as individuals. And so I posed the question that I would like to ask anybody who is contemplating such a use: "Do you really believe that you would be able to treat that new person as an individual?" And that, I think, is the limit of my responsibility.

KC: Human cloning—obviously this was bound to raise its head. Personally and medically, I see no reasons for cloning humans to term. I am against using nuclear transfer to produce humans. Full stop!

AJK: Is a risk of producing deformed humans a primary moral consideration in your objections to human cloning?

GB: There is no doubt about it. This is a sixth reason that you wouldn't want to clone humans. It is true that if you did want to apply it to humans, which we don't, the abnormalities would be a major problem.

KC: As I am morally against cloning humans, potential abnormalities are not a factor. As we understand more of the processes involved in creating animals by nuclear transfer, then I am sure that we will be able to remove these abnormalities.

AJK: What about the use of the technique of nuclear transfer to help an infertile couple? If the man was unable to produce any sperm at all, do you think that it would be ethically suspect to clone either the man or the woman and enable them to have a child?

IW: If the situation existed wherein the woman is potentially fertile but her husband is not, I would regard it as more appropriate to create a different individual by using artificial insemination with donated sperm. With the technique of nuclear transfer applied in this situation, you would make a copy of one of the parents. There would inevitably be distortions and pressures in the relationship if one produced a copy of the husband or the wife. [The copy of course would be an infant.]

AJK: So in your view the advantage of not bringing an outside party into the relationship as a genetic parent would be outweighed?

IW: Yes.

AJK: With respect to therapeutic cloning[9]—human cloning that produced embryos, not persons—please tell us about the medical implications of the techniques developed at Roslin.

IW: One potential use for this technique would be to take cells—skin cells, for example—from a human patient who had a genetic disease. These cells inevitably have a limited potential to do other things. You take these and get them back to the beginning of their life by nuclear transfer into an oocyte to produce a new embryo. From that new embryo, you would be able to obtain relatively simple, undifferentiated cells, which would retain the ability to colonize the tissues of the patient. Now that means that once they are in the laboratory, there would be the ability to make genetic changes or even add a gene, and so help the patient to deal with the particular condition.

AJK: Do you think that society should allow cloning human embryos because of the great promise of medical benefit?

KC: Yes. Cloning at the embryo stage—to achieve cell dedifferentiation—could provide benefits that are wide ranging, and have significant effects on the lives of many people. I feel quite sad that the world media decided to concentrate on many of the misuses. Positive education on the use of this technology would help to educate the public about the benefits. [Editor's note: About one and one half years after this interview was conducted, Dr. James Thomson and his group at the University of Wisconsin, Madison, announced that they had isolated stem cells from human embryos and grown them into 5 immortal cell lines (Thomson J, Iskovitz-Eldor, J. Shapiro S. et al. Embryonic stem cells lines derived from human blastocysts *Science*, 1998: 282: 1145–1147).

Dr. John Gearhart and his team at Johns Hopkins University in Baltimore, Maryland announced that they had isolated embryo germ cells from the primordial reproductive cells of the developing fetus (Shamblott M., Axelman. J. Wang S et al. "Derivation of pluripotent stem cells from cultured human primordial germ cells." PNAS: 1998:95:13726–13731). Drs. Gearhart and Thomson, along with other researchers, have demonstrated the therapeutic potential of stem cells. Therapeutic cloning for the derivation of immunologically compatible embryonic stem cells is described in chapter 27.]

AJK: What is the likelihood of using your technique to actually grow human organs? Will this be a possibility?

KC: At present, we can hypothesize or suggest that we may be able to dedifferentiate somatic cells using nuclear transfer, to preimplantation stages of development, and then produce specific cell lineages. However, the production of organs requires an interaction between different cell types. At present, in vitro, this is a dream, in my opinion.

AJK: Do you think that the cloning of human beings by nuclear transfer from a differentiated cell will someday be possible?

GB: I don't know. Ian tells me that the embryology of humans is somewhat different [from that of sheep.]. We don't know that it can be done in humans. Incidentally, we don't know that it can be done in pigs or mice yet. Their embryology is also different. With cattle, the embryology is very similar to sheep so it might be more likely to be able to do it in cattle. [Editor's note: Since this interview was conducted, mice, cattle, and pigs have been cloned.]

AJK: Would there be major logistic obstacles to the cloning of humans?

GB: With respect to humans, there would be enormous problems in getting the approximately 1,000 oocytes [eggs] that would be required. I have been told that we only have 1,000 oocytes for all the human IVF work that is done in the United Kingdom in a year. So you'd probably need all that number and twenty to fifty healthy women to be surrogates, just to clone one human being. And, in current conditions, a large number of the embryos may turn out to be abnormal. So the situation is absolutely untenable. I can't imagine any civilized country doing this sort of research.

AJK: What do you think it would be like to be human clone?

GB: I have no idea. The person you ought to ask about this—who is always going on about how fascinating it would be to watch a clone of yourself grow up—is Richard Dawkins. And you could, of course, ask an identical twin.

AJK: Could it be that it will never be scientifically possible to clone a human?

KC: I think that very few things are clearly impossible. I think that it will be possible to clone a human. However, it may be more difficult to clone a human than to clone a sheep.

AJK: Does it appear to you that the public has been educated by the media and the various government entities that have examined cloning? Is there now more understanding of the nature and the limitations of what you did with respect to human applications?

IW: Not a lot. No. An extreme example of this was a request brought to us by a woman wishing to produce a copy of her father. She did not wish to give birth to him, but she wanted a copy. This person had simply not thought about what would have happened. I judged that her father was about 70. She sees him as he is at present and does not think about the fact that if we made a genetically identical copy it would be born as baby and the relationship would be completely different. Nor does she think about the fact that, because the baby would be born 70 years later to a different mother, to a different family, with all of the expectations that would fall on him, the personality that developed would be completely different. In these cases where people wish to produce a copy of an individual, it simply cannot be done. And people have still not learned that.

NOTES

This interview was conducted during Spring and Summer, 1997.

1. The cloning—or copying—of human cells or DNA has been going on for some time. The discussion in this interview is concerned with cloning to produce a newborn person or animal or—short of that—an embryo.

2. Gould S J. On common ground: individuality: cloning and the discomfitting case of Siamese twins. *The Sciences 1997*; Sept–Oct: 14.

3. For a full discussion and analysis of the media, public, and political reaction in the United States, please see chapter 11.

4. Greenberg, D S. Much ado about mutton? *The Lancet 1997*; 349: 1850.

5. Wilmut, I: Schneike, A E, McWhir, J; Kind, A J, Campbell, K H S. Viable offspring derived from fetal and adult mammalian cells. *Nature 1997*; 385:810–13-.

6. Wolpert I. Richards. A. *A Passion for Science*. Oxford: Oxford University Press, 1988:5, 7.

7. Watson J L. *The Times Higher Education Supplement* 1977; March 14:15.

8. Wilkie T. *Perilous Knowledge*. London: Faber and Faber, 1993.

9. For a full discussion of therapeutic cloning, see chapter 27.

2

Mammalian Cloning by
Nuclear Microinjection

Anthony C. F. Perry

In the cloning of whole animals, genetic information from a *single* cell programs the development of a new individual; it is a form of asexual reproduction. Consequently, cloning circumvents the requirement for a genetic contribution from *two* cells (spermatozoon and egg)[1] exhibited in sexual reproduction. Cloning provides a method of producing multiple offspring from a single individual, embryo, or cell line, and cloned animals share much the same genetic relationship as identical twins. In practice, whole animal cloning (also referred to as "reproductive cloning") is often defined euphemistically as a process in which the *nuclear genome* of a single cell programs full development;[2] this is the definition we will adopt here. While the possibility that mammals might one day be cloned by nuclear transfer from adult cells has evoked enormous interest, some believed it impossible.[3] This skepticism has been abandoned following advances that include the first reproducible method of cloning mice from the nuclear genetic material of an adult cell. The method was developed by Dr. Teruhiko Wakayama and is here termed *c*loning by *n*uclear *m*icroinjection (CNM).[4] In CNM, genetic (chromosomal) material from the nucleus of one cell is transferred into a chromosomally devoid, unfertilized egg using a micropipette. Cloning from adult somatic cells[5] as exemplified by CNM provides biologists with a powerful new device for studying fundamental aspects of development and aging. Its applications likely have wide-ranging benefits but also potential pitfalls, and it is reasonable to imagine that it represents a milestone in biology, medicine, agriculture, biotechnology, and bioethics. This article seeks to describe CNM, its background and potential.

TOWARD MAMMALIAN CLONING

In perhaps the first report of cloning mammals, Illmensee and Hoppe[6] described a method with features common to that of cloning amphibians,[7] through which the general characteristics of cloning had earlier been defined and shown to be achievable. This generic method involves the transfer of genetic information (chromosomes, normally housed within the nucleus) from a nongamete cell (the nucleus donor) into an egg or early embryo whose native chromosomes have been removed (an enucleated recipient cell). The result is a reconstituted cell that can become an embryo and develop fully if it is transferred into the womb of a pseudo-pregnant surrogate mother. The experiment reported by Illmensee and Hoppe used microinjection to supplant the chromosomes of 1-cell mouse embryos with those of early embryonic mouse cells, to generate three cloned mice.[6] But their findings triggered controversy when others could not repeat them, and prompted the contemporaneous conclusion that "the cloning of mammals by simple nuclear transfer is biologically impossible."[3,8]

Parallel experiments from the early 1980s suggested a potent scientific argument against the suitability of early mouse embryos as recipients in cloning. This hinges on a critical switch in the very early embryo from egglike to embryolike gene activity: the zygotic switch. In the mouse, the zygotic switch was thought to occur rapidly after fertilization relative to that in other species.[11–17] This rapid onset of the zygotic switch is clearly not a problem in fertilization, where gamete-derived chromosomes are biologically equipped for it. But in a cloning procedure, where chromosomes are transferred from a nucleus donor cell into the artificial environment of the recipient egg, they may need time to undergo preparatory changes necessary for the switch. The umbrella term for these preparatory changes, which are ultimately responsible for the new role of the incoming chromosomes in directing embryonic development, is "reprogramming".[18,19] It was reasoned that in the mouse, there was insufficient time to allow reprogramming in a cloning procedure before the zygotic switch occurred, and many workers therefore turned to different species, such as sheep and cattle, in which the zygotic switch was thought to be relatively delayed after fertilization.[11] Reports of mouse cloning from the 1980s to the late 1990s were sparse and attempts generally unsuccessful, regardless of whether the recipient cell was a (fertilized) zygote or an (unfertilized) oocyte.[20–22]

By contrast, an archetypal report by Steen Willadsen in 1986 described how a single cell (blastomere) from an 8-cell sheep embryo could be fused with an enucleated sheep oocyte to produce offspring; fetuses were produced from the blastomeres of 16-cell embryos.[23,24] This paved the way for further developments; cell fusion techniques produced clones from more differentiated nucleus donor cells and from different species; cattle and rabbits.[25–30] One case report described how the use of an adult, mammary-derived cell led to the birth of a living offspring, the sheep Dolly,[27] this 1997 report was the first of mammalian cloning using an adult cell as nucleus donor and has been followed by reports of cloning by fusion from the cells of adult cattle.[28,29] The surge of activity in the late 1990s included reports of lambs cloned by all fusion methods from cells cultured for a relatively short term in the laboratory[31–33] and cloning of additional species, the goat[34] and rhesus macaque.[35]

Notwithstanding the supremacy of the mouse as a research model compared to these larger species, until recently the biological sciences lacked what they greatly needed: a method to clone mice. This implicit shortcoming was overcome by Dr. Teruhiko Wakayama, then in Hawaii, and reported in a 1998 paper describing a distinctive method of cloning: cloning by nuclear microinjection (CNM).[4] CNM is now described, followed by a brief consideration of its potential applications.

A COMPARATIVE DESCRIPTION OF CNM

The CNM method reported by Teruhiko Wakayama and his colleagues[4] is the first report of mouse cloning from adult, somatic cells. The report describes how nuclear material from three adult-derived somatic cell types programmed development in vitro and to varying degrees in vivo. Remarkably, offspring were produced using nuclear material from cumulus cells. This reproducible method enabled clones to be generated from the cumulus cells of clones. Indeed, it is the first report for any organism of adult somatic cells producing cloned clones.[36] The first cumulus-cell-derived clone generated by this method to survive to adulthood was named Cumulina to reflect her provenance.[4] Cumulina was born on October 3, 1997, and died on May 5, 2000, at age 2 years and 7 months (old age for a mouse). The method that produced Cumulina can be described in two parts.

Cell Reconstitution

The CNM method used to produce Cumulina employs (unfertilized) mouse eggs as nucleus recipients; the egg chromosomes are removed with a micropipette (also known as microinjection needle). Removal of chromosomes from the egg is termed "enucleation." Enucleation is typically enabled by compounds that destabilize microfilaments (important structural components of cells) so that the chromosomes can be gently removed from each oocyte without killing it. Batches of 10–20 oocytes are enucleated in this way before being returned to an incubator; an exceptional operator like

Teruhiko Wakayama generally enucleates 140–200 oocytes per 1-hour session. Use of a piezo-actuated micropipette[37] in this procedure means that penetration of the egg-protecting layer (the zona pellucida) and subsequent chromosome removal can be accomplished relatively quickly using the same micropipette. The resulting eggs, now devoid of their native chromosomes (i.e., enucleated), are ready for the introduction of genetic material from the nucleus donor cell to produce a reconstituted cell.

In CNM, removal of the nuclear genome from the donor cell and its insertion into the recipient cell are also typically performed with a piezo-actuated micropipette. The steps of nucleus removal and insertion in CNM are spatiotemporally separate. In the first step, a nucleus is removed from a donor cell. The microinjection needle (which is usually different from the one used for enucleation) is moved into a suspension of nucleus donor cells (such as cumulus cells), and a cell selected and aspirated in and out of the needle. If the internal diameter of the needle is smaller than that of the donor cell, this action usually breaks the outer (plasma) and occasionally the nucleus-delineating membranes of that cell. The process is sometimes assisted by the application of a small number of low-intensity piezo pulses.[37] This method of nuclear selection means that the chromosomal aggregate can swiftly be washed free of much of its surrounding cellular material; it is conceivable (although not yet known) that this increase in the purity of the introduced material is advantageous to developmental outcome.[38] The chromosomal complement contained within several nuclei can quickly be harvested such that the microinjection pipette contains a "queue" of nuclear boluses. These are then moved via the injection needle to a different droplet (on the microscope stage) containing a batch of enucleated oocytes in preparation for the second step in the transfer process: donor nucleus insertion. Once again, by piezo-actuation of the micropipette it is possible rapidly to pierce the zona pellucida and then inject a single nuclear bolus deep inside each recipient egg. The time between nuclear removal and insertion is usually no more than 5 minutes. The relative speed of manipulation afforded by piezo-actuated micropipettes (as compared with conventionally mounted ones) is likely to contribute to the efficacy of CNM.[37]

The CNM method outlined in *Nature*[4] is distinctive compared to the cell fusion archetype in vogue at the time it was being developed.[23,26–28] CNM selected and removed the chromosomes of readily identifiable cells from an adult mouse. Fusion methods juxtaposed the entire nucleus donor cell with an enucleated recipient cell and caused their membranes to fuse and become continuous. Fusion is usually achieved by challenging the adjacent cells with an electric pulse, although agents such as fusogenic viruses can be used.[23,26–35] In fusion methods, there is no spatiotemporal separation between donor nucleus isolation and insertion; there is essentially a single step, in which the entire contents of one cell are caused to mix with the contents of another, melding to generate a reconstituted cell. Fusion methods therefore lack the inherent precision of CNM; in fact they are really *cell* (not only nuclear) transfer methods. They are further limited by technical constraints on the size of the nucleus donor cell relative to that of the enucleated egg; such constraints do not apply to CNM. CNM proffers greater control through the selective removal of extraneous material from input chromosomes during micromanipulation and, perhaps more significantly, through the potential to introduce agents in addition to the chromosomes of the donor cell nucleus.[38] This would be an extremely valuable feature of CNM were the extra agents to improve developmental outcome. A further drawback to fusion-based methods is that, unlike CNM, they often require the removal of the first polar body, a byproduct of the in vivo process that normally generates eggs ready for fertilization.[39] Since many fertilizable eggs possess an extant first polar body (containing chromosomal remnants) attached to them, any fusion procedure without prior polar body removal potentially results in reconstituted cells with an abnormal chromosome number (hyperploidy). Such hyperploid cells are unable to support full embryonic development.

Development of the Reconstituted Cell

At fertilization, the signal to initiate embryonic development is provided by the fertilizing sperm. However, since cloning protocols are devoid of spermatozoa, reconstituted cells must be artificially triggered to activate embryonic development. CNM is able to exploit a variety of activating protocols, such as those involving electrical discharges, or chemicals such as ethanol or strontium chloride.[4,40] Strontium chloride mimics the signal supplied by spermatozoa—for example, it causes a similar mobilization of an important intracellular secondary messenger, the calcium ion, Ca^{2+}. Hence, following microsurgery, reconstituted cells are transferred to media containing strontium chloride. In mimicking normal fertilization, this step could (depending on the timing of the signal) result in the expulsion of chromosomes in a structure resembling a second polar body.[39] Although extrusion of the second polar body is a programmed feature of fertilization, chromosome loss would sometimes be disastrous in CNM, and it is therefore prevented by coinclusion (with the strontium chloride) of a microfilament-destabilizing compound.[39] In this situation, the reconstituted cell initiates development without polar body extrusion and therefore without chromosome loss. By contrast, the electric discharge used in cell fusion methods such as the one that produced Dolly,[23,27] often concomitantly provides the signal to begin development, although it is possible to employ separate electrical discharges such that one causes fusion and another triggers development.

Simultaneous nuclear introduction and initiation of development was not a feature of the initial report of CNM.[4] It was reported that a latent period of 1–6 hours between chromosome insertion and the chemical activation of development favored embryonic outcome. The latent period was held consistent with the requirement by incoming chromosomes (from the donor nucleus) for an adjustment time to be reprogrammed in their new cellular environment, to enable them to undergo the zygotic switch and to issue the correct instructions to begin embryonic development.[11,18,19] CNM also succeeds with latency periods of less than 1 hour, suggesting that any reprogramming can occur rapidly.[10,11] One corollary of this work is that 1-cell embryos (zygotes) are *not* able to effect reprogramming when they are used as nuclear recipients in CNM; Illmensee and Hoppe, however, reported success in such experiments.[6,8,10]

Embryos generated by CNM are transferred to surrogate mothers to permit placentation and development in vivo. As part of a rigorous check, white-colored mice are chosen as surrogate mothers. Since the oocyte donors are generally black and the nucleus donors coffee-colored, cloned offspring are invariably coffee-colored, consistent with the provenance of their chromosomes. At its inception in 1997, CNM produced offspring with an efficiency of about 1 pup per 100 enucleated eggs (1%).[4] Factors predisposing to the success of CNM had therefore evidently not all been elucidated. For example, the influence of trace components in the distilled water and mineral oil required for embryo culture, the temperature and relative humidity during micromanipulation, and the duration of micromanipulation were among the manifold parameters which have not been standardized. While the efficiency of CNM is still only about 1%, other advances have been swift to follow its advent. Different methods of activation of embryonic development have been employed,[40] different workers have generated offspring in different laboratories,[41] and males have also been cloned from adult-derived cells.[42] Among the most poignant demonstration of the versatility of CNM came with a report[43] in 1999 that it had been used to clone mice from well-established, cultured cells, derived from embryos (embryonic stem cells) in one case around 13 years earlier.[43–45]

SOME FUTURE DIRECTIONS FOR CNM

The elucidation of a cloning method that works for mice will impact greatly on science and medicine. The mouse—itself a mammal—has become a supreme model for mammalian development and pathology, not least because it is relatively inexpensive and has a short life cycle; it has a gestation period of 19.5 days and can reproduce within 6 weeks of birth. There exists unparalleled information on mouse husbandry, genetics, development, embryology, and gamete biology.

With the advent of CNM, key biological questions can be addressed in ways that would have been prohibitive in larger species. It is poised to unlock some of the mysteries of aging, and the way cells specialize (differentiate) within an animal as it develops. CNM will be used to ask the meta-questions "What, if any, are the limitations of CNM and how may they be overcome?" It will allow us to assess the safety of cloning per se and to make improvements. On a molecular level, it can be used to address questions about changes in the molecular architecture of chromosomes during reprogramming,[18,19] about the biochemical "erosion" of chromosome ends (telomeres) during the life of cells in adults,[46-49] about an important phenomenon in some human diseases called imprinting,[50-52] and about the contribution of the enucleated oocyte to these processes.

The study of aging provides an illustrative paradigm. One of the themes in research on aging holds that an organism ages in a programmed manner, with the timing mechanism—the "clock"—corresponding somewhat to the number of divisions the cells in a given individual have undergone.[46-49,53,54] This clock is presumably reset either during the process that generates gametes or very soon after normal fertilisation, so that the resulting embryonic cells start life at time zero. Yet with asexual reproduction using adult cells, that reset cannot have occurred in the manner envisaged for gametes. Therefore, the finding that clones can be produced from adult clones is suggestive, as is the extension of this finding that the process is reiterative beyond two cloned generations.[4,36] Are the cells of a cloned cloned clone biologically four times older than those of a noncloned littermate? This and similar questions posed by CNM have to be accommodated by prevailing ideas on aging; either the hypothetical clock does not exist after all in vivo or it does not need to be reset at the time when gametes are synthesized (gametogenesis), or at all. CNM will also enable scientists to examine the molecular machinery that they postulate might be key to the aging process, and do so in the context of whole animals such as mice. Reiteration of the cloning process can, in principle, be readily extrapolated in the mouse so that one might address the question "Will the 50th successive generation of cloned mice be viable?" In the mouse, this experiment would take around 6 years; in cattle, it could take a century or longer.

Using CNM as a research tool, scientists will doubtless seek to improve the cloning efficiency and to extend its applications. For example, mouse CNM will likely be used to address the possibility that rather than giving rise to an entire individual (reproductive cloning), cell differentiation following cloning might be steered to favor the genesis of certain cell types, tissues, or even organs ("therapeutic cloning"). Such applications of the cloning technology could dovetail with technologies such as tissue engineering to provide perfectly immunologically matched tissue for transplantation.[55-62] Ultimately, it may be possible to induce the controlled reprogramming of any nucleus in situ to give rise either to offspring or to the programmed differentiation of specified tissues. Even if possible, such obviation of the need for nuclear transfer will require that we attain a far greater understanding of the contribution of the egg in nuclear reprogramming during CNM and other nuclear transfer methods.

The clonal replication of genetically modified mice should accelerate the development of medicines ipso facto,[63] but a major application will likely be to clone other species. This area of biology is today not a theoretical subject and will require experimentation. Although the mouse is an excellent starting point for generalizations about mammals, there are, of course, differences between any given aspect of mouse biology and its counterpart (if there is one) in other mammalian species.[64] For several of the larger species experiments will likely be justified by the manifold advantages that successful cloning would yield. The farming community could elect to replicate choice animals in a controlled manner, making it possible to produce a field containing many of what are, in effect, identical 'twins' of the animal, some born at least one generation apart. Cattle might be bred from genetically redesigned stock that are refractory to bovine spongiform encephalopathy (BSE, also known as "mad cow disease") or other diseases, or a heffer producing high milk yield in unfavorable terrains.

Larger cloned animals could also have great utility in medicine, facilitating both the production of therapeutic agents in a technology referred to by the unfortunate, phonetically ambiguous term "pharming," and the generation of "immunologically less visible" organs for transplantation (xenotransplantation) into the terminally ill.[65,66] The generation of large animals for pharming, xenotransplantation and other applications is currently slow and expensive. It requires genome manipulation (such as transgenesis) that is inefficient in these larger species (relative to the efficiency in the mouse) and usually necessitates unavoidably lengthy breeding programs to replicate the genetic changes. CNM promises to circumvent the long breeding programs required by existing methods.[63] If the technology is permitted any significance, one manifestation might be hospital farms whose genetically redesigned livestock are universal tissue donors for xenotransplantation to human patients such as those in renal failure.[65,66]

CONCLUDING COMMENTS

At the inception of CNM, the vogue technology for cloning whole animals used cell fusion. In contrast, CNM is a microinjection-based nuclear transfer method which enables the cloning of mice from adult-or embryo-derived cells and cell lines. This development will allow a rigorous analysis of cloning itself, and is poised to benefit agriculture and both human and veterinary medicine. It behooves those interested in seeing such developments to ensure that debates concerning cloning are well-informed and balanced, so that the promise of cloning technologies can be harnessed for the benefit of all.

NOTES

1. For the purposes of this text, "sperm" is interchangeable with "spermatozoon" (singular) and "spermatozoa" (plural). "Egg" and "oocyte" refer to a mature female gamete that has not been fertilized. "Fertilization" (resulting from the union of sperm and egg) produces a zygote, also known as "1-cell" embryo.

2. Formally speaking, the cloning methods described here (nuclear and cell transfer methods) refer to an embodiment of cloning in which the nuclear genome is derived from one individual and the mitochondrial genome from another, or both (mitochondria are cellular components that have their own mini-genomes). This is because the nucleus is contributed by one cell (the nucleus donor), and the mitochondria (many of them) from another (the oocyte). This common euphemism for cloning—also adopted throughout this article—is arguably justified by the fact that the nuclear genome constitutes 99.9994 % of the total.

3. McGrath, J., and Solter, D. (1984). Inability of mouse blastomere nuclei transferred to enucleated zygotes to support development *in vitro*. *Science* **226**, 1317–1319.

4. Wakayama, T., Perry, A. C. F., Zuccotti, M., Johnson, K. R., and Yanagimachi, R. (1998). Full-term development of mice from enucleated oocytes injected with cumulus cell nuclei. *Nature* **394**, 369–374. This is the first report of CNM; of cloning by microinjection with adult-derived cells; of cloning mice from adult-derived cells; and of generating a clone from a cloned animal. Cumulina, the first adult mouse clone to survive (October 3, 1997–May 5, 2000) was in fact the third to be cloned; the other two perished soon after birth of unknown causes.

5. Somatic cells are those that are of neither sperm nor egg lineages—that is, the vast majority.

6. Illmensee, K., and Hoppe, P. C. (1981). Nuclear transplantation in Mus musculus: developmental potential of nuclei from preimplantation embryos. *Cell* **23**, 9–18. This paper reports the cloning of two females and one male mouse following transfer of blastocyst inner cell mass nuclei into zygotes whose native chromosomes were removed *following* nuclear insertion. Its findings were subsequently questioned (McGrath and Solter; note 3).

7. Briggs, R., and King, T. J. (1952). Transplantation of living nuclei from blastula cells into enucleated frog's eggs. *Proc. Natl. Acad. Sci.* USA **38**, 455–461.

8. The mouse nucleus donor cells used by Illmensee and Hoppe were from the inner cell mass. However, as McGrath and Solter (note 3) document, an attempt to reproduce this work failed even when formally less differentiated cells (from 2-cell embryos) were used as nucleus donors. As if to add to the confusion, a different (fusion-based) method succeeded, although the report (see note 9) suggested that all of the blastocysts obtained were capable of developing to term. In summary, this and subsequent work by Teruhiko Wakayama et al. (note 10) supported the notion that the late-twentieth-century understanding of zygote biology was insufficient to permit zygotes to be harnessed as nuclear recipients in reproductive cloning when the nucleus donor cell was from a developmental stage more advanced than an 8-cell embryo.

9. Tsunoda, Y., Yasui, T., Shioda, Y., Nakamura, K., Uchida, T., and Sugie, T. (1987). Full-term development of mouse blastomere nuclei transplanted into enucleated two-cell embryos. *J. Exp. Zool.* **242**, 147–151.

10. Wakayama, T., Tateno, H., Mombaerts, P., and Yanagimachi, R. (2000). Cloning mice by nuclear transfer into zygotes? *Nature Gen*, **24**, 108–109.

11. In the zygotic switch (or "maternal-to-embryonic transition"), the subset of genes that are active in an egg switches to a new subset of genes whose activity is necessary for embryonic development. The zygotic switch in the mouse was long thought to be initiated by the mid-2-cell stage of embryonic development. More recent studies (Latham, 1999; note 13) have suggested that the zygotic switch in mice may occur even earlier. The relative delay of the zygotic switch in other species was reported for cattle, humans, and sheep. Later work suggested that the zygotic switch in these other species may, after all, occur earlier than at first suspected and perhaps mirror that of the mouse. See Memili and First, note 17.

12. Flach, G., Johnson, M. H., Braude, P. R., Taylor, R. A. S., and Bolton, V. N. (1982). The transition from maternal to embryonic control in the 2-cell mouse embryo. *EMBO J.* **1**, 681–686.

13. Latham, K. E. (1999). Mechanisms and controls of embryonic genome activation in mammalian embryos. *Int. Rev. Cytol.* **193**, 71–124.

14. Kopecny, V., Flechon, J. E., Camous, S., and Fulka, J. (1989). Nucleogenesis and the onset of transcription in the eight-cell bovine embryo: fine-structural autoradiographic study. *Mol. Reprod. Dev.* **1**, 79–90.

15. Braude, P., Bolton, V., and Moore, S. (1988). Human gene expression first occurs between the four and eight-cell stages of preimplantation development. *Nature* **332**, 459–461.

16. Crosby, I. M., Gandolfi, F., and Moor, R. M. (1988). Control of protein synthesis during early cleavage of sheep embryos. *J. Reprod. Fertil,* **82**, 769–775.

17. Memili, E., and First N.L. (1998). Developmental changes in RNA polymerase II in bovine oocytes, early embryos and effect of alpha-amanitin on embryo development. *Mol. Reprod. Dev.* **51**, 381–389.

18. Chromosomes which only moments before nuclear transfer were issuing one set of instructions (for instance, those required to produce a cumulus cell) were thought to require time to become "reprogrammed" such that they started to issue the distinct set of instructions necessary to initiate embryonic development.

19. *Reprogramming* is an umbrella term describing changes undergone by chromosomes transferred from one nuclear environment into another (such as an enucleated egg). Reprogramming can be thought of functionally as it accounts for changes in gene activity (expression) and probably involves architectural remodeling of chromosomes, although the molecular events responsible are unknown.

20. Progress in mouse cloning was slow relative to that in larger mammals. (See note 9.)

21. Kono, T., Ogawa, M., and Nakahara, T. (1993). Thymocyte transfer to enucleated oocytes in the mouse. *J. Reprod. Dev* **39**, 301–307.

22. Cheong, H. T., Takahashi, Y., and Kanagawa, H. (1994). Relationship between nuclear remodelling and subsequent development of mouse embryonic nuclei transferred to enucleated oocytes. *Mol. Reprod. Dev* **37**, 138–145.

23. Willadsen, S. M. (1986). Nuclear transplantation in sheep embryos. *Nature* **320**, 63–65. This landmark paper is the first description of virtually and electrically mediated cell fusion in mammalian cloning by nuclear transfer; this method is the prototype of mammalian cloning by cell fusion.

Cloning by splitting embryos (rather than by transferring nuclei between cells) was earlier described by Steen Willadsen (see note 24).

24. Willadsen, S. M. (1979). A method for culture of micromanipulated sheep embryos and its use to produce monozygotic twins. *Nature* **227**, 298–300.

25. The process by which functionally specialized cells arise from progenitor populations is known as differentiation. In the 1-, 2-, and 4-cell embryos of most species, cells are so undifferentiated that they can each give rise to a complete individual; each cell is said to be totipotent. Highly ordered differentiation occurs cumulatively as an embryo develops. Certain cells of post-4-cell embryos may give rise to an increasingly restricted plurality of cell types, but not necessarily all, and are thus said to be pluripotent. The more highly differentiated a cell, the less it resembles the cell of an early embryo and, it was therefore argued, the less likely it would be to support the development of all cell types required by cloning. Many adult-derived cell types may be considered irreversibly and maximally specialized (terminally differentiated), including cumulus cells, yet they are amenable to CNM. Cloning has thus challenged previous views of totipotency and produced the notion of "nuclear totipotency": the ability of a *nucleus* to direct full embryonic development. Biology has now to determine whether nuclear totipotency is universal among cell types, and if not, why not.

26. Sims, M., and First, N. L. (1994). Production of calves by transfer of nuclei from cultured inner cell mass cells. *Proc. Natl. Acad. Sci. USA* **91**, 6143–6147.

27. Wilmut, I., Schnieke, A. E., McWhir, J., Kind, A. J., and Campbell, K. H. S. (1997). Viable offspring derived from fetal and adult mammalian cells. *Nature* **385**, 810–813. This paper reports the cloning of Dolly the sheep.

28. Kato, Y., Tani, T., Sotomaru, Y., Kurokawa, K., Kato, J., Doguchi, H., Yasue, H., and Tsunoda, Y. (1998). Eight calves cloned from somatic cells of a single adult. *Science* **282**, 2095–2098.

29. Wells, D. N., Misica, P. M., and Tervit, H. R. (1999). Production of cloned calves following nuclear transfer with cultured adult mural granulosa cells. *Biol. Reprod.* **60**, 996–1005.

30. Stice, S. L., and Robl, J. M. (1998). Nuclear reprogramming in nuclear transplant rabbit embryos. *Biol Reprod.* **39**, 657–664.

31. Campbell, K. H. S., McWhir, J., Ritchie, W. A., and Wilmut, I. (1996). Sheep cloned by nuclear transfer from a cultured cell line. *Nature* **380**, 64–66.

32. Schnieke, A. E., Kind, A. J., Ritchie, W. A., Mycock, K., Scott, A. R., Ritchie, M., Wilmut, I., Colman, A., and Campbell, K. H. S. (1997). Human factor IX transgenic sheep produced by transfer of nuclei from transfected fetal fibroblasts. *Science* **278**, 2130–2133.

33. Cibelli, J. B., Stice, S. L., Golueke, P. J., Kane, J. J., Jerry, J., Blackwell, C., Abel Ponce de León, F., and Robl, J. M. (1998). Cloned transgenic calves produced from nonquiescent fetal fibroblasts. *Science* **280**, 1256–1258.

34. Baguisi, A., Behboodi, E., Melican, D. T., Pollock, J. S., Destrempes, M. M., Cammuso, C., Williams, J. L., Nims, S. D., Porter, C. A., Midura, P.,

et al. (1999). Production of goats by somatic cell nuclear transfer. *Nature Biotech.* **17**, 456–461.

35. Meng, L., Ely, J. J., Stouffer, R. L., and Wolf, D. P. (1997). Rhesus monkeys produced by nuclear transfer. *Biol. Reprod,* **57**, 454–459.

36. Wakayama, T., Shinkai, Y., Tamashiro, K. L., Niida, H., Blanchard, D. C., Blanchard, R. J., Ogura, A., Tanemura, K., Tachibana, M., Perry, A. C. F., Colgan, D. F., Mombaerts, P., and Yanagimachi, R. (2000). Cloning of mice to six generations. *Nature* **407**, 318–319. This article describes a sixth generation cloned mouse (a clone derived from an adult clone derived from an adult clone derived from an adult clone derived from an adult clone derived from an adult clone of a founding nucleus donor) whose great great great great great grandmother was essentially its identical twin sister.

37. Microinjection needles (typically of 5–7 μm internal diameter) used in CNM are housed in a piezo micromanipulation unit, permitting piezo-actuated microinjection. The unit transmits the piezo-electric effect in a friction-dependent manner to the microinjection needle tip, causing it to move a short distance (approximately 0.1 μm) in a short time (microseconds). The frequency and intensity of piezo pulses applied to the microinjection needle are highly reproducible and can be varied according to empirically determined optima. This gives the operator great control over the needle tip during microinjection. The control afforded by piezo-actuated microinjection enables the efficient manipulation of mouse eggs where other methods fail. It assists in penetrating the zona pellucida, in nonlethally puncturing the sensitive plasma membrane of the egg and in breaking the membranes of nucleus donor cells.

38. Being a microinjection-based technique, CNM allows microsurgical fractionation of cellular components; harvested nuclear components can briskly be washed free of extranuclear (and sometimes nuclear) debris by micropipetting. Perhaps of greater utility, this feature of CNM allows the *addition* of cloning adjuncts. The precisely controlled delivery of components during CNM is difficult in cell fusion methods.

39. This is because immediately after fertilization, the zygote briefly contains three sets of chromosomes (it is thus said to be 3n) but soon expels n chromosomes in a structure known as the second polar body, so that the zygote retains 2n chromosomes (n each from mother and father). 2n is the complement of chromosomes normally per somatic cell and is the number of sets of chromosomes per cell obligatory for a nonchimaeric early embryo to develop fully. Any loss of chromosomes from the 2n somatic cell nucleus after transfer into the enucleated oocyte would thus preclude development. Exposure to the fungal metabolite cytochalasin B prevents the critical loss of chromosomes by abrogating the formation of a polar bodylike structure. Although loss of any of the chromosomes would preclude embryonic development following the transfer of a 2n somatic cell nucleus, Teruhiko Wakayama and colleagues have shown that the nuclei of cells containing twice the normal content of genomic DNA (equivalent to 4n) can be used in CNM.

40. Kishikawa, H., Wakayama, T., and Yanagimachi, R. (1999). Comparison of oocyte-activating agents for mouse cloning. *Cloning* **1**, 153–159.

41. The list includes (in chronological order of producing offspring by

CNM): Teruhiko Wakayama, Hidefumi Kishikawa, Tsuyoshi Kasai, Atsuo Ogura, Yukiko Yamazaki.

42. Wakayama, T., and Yanagimachi, R. (1999). Cloning of male mice from adult tail-tip cells. *Nature Gen.* **22**, 127–128. This application of CNM is the first well-documented description of the cloning of a male mammal from an adult-derived cell.

43. Wakayama, T., Rodriguez, I., Perry, A. C. F., Yanagimachi, R., and Mombaerts, P. (1999). Mice cloned from embryonic stem cells. *Proc. Natl. Acad. Sci. USA.* **96**, 14984–14989. This paper reports that cells maintained in the laboratory over a period of many years (by a combination of culture and long-term freezing at 195°C) can be used to generate offspring via CNM; the cells were derived in February 1986. In human terms, this is akin to cloning an individual who was conceived in 600 A.D. The paper also suggests that donor nuclei with the equivalent of 4n chromosomes can be utilized. In this adaptation of the CNM protocol, compounds with effects analogous to that of cytochalasin B are omitted so that the equivalent of 2n chromosomes are lost into a polar-body-like structure, thereby effectively restoring the 2n chromosome number of the cell. This provided strong evidence that the cell cycle is not a critical parameter in cloning at a time when many workers believed otherwise. The cultured cells were embryonic stem (ES) cells of the types used for gene targeting; indeed, the paper described cloning of the first live-born, gene-targeted animal.

44. The cells are called embryonic stem (ES) cells and are derived from the inner cell mass of blastocysts. ES cells can be grown in vitro over extended periods and have great utility, supporting the generation of targeted genetic changes. This means that it is possible precisely to alter any one of the 3 billion or so nucleotides that comprise the genetic makeup of a mouse, so that every cell in the resulting mouse contains the specified alternation.

45. Capecchi, M. R. (1989). Altering the genome by homologous recombination. *Science* **244**, 1288–1292.

46. C. B. Harley, who presents one of the several models that casually links the number of cell divisions with aging, envisages a role for the programmed shortening of chromosome ends (telomeres)4; see note 47. Indeed, the relationship between telomere loss and viability in mice has been described (see note 48). It is perhaps paradoxical, then, that the telomeres of cloned cattle are not shorter than those of age-matched controls (as shown by Lanza et al., note 49), and those of successive generations of cloned mice may actually get longer (note 36).

47. Harley, C. B. (1991). Telomere loss: mitotic clock or genetic time bomb? *Mutation Res.* **256**, 271–282.

48. Lee, H. W., Blasco, M. A., Gottlieb, G. J., Horner II, J. W., Greider, C. W., and DePinho, R. A. (1998). Essential role of mouse telomerase in highly profilerative organs. *Nature* **392**, 569–574.

49. Lanza, R. P., Cibelli, J. B., Blackwell, C., Cristofalo, V. J., Francis, M. K., Baerlocher, G. M., Mak, J., Schertzer, M., Chavez, E. A., Sawyer, N., Lansdorp, P. M., and West, M. D. (2000). Extension of cell life-span and telomere length in animals cloned from senescent somatic cells. *Science* **288**, 665–669.

50. The vast majority of genes are present in two copies per genome per cell; one each inherited from mother and father. For approximately

0.2% of the total number of genes, it is important which copy of the gene was inherited from the mother and which from the father. Such genes are said to be "imprinted," because each apparently carries a parent-specific imprint (or "mark") that determines its activity. The molecular nature of the imprint is not known, but it does not reside in the gene sequence per se, and is therefore said to be epigenetic. Alterations in the activity of imprinted genes can result in embryonic lethality. Among the first to recognize imprinting phenomena were Azim Surani and Davor Solter and colleagues (see notes 51 and 52).

51. Surani, M. A. H., Barton, S. C., and Norris, M. L. (1984). Development of reconstituted mouse eggs suggests imprinting of the genome during gametogenesis. *Nature* **308**, 548–550.

52. McGrath, J., and Solter, D. (1984). Maternal T^{hp} lethality in the mouse is a nuclear, not a cytoplasmic effect. *Nature* **308**, 550–551.

53. Other possible prescriptive meters of biological aging include: progressive DNA modification (demethylation), protein modification (e.g., dephosphorylation of the retinoblastoma protein), and an accumulation of mutations. As Ikram, Norton, and Jat discovered (see note 54), irrespective of the mechanism by which the putative aging clock is read, it can apparently be suspended without resetting the time to zero so that although cells "know" that it is time to die, they continue to live. It remains to be seen whether a hypothetical suspension of clock function is necessary for cloning from adult cells.

54. Ikram, Z., Norton, T., and Jat, P. S. (1994). The biological clock that measures the mitotic life-span of mouse embryo fibroblasts continues to function in the presence of simian virus 40 large tumor antigen; the clock is over-ridden. *Proc. Natl. Acad. Sci. USA* **91**, 6448–6452.

55. There is no known cell type that fails to yield blastocysts when used as nucleus donor in CNM (T. Wakayama, personal communication). The inner cell mass of such blastocysts might be used as a source of stem cells. This has at least two implications. First, that cloning does not require a preexisting embryo generated *de novo* by fertilization. Second, such stem cells might be cultured in vitro and induced to differentiate along predetermined pathways. This application is known as "therapeutic cloning" (as opposed to reproductive cloning, in which embryos are allowed to placentate and may produce an entire individual). Success in prescriptive differentiation in vitro has been greatest in glial, neuronal, cardiac, and blood cells. (See notes 56–59 for example.) The growth of clonally derived cells on matrices in vitro could provide a link between the technologies of cloning and tissue engineering, as Kaihara and Vacanti showed (note 60). Therapeutic cloning is reviewed, in Lanza, Cibelli, and West, "Human therapeutic cloning" (note 61)

56. Brustle, O., Jones, K. N., Learish, R. D., Karram, K., Choudhary, K., Wiestler, O. D., Duncan, I. D., and McKay, R. D. (1999). Embryonic stem cell-derived glial precursors: a source of myelinating transplants. *Science* **285**, 754–756.

57. Bain, G., Kitchens, D., Yao, M., Huettner, J. E., and Gottlieb, D. I. (1995). Embryonic stem cells express neuronal properties in vitro *Dev. Biol.* **168**, 342–357.

58. Klug, M. G., Soonpaa, M. H., Koh, G. Y., and Field, L. J. (1996). Ge-

netically selected cardiomyocytes from differentiating embryonic stem cells form stable intracardiac grafts. *J. Clin. Invest.* **98**, 216–224.

59. Wiles, M. V., and Keller, G. (1991). Multiple hematopoietic lineages develop from embryonic stem (ES) cells in culture. *Development* **111**, 259–267.

60. Kaihara, S., and Vacanti, J. P. (1999). Tissue engineering: toward new solutions for transplantation and reconstructive surgery. *Arch. Surg.* **134**, 1184–1188.

61. Lanza, R. P., Cibelli, J. B., and West, M. D. (1999). Prospects for the use of nuclear transfer in human transplantation. *Nature Biotechnol.* **17**, 1171–1174.

62. Lanza, R. P., Cibelli, J. B., and West, M. D. (1999). Human therapeutic cloning. *Nature Med.* **5**, 975–977.

63. The new area of functional genomics seeks to elucidate the function of genes, usually as part of genomewide analyses, and likewise pharmacogenomics the therapeutic potential of gene products as pharmacological targets. In both, it is necessary to generate mice that contain desired targeted alterations of their genetic information. Even in mice, this process is inefficient, the requisite breeding program time-consuming, and the offspring often infertile. Reproductive cloning promises to alleviate these problems via rapid, asexual reproduction. Moreover, the interpretation of experimental data from cloned offspring would be more reliable since they are genetically near identical.

64. Assertions such as "It is possible to clone species x, therefore it is possible to clone species y" are today unsupported and not axiomatic. Perhaps they are akin to "Because we have invented the bicycle allowing us to travel at 25 miles per hour, it is thus possible to travel at speeds near that of light (186,000 miles per second at temperatures permissive for cycling)." For example, notwithstanding considerable effort, it took until after 2000 for pigs to be cloned from differentiated cells. Even if offspring of a given species can be clonally generated; this is not necessarily enough. A mouse model provides a powerful means by which to catalog, analyse, and rectify any features of the cloning process that produce abnormalities (T. Wakayama, personal communication).

65. This is an example of cross-species organ or tissue transplantation called xenotransplantation.

66. Platt, J. L. (1998). New directions for organ transplantation. *Nature* **392**, 11–17.

67. The author thanks Dr. Teruhiko Wakayama for communicating data prior to publication.

3

On Recent Developments in Mammalian Nuclear Transplantation and Cloning

Steen M. Willadsen

I have been asked to compare the report by Wilmut, Schnieke, McWhir, Kind, and Campbell published in *Nature* in February 1997,[1] about the experiments that resulted in the birth of Dolly, "the first mammalian clone," with that published more recently in the same journal by Wakayama, Perry, Zucotti, Johnson, and Yanagimachi,[2] which describes the production of a whole swarm of cloned mice—and to add my own comments. It is with some reluctance that I have accepted the commission, for although I have been, and in a way still am involved in precisely the type of work these articles deal with—nuclear transplantation as a means of cloning mammals—I cannot pretend to have any precise, let alone complete knowledge of the underlying biology, especially not the molecular part, which is becoming increasingly important in the discussion. But then, it is doubtful whether any single person has such complete knowledge or understanding at this point. However, although our insight is far from complete, nuclear transfer experiments involving mammalian eggs and embryos have been surprisingly, to some worryingly, successful of late.

Active interest in mammalian cloning by nuclear transplantation dates back at least to the late sixties[3] and was originally inspired by the successful cloning of certain amphibians by transfer of embryonic cell nuclei into activated unfertilized eggs whose own nuclei had been removed or neutralized.[4] However, in mammals the beginnings were rather unpromising, despite discreet coaching by amphibian cloners. A number of major technical hurdles had to be negotiated before it was possible even to conduct such experiments on a sound basis, and although several attempts[5] and false starts were made during the seventies,[6] it was not until the early eighties that sufficiently effective methods of micromanipulation, nuclear transplantation, and culture had been developed for mammalian embryos and eggs. The central biological questions could now be addressed experimentally; but of course, these technical advances did not in themselves imply that cloning of mammals by nuclear transplantation was possible, let alone imminent.

Indeed, it is interesting to note that the main conclusion drawn from the very first reliable report on nuclear transplantation experiments aimed directly at cloning with mouse embryonic cells was that it was probably biologically impossible to clone mammals by simple nuclear transplantation.[7] However, a few months later, in the spring of 1984, I successfully cloned sheep embryos by nuclear transplantation.[8] The nucleus donors were 8-cell embryos, the recipients enucleated, slightly overripe unfertilized eggs. The first genetically identical nuclear transplant lambs were born in July-August of that same year.

While these sheep embryo cloning experiments were inspired by the successful cloning of amphibian embryos, the procedure used, except for the actual transfer of the donor nucleus, depended critically on sheep embryo micromanipulation and-culture techniques that I had developed in the late seventies.[9] These techniques had proved extremely effective for producing genetically identical animals and chimaeras from cells isolated from very young embryos not only in sheep, but in a number of other mammalian species as well.[10] Together with a standard cell fusion technique[3] for the actual nuclear transfer (and a new egg enucleation technique which I developed in the course of the experiments) they formed the basis of the complete embryo cloning procedure. However, even with highly efficient techniques my attempts to clone sheep embryos were unsuccessful until I switched from the enucleated *fertilized* eggs used by would-be mouse cloners, as recipient cells for the donor nuclei, to enucleated *unfertilized* eggs. With the enucleated unfertilized eggs as recipients, success was immediate and reproducible. Over the next couple of years I further refined and simplified this procedure and also adapted it for use both in other mammalian species, most notably the cow, and with other nucleus donor and recipient cells than those used in the original experiments.[10] Once this had been accomplished, a biotechnical framework had been created within which the cloning of livestock might be pursued—which had been the practical objective of the work all along. At the same time, a much larger research area concerning the functional relationship between nucleus and cytoplasm in mammalian eggs and embryos had become accessible to radical experimentation and,—potentially,—exploitation. Since interactions between nucleus and cytoplasm are of crucial importance in regulating function, reproduction, and differentiation in all nucleated cells, the results of nuclear transplantation experiments involving eggs and embryos are often of more general biological significance. Dolly the cloned sheep produced by Wilmut and company is the supreme example so far.

From the first successful sheep embryo cloning experiment it has been evident that enucleated mammalian eggs have an amazing ability to control and modify transplanted nuclei. Enucleated eggs have therefore remained at the very center of nuclear transplantation research. The term "egg" is used here to cover the whole succession of developmental, maturational, and immediately postfertilizational (or rather post*activational*) stages of the female germ cell. There are a number of reasons why the egg is almost uniquely suited for such studies, including the following:

1. Maturation of the egg and activation of embryonic development, which together span three cell divisions with their associated nuclear and cytoplasmatic processes, are strictly controlled and precisely timed events that may be triggered and at any point temporarily interrupted in the laboratory.
2. The egg is relatively large and therefore more easily micromanipulated and analyzed than most other mammalian cells; the huge cytoplasm volume also tends to overwhelm any disturbing effect of the small amount of cytoplasm that almost inevitably is imported along with the transferred nucleus.
3. Cytoplasm from the same egg may even be divided into smaller fragments, which may be used or analyzed individually.
4. The eggs of a number of species are by now relatively easy to maintain in culture in the laboratory.
5. The eggs has a huge developmental potential, which may be tested and analyzed in considerable detail.

The basic embryo manipulation and culture techniques have probably only been improved marginally since the mid-eighties, despite various refinements in instrumentation and media. However, the introduction of molecular techniques—and talent—has facilitated a more discriminating and analytical approach, so that experiments may now be designed to answer increasingly pointed and specific questions.

The two studies referred to in the introductory paragraph represent this second, or perhaps even third generation of nuclear transplantation experiments involving eggs or embryos. Both focus on the ability of the enucleated mature unfertilized egg to synchronize and, more importantly, reprogram a nucleus taken from a cell that does not normally give rise to an egg or an embryo, so that the nucleus may function not as the original egg nucleus, but as that of a 1-cell embryo (or zygote) which normally derives half of its chromosomes from the unfertilized egg while the other half is supplied by the fertilizing sperm cell. On top of that, and most important in the present context: the donor nuclei used in both studies were not from relatively undifferentiated embryonic or fetal cells, but from more highly differentiated cells from mature animals. It all adds up to a very tall order, at least according to conventional, premammalian nuclear transfer wisdom, which predicted failure of any such attempt.[11] Until the birth of Dolly, this prediction was without a doubt the single most effective deterrent against mammalian cloning at the operational level: why bother with it if it could not even be done in frogs?

Very few mammalian embryologists bothered or maintained an active interest in cloning by nuclear transplantation, and by the time Dolly was born, this interest had if anything dwindled due partly to the difficulties encountered in implementing embryo cloning in commercial cattle breeding. Against the latter sea of trouble, cloning with somatic cell nuclei was not the arm I myself reached for. For although there was no hard experimental evidence in the sheep and cattle work to exclude the possibility of successful cloning with cell nuclei from mature animals (rather, the results and common sense kept this possibility open),[10] it would clearly be a long shot from a practical point of view, with no guarantee of success, and even if such an attempt was successful, it would almost certainly entail a lower efficiency than that obtainable with nuclei from early embryos. Besides, alternative strategies were already in place that would, in effect, allow the same practical result to be achieved, namely the production of animals that were genetically identical to an already existing animal deemed to be of outstanding quality. In fact, this had already been achieved in sheep in 1979 by combing embryo splitting and the freezing of one of the resulting half-embryos.[12] A similar approach was later used successfully with cloned cattle embryos. So much for practice!

But there is science too! One could argue—as I would—that one good starting point for experimental research is an objective that has been declared impossible by authorities in the field. It is to Ian Wilmut and his workers' lasting credit that they chose, or accepted, such an objective and stuck with it.

Both Wilmut and company and Wakayama et al. assume that reprogramming of the transplanted nucleus is most readily achieved if the latter comes from a cell in a particular—relatively inactive—phase of the cell cycle: the G0 phase. The exact nature of the reprogramming is still unknown, and it is still unclear which cells from a mature mammalian organism can serve successfully as nucleus donors in cloning. One suspects that the current ultra-short list will grow—perhaps even exponentially so—in the next few years. Wilmut and company chose cells originating from a sheep's udder, but propagated in the laboratory, while Wakayama et al. and used uncultured cumulus cells—the cells that normally envelop the maturing egg. It is tempting to speculate that there is a connection between the greater rate of success experienced by Wakayama et al. and their use of nucleus donor cells which had, at least indirectly, been involved in the programming of an egg nucleus, and might therefore already have experienced some of the very same influences as those to which a maturing egg is exposed. On the other hand, no study so far—and there have been quite a few of them—in which the mammalian egg nucleus itself and on its own was encouraged to serve as a complete genetic blueprint for an animal has produced a viable conceptus. This may well change in the future. It is also reasonable to predict that cloning of amphibians from somatic cells of mature animals will be revisited.

Perhaps the most notable difference between the two studies, apart from the choice of donor cells and experimental animal, is in the timing of activation of development. Whereas in Wilmut and company's experiments the composite eggs were immediately activated to behave like embryos, those of Wakayama et al. were maintained for a few hours in the arrested state (metaphase of the second meiotic division) characteristic of mature unfertilized eggs before activation. In other words: Wilmut and company's tactic was to start with producing the equivalent of *1-cell embryo*, whereas Wakayama et al's tactic was to start with producing the equivalent of *a mature unfertilized egg* which was only subsequently encouraged to develop into an embryo, *parthenogenetically*, so to speak. In both instances the mosaic cell resulting from the nuclear transplantation contained the correct number of chromosomes for a 1-cell embryo about to begin its development, although the states of the chromosomes differed. The mature egg normally expels half of its chromosomes in a small cell (the second polar body) as soon as it has been activated whether by a penetrating sperm cell or artificially. This tendency (although only under certain circumstances with the correct segregation of the chromosomes) is maintained even in eggs produced by nuclear transplantation unless they are activated immediately. To ensure that all the chromosomes of the transferred nucleus would remain within the egg after activation, Wakayama et al. included a special compound (Cytochalasin) which prevents division of the cell, essentially without interfering with nuclear events, in the activation medium. Wakayama and his colleagues think that there is a connection between the particular state the chromosomes (are forced to) assume in the cytoplasm of the mature nonactivated egg and the successful reprogramming of the transferred nucleus. That is possible, but under the circumstances, the postmaturation age of the recipient egg at activation, rather than the timing of activation relative to the nuclear transfer, might be invoked as an equally decisive factor in the successful reprogramming of the nucleus.

However, the deliberate delay of activation *is* interesting. This approach was in fact used by Wilmut and his colleagues in a previous study—and the Roslin Institute, where the work took place, filed a patent application concerning the use of unactivated oocytes as cytoplast recipients for nuclear transfer in 1996. So did I—in 1992. Others may very well have developed the same or closely similar ideas and procedures, indeed, it would have been hard for anyone working intensely with nuclear transplantation involving eggs and embryos for any length of time not to. For once the basic techniques of oocyte and embryo culture, micromanipulation and nuclear transplantation are mastered, a number of interesting experimental possibilities and potential practical applications immediately present themselves, among which are the above. This is not intended to detract from the achievements of Wakayama et al., or indeed those of Wilmut and company, but rather to emphasize that it is not so much the originality of the ideas, but their realization, that counts at this stage. And perhaps rightly so, for the road to realization is almost invariably rocky and tortuous, while the basic idea is usually straightforward and well-trodden.

With regard to the techniques used by the two groups, there are some additional differences even when one allows for those that are necessitated by the choice of experimental animals. For instance, Wilmut and company used electroporation for simultaneous fusion of donor cell and recipient egg and activation of the egg, and they employed a sheep as a temporary incubator before the nuclear transplant embryos were transferred to definitive foster mothers. This is the procedure originally developed for cloning of sheep embryos.[8] Wakayama et al., on the other hand, used a nuclear transplantation technique originally developed for injection of sperm directly into unfertilized eggs.[13] The eggs were later activated by exposure to a medium containing strontium, followed by culture of the resulting embryos in the laboratory for a varying period of time before transfer into foster mothers. It is impossible to precisely assess the effect of these differences and others that may have existed, but are not obvious from the written reports. However, they probably played a relatively minor role in the qualitative results.

In many ways the two studies are more remarkable for their similarities than for their differences. The most central and important similarities concern the principles employed and the outcome—the demonstration, in both instances, that it is possible to produce viable embryos by combining nuclei from certain somatic cells derived from a mature mammalian organism (kelp) with enucleated mature, unfertilized eggs of the same species, and thus, to all intent and purpose, to clone that mature mammal. It is fitting and thought provoking, although probably coincidental, that the first mature mammals to be cloned in this way should be female—although, alas, dead ones—for the results could be seen also as having opened the way to parthenogenesis in mammals.

Wilmut and company, of course, hold priority with regard to demonstrating that a fully viable embryo can be produced by transplantation of a nucleus from a somatic cell derived from a mature mammal. The importance of this demonstration is well known, if not fully understood. However, Wakayama et al. deserve credit, not only for being the first to confirm that general and very significant observation, but also for delivering the confirmation in what appears to be a more repeatable form—and in a species that until then had been considerably less cooperative as regards cloning than the meek sheep. Above all, the two studies, and many others in recent years, serve to remind us what fascinating cells mammalian eggs are and what dynamic and powerful, yet reliable and forgiving subjects of exploration and experiment, they have become.

With regard specifically to cloning: there are several other ways in which mammals may be cloned more effectively and reliably than was the case in the two studies dealt with here.[10] Furthermore, as pointed out by a number of authors, nuclear transplantation does not generally lead to cloning in the strictest sense. In order to achieve the latter, one would have to use a recipient egg from a female whose extranuclear DNA—which represents a relatively small number of genes concerned primarily with energy metabolism and located in the mitochondria—was identical to that of the nucleus donor (ideally the nucleus donor itself, its mother or another member of the same female line). While the recipient oocytes were often from a single known oocyte donor in the early livestock embryo cloning experiments, so that any offspring produced would be genetically identical also with respect to extranuclear DNA, that is now rarely the case. On the other hand, with the approach used by Wakayama et al. the same female might easily be used as the donor of both nuclei and recipient eggs. Of course, these considerations do not affect the general significance and importance attached to the birth of Dolly the sheep and the swarm of cloned mice.

Although things have progressed spectacularly since the early eighties, it is worth keeping in mind that the original sheep embryo cloning procedure, on which one may fairly say that most current mammalian cloning work is based, is by far the most efficient way of cloning mammals by nuclear transplantation even today. It uses very young embryos (usually between the 8-cell and the 64-cell stage) as nucleus donors. The nucleus donor cells may constitute only a sample taken from a parent embryo without overtly affecting the subsequent development of the latter. It was also shown very early on that embryos that are themselves the products of cloning by nuclear transplantation may be used as nucleus donors.[8,10] Whatever the precise history of the embryonic nucleus donor cells is, it is assumed that they are relatively undifferentiated and therefore require minimal reprogramming. Although the success rate varies widely, it can reach 100 percent. By that is meant that all the nuclear transplant embryos produced from one particular donor embryo may be viable. That happens rarely, and there are other more worrying problems that plague either the procedure as a whole, or its constituent parts. Nevertheless, it would probably be scientifically rewarding at this point to redirect some of the excellent efforts and talents currently expended in attempts to clone with more highly differentiated cells back to these undifferentiated embryonic cells. It would also serve some very sensible practical purposes.

Mammalian cloning has come a long way from its (relatively) innocent beginnings, whether one considers them scientific or pastoral. Other interests and masters are being served now than at the outset. As we have been told repeatedly, the primary aim is no longer selective breeding of livestock, but production of genetically engineered animals for the pharmaceutical industry, and of replacement tissues, even entire organs—for all of us, presumably, should we need any: the mind boggles! It will take time and ingenuity to realize these new objectives; but they are by no means over-ambitious from a scientific point of view. Nor is it probable that their realization will pose any real threat to individuality or humanity, at least not compared to developments already implemented or under way in other areas. They might even be of some benefit. Yet they are somehow laughable—as are many other preoccupations of ours. More sinister, perhaps, is the suggestion that since many of the basic mechanisms currently being harnessed are of a general nature, they ought to be usefully employed across species boundaries. For instance, by transfer of somatic nuclei into enucleated cow eggs it might be possible to produce patient specific human stem cells from which replacement cells, tissues, and organs could be engineered. This may give most of us pause—but probably not for too long! Indeed, why stop at cow eggs? What is wrong with frog eggs (several reports on the use of frog eggs as recipients of human nuclei were in fact published back in the mid-seventies [see for instance Gurdon et al.][12] without causing any noticeable reaction on the part of the mass media or, for that matter, anybody else outside their scientific readership) or insect cells? And so on via Protista to Planta! Although these possibilities may seem bizarre and outlandish when presented out of a reasoned scientific context, the future may well provide such a context, and a reasonable and perfectly acceptable one, not only for them, but also for others as yet undreamed of. Here, as elsewhere, it is important to keep an open mind.

One thing seems certain: The colossal efforts and resources that are currently being devoted to molecular genetics and molecular biology generally make the ultimate biotechnical success of mammalian cloning in various forms by nuclear transplantation a near certainty. Already a more differentiating terminology is beginning to appear in discussions concerning the future of human cloning. For instance, a distinction is now made between "reproductive" and "therapeutic" cloning. [Editor's note: See chapter 27.] At present such a distinction, based on intent rather than substance, may be somewhat questionable. However, the miniaturization and refinement of technique which has allowed the leap from experiments with amphibian eggs to similarly individualized experiments with their mammalian cousins are far from having reached their ultimate limits, and besides, the factors and processes which bring about differentiation and dedifferentiation are molecular in nature, and there is no particular reason to believe that any of them are, let alone will remain, exclusive to eggs and embryos. Therefore, it is not too difficult to envisage human cloning of the "therapeutic" variety being carried out at an even more refined and sophisticated level, with cells and cell components that do not exhibit developmental totipotency as glaringly and controversially as do eggs and embryos. Indeed, future human cloning of the "therapeutic" variety may not involve eggs or embryos at all. Be that as it may, with continued research, more fascinating insights will be gained, and more interesting, wonderful, laughable, or frightening possibilities will present themselves. It is only to be expected that at this relatively early stage confusion and uncertainty should reign with regard to the potential and proper uses of the new reproductive and genetic discoveries and technologies. At least until now, embryologists have on the whole been remarkably reluctant to engage in human cloning of any kind, although the biotechnical possibility existed virtually from the beginning of the human in vitro Fertilization in the late seventies.[13] That is not to argue that human cloning will not occur in the future. On the contrary, I think this is very likely to happen sooner or later, in one form or another. But apart from this, the major decisions concerning the future of mammalian cloning are now in the hands of politicians—a great comfort, I am sure!

We are, of course, under no obligation to use cloning for anything practical or controversial whatsoever, but I think most people, once they have recovered from the initial shock, will recognize that the insights gained so far from mammalian cloning experiments have contributed in a positive way to our intellectual universe, whatever new shocks may await us. Whether in the long run cloning will alter the human condition quite as drastically as anticipated by some or feared by many, remains to be seen. I have my doubts.[14]

NOTES

1. Wilmut, I., Schnieke, A. E., McWhir, J., Kind, A. J., & Campbell, K. H. S. Viable offspring derived from fetal and adult mammalian cells. *Nature* 385, 810–813 (1997).

2. Wakayama, T., Perry, A. C. F., Zucotti, M., Johnson, K. R., & Yanagimachi, R., Full term development of mice from oocytes injected with cumulus cell nuclei. *Nature* 394, 369 (1998).

3. Graham, C. F., Virus assisted fusion of embryonic cells. *Karolinska Symposium on Research in Reproductive Endocrinology*, 154–165 (1971).

4. Briggs, R., & King, T. J., Transplantation of living nuclei into enucleated frog eggs. *Proc. Nat. Acad. Sci. U.S.A.* 38, 455–463 (1952).

5. Bromhall, J. D., Nuclear transplantation in rabbit eggs. *Nature* 258, 719–721 (1975).

6. Illmensee, K., & Hoppe, P. C., *Cell* 23, 9– (1981).

7. McGrath, J., & Solter, D., Nuclear transplantation in the mouse embryo by microsurgery and cell fusion. *Science* 22, 1300–1302 (1983).

8. Willadsen, S. M., Nuclear transplantation in sheep embryos. *Nature* 320, 63–65 (1986).

9. Willadsen, S. M., A method for culture of micromanipulated sheep embryos and its use to produce monozygotic twins. *Nature* 277, 298–300 (1979).

10. Willadsen, S. M., Cloning of sheep and cow embryos. *Genome* 956–962 (1989).

11. Gurdon, J. B., Partington, G. A., & Robertis, E. M., Injected nuclei in frog oocytes: RNA synthesis and protein exchange. *J. Embryol. exp. Morph.* 36, 541–553 (1976).

12. Willadsen, S. M., The viability of early cleavage stages containing half the normal number of blastomeres in the sheep. *J. Reprod. Fertil.* 59, 357–352 (1980).

13. Kimura, Y., & Yanagimachi, R., Intracytoplasmic sperm injection in the mouse. *Biol. Reprod.* 52, 709–720 (1995).

14. This chapter was written in the autumn of 1998. Since then several more laboratories in the United States and elsewhere have reported cloning of mammals from somatic cells. Two or three more species have been added to the list, and two or three more cell types. More will surely follow. I have decided to leave the chapter as it was, even though the field is developing at a fair pace. So far, neither man's closest relatives nor his best friends have been cloned, although for both groups the matter has progressed well beyond mere declarations of intent. As for ourselves, the near completion of the human genome project and the predictable shift in research interest and effort from gene structure to gene function has helped to boost the interest in human embryonic stem cells and nuclear transplantation, and hence in cloning (*therapeutic* cloning, that is). It would be beyond the scope of the original commission and the capabilities of this author to deal comprehensively with mammalian cloning in the light of these developments. Suffice it to say that we are unlikely to run out of exciting developments and heated debate in that area for some years to come.

4

Dolly Mice

Anne McLaren

Dolly the cloned sheep was derived by nuclear transfer to an enucleated oocyte, using a nucleus from a cell culture grown from mammary gland tissue taken from a 6-year-old pregnant ewe.[1] Doubts expressed by a few scientists as to the provenance of the progenitor nucleus have recently been allayed by microsatellite analysis and DNA fingerprinting of Dolly's DNA and that of the donor ewe.[2,3] Dolly herself appears healthy and sociable, with no reported signs of premature aging. She has given birth to six lambs conceived by entirely conventional means.

One solitary sheep could never have sparked such a hubbub of ethical and social speculation had the clone donor not been an *adult* animal. Cloning by nuclear substitution using nuclei derived from early embryos had been successfully carried out in a number of mammalian species for several years. From the scientific point of view, the cloned lambs derived from nuclei taken from cultures of differentiated *fetal* fibroblast cells, reported along with Dolly,[1] as well as the subsequent U.S. claim of two calves similarly derived from fetal fibroblasts[4], were as important as Dolly herself. The nucleus from any differentiated cell has a characteristic pattern of gene expression, with some genes switched on and others switched off. It is the abolition of this pattern by transfer into an enucleated oocyte, and the reprogramming of the nucleus to support a whole new cycle of embryonic development, that biologists find so unexpected and intriguing. But from the point of view of human interest, the clone of a fetus could never compete with the clone of a long-dead 6-year-old ewe.

But why only one Dolly, the public asked. Why was the work not repeated, in sheep or in some other species? Was Dolly destined forever to stand alone, the product of some miraculous never-to-be-repeated concatenation of circumstances? Some scientists took the view that the work should never have been accepted for publication until it had been repeated. Those who were disappointed that the findings were not instantly replicated in other labs had little understanding of the time and effort that has to intervene between the initiation of a scientific project and its eventual publication.

OF MICE AND SHEEP

A mere 17 months after the Dolly paper, another paper appeared in *Nature*, reporting the successful cloning of mice by nuclear transfer, again using nuclei derived from adult cells.[5] A few died at or soon after birth (this is not unexpected, since for technical reasons all were delivered by Caesarian section), but more than twenty "Dolly mice" developed into healthy fertile adults. Many more have been produced since the paper was submitted, including male mice cloned from the tail tip of an adult.[6] Successes have also been reported in cattle, including a report of eight calves cloned from a single adult cow,[7] and more recently in goats[8] and pigs.[9] Thus the reproducibility of cloning adult mammals by nuclear transfer is no longer in question, although the success rate for all species remains low.

There are important differences between the methods used to clone sheep and the methods used to clone mice, as follows:

1. In sheep, the adult mammary gland cells as well as the fetal fibroblasts were first used to grow primary cell cultures. Nuclei for cloning experiments were derived from these cultures, with the cells rendered quiescent by reducing the concentration of serum in the culture medium. The main objective of the sheep program was not cloning itself, but rather devising a technique for genetic manipulation of farm animals, in particular gene targeting. For this reason a cell culture strategy was essential. In mice, an effective technique for genetic manipulation and gene targeting already exists, so that the motivation was entirely to explore the basic biology of the nuclear reprogramming that must accompany cloning by nuclear substitution. Nuclei for cloning were therefore taken directly from cumulus cells. These cells surround each developing egg in the ovary and are shed with it at ovulation. After ovulation, more than 90% of the cumulus cells are already in the quiescent stage (G0/G1) of the cell cycle.

2. In sheep, the donor nucleus was then inserted into the enucle-

ated unfertilized egg using an electrical pulse to fuse the small cultured cell with the large egg. In this way the entire contents of the donor cell, including the mitochondria that are involved in energy metabolism, were transferred to the egg. In mice, a piezo-electric pipette was used for the injection of the donor nucleus, which had previously been removed from the cumulus cell. Using this method, the transfer of donor mitochondria and other cytoplasmic components into the egg was minimal.

3. In sheep, the same electrical pulse that induced fusion of the donor cell with the egg induced simultaneous activation of the egg, initiating embryonic development. In mice, the eggs were activated chemically. The best results were obtained when the activation stimulus was given not at the time that the nucleus was transferred, but 1–3 hours later. This interval may have been helpful in allowing more time for the donor nucleus to be reprogrammed before embryonic development began.

WHY ARE DOLLY MICE SO EXCITING?

As a mouse worker for many years, I must declare a vested interest. But few would question that the mouse is the animal par excellence for the study of the molecular basis of development.

- The Mouse Genome Project is not far behind the Human Genome Project.
- Using embryonic stem cell technology, genes in mice can be added, removed, or replaced. Nuclei from mouse ES cells, including targeted ES cells, have recently been used to produce cloned mice.[10,11]
- The time and place of gene expression can be experimentally manipulated.
- Mouse model systems have contributed extensively to both agricultural and medical research, not least by providing experimental models for human diseases.
- As mammals go, mice have a short generation time and low maintenance costs.
- Background knowledge of their reproductive biology, genetics, and embryology is unrivaled.

WHAT HAVE THE CLONED MICE SHOWN US SO FAR?

- They were the first to show that the feasibility of somatic cell nuclear transfer cloning from adult mammals is not confined to one species, nor to one sex.

- In sheep, the embryonic genome is not required to direct development until the 8–16 cell stage, but in mice it begins to direct
development already at the 2-cell stage. The shorter interval for
reprogramming after nuclear substitution appears adequate to
support the development of cloned embryos.
- In terms of adult mice or sheep produced, viewed as a proportion
of the number of unfertilized eggs used for nuclear substitution,
the mouse procedure is more efficient (0.89% versus 0.36%), perhaps because the piezo-electric technology allows the nuclear
substitution to be carried out more rapidly. In terms of adults
surviving as a proportion of embryos transferred, the mouse procedure is less efficient (1.59% versus 3.45%), but since the sheep
figure is based on a single animal (Dolly), the comparison is not
very meaningful. Given the speed with which variant procedures
can be assessed, efficiency might be expected to improve more
rapidly with the mouse system.
- Some of the mouse clones derived from cumulus cell nuclei
taken from females that were themselves clones. The success
rates for producing first and second generation clones were similar. This suggests that clones do not undergo changes that affect
a subsequent cloning procedure for either better or worse.
- Cumulus cells are terminally differentiated: they do not divide
further, nor develop into any other cell type. The cultured mammary gland cell responsible for Dolly may not have been terminally differentiated: it could for example have been an epithelial
stem cell. Thus the Dolly mice were the first proof that the terminally differentiated state can still be amenable to successful
reprogramming by unfertilized egg cytoplasm.
- Neither neuronal nuclei nor Sertoli cell nuclei proved successful
as nuclear transfer donors.[5] Since both neurons and Sertoli cells
are in the G0/G1 stage of the cell cycle, it appears that quiescence
is not a sufficient criterion to ensure that the nucleus is open to
reprogramming.

WHAT MIGHT THEY SHOW US IN THE FUTURE?

Virtually all of the basic biological questions raised by Dolly will be more
readily investigated in mice than in sheep. For example:

- Do nuclei taken from different adult cell types really differ in
their capacity to be reprogrammed by transfer to an enucleated
unfertilized egg? If so, why?
- How does the cell cycle stage of the donor nucleus affect its capacity to be reprogrammed? Is quiescence essential? Why?

- How is the shortening of the telomeres at the ends of the chromosomes that accompanies cell division in normal development repaired by transfer to an enucleated unfertilized egg? Dolly's telomeres appear[12] to be shorter than expected for a sheep of her age, perhaps reflecting the fact that her progenitor was already six years old. On the other hand calves cloned from nuclei taken from fetal fibroblasts cultured to senescence are reported to have unusually long telomeres.[13] Will cloned mice, or clones of clones of clones, exhibit shorter or longer telomeres? Will they undergo premature aging, or perhaps have an extended life-span? When Cumulina, the first cloned mouse, died in May 2000 she was more than two years old, a ripe old age for a mouse.
- Will somatic mutations or chromosome damage in the donor nucleus result in cancer or other abnormalities in the cloned animal? Will these be more common in the clones of older progenitors?
- What genes are switched off and which are activated during reprogramming? How does chromosome conformation change?
- What elements in the egg cytoplasm are required to achieve this reprogramming, and how does it work?
- Would immature eggs, matured in vitro, be equally effective in bringing about nuclear reprogramming?
- Would eggs of another species achieve nuclear reprogramming? If so, how distant a species could be used?
- Will an increased understanding of nuclear reprogramming throw light on the transformation of malignant cancer cells?
- Will genomic imprinting, in which the expression of genes during development is influenced by their recent exposure to gametogenesis, be affected by absence of any recent exposure to gametogenesis?
- If mitochondria are transferred along with the nucleus, will the mix of mitochondria be maintained or will one or other population win out? The mitochondria of Dolly herself, and another 9 sheep cloned from fetal fibroblasts, are reported to be derived exclusively from the host oocytes.[14]
- Is the low efficiency and high embryonic mortality that at present characterizes nuclear transfer cloning a result of incomplete nuclear reprogramming? If not, what is its cause? Can it be avoided?
- How similar are clones to each other and to their progenitor? In their immune responses? In their behavior?

IMPLICATIONS FOR HUMAN CLONING

There were those who believed that cloning of adult animals by nuclear substitution would work only in sheep and perhaps cows, because of the relatively late stage at which the embryonic genome starts to control development in ruminants. There were those who believed that it would be unlikely to work in mice; and there were those who hoped or feared that it would never work in humans.

The fact that it has now been shown to work in mice, as well as in cattle, goats and pigs, increases somewhat the likelihood that it might work in other species, including humans. The Texas multimillionaire who is said to have promised a laboratory 2.3 million dollars if it succeeds within two years in cloning his pet dog, Missy the mongrel, may have to pay up. However, the efficiency is still exceedingly low, and many of the cloned mouse embryos die at or before birth. Future research on mice is likely to reveal the causes of this high failure rate, whether it can be ameliorated or whether it is inherent in the nuclear reprogramming that lies at the heart of the procedure.

On balance the emergence of Dolly mice is encouraging to those who look forward to the possibility of growing immunologically compatible cell lines for people whose tissues or organs have been damaged by trauma or by degenerative disease. The *in vitro* research that would be required to realize this possibility could be carried out only in countries where human embryo research for therapeutic purposes was not prohibited. Such research would not involve reproductive cloning, that is, the cloned embryo would not develop into a fetus or baby.

Future research on mouse cloning may include investigations into the properties of aggregation chimeras made with cloned embryos. When two genetically distinct 8-cell embryos are aggregated, their cells mingle, and the two components contribute in differing proportions to every part of the body.[15] Each chimera is therefore different from every other chimera. If cloned embryos showed this same pattern of development after aggregation, the resulting "cloneras" would of course not themselves be clones, since they would not be genetically identical to any other clonera, nor to either of their progenitors. They would exhibit a mix of the characteristics of the two progenitors, with new characteristics appearing through interaction between the two genetically different components, in an analogous way to normal postfertilization development. Many of the present ethical objections to human reproductive cloning would therefore not apply to human cloneras. Unfortunately, however, even if problems of safety and efficiency were eventually to be overcome, the approach could be used only for progenitors of the same sex, since chimeras made between male and female mouse embryos show numerous abnormalities of the reproductive system.

How are the prospects of human reproductive cloning, in the sense of cloning that produces a cloned fetus or baby, affected by the advent of Dolly mice? In those parts of the world where human reproductive cloning is already prohibited by law or by regulation, the recent findings are irrelevant. Elsewhere, however, there is now an increased need for society and professional and governmental decision makers to determine their stance on human reproductive cloning, and to take appropriate regulatory measures.

NOTES

The author gratefully acknowledges support from the Wellcome Trust.

1. Wilmut, I., Schnieke, A. E., McWhir, J., Kind, A. J. & Campbell, K. H. S. (1997). Viable offspring derived from fetal and adult mammalian cells. *Nature* **385**, 810–813.

2. Ashworth, D., Bishop, M., Campbell, K., Colman, A., Kind, A., Schnieke, A., Blott, S., Griffin, H., Haley, C., McWhir, J. & Wilmut, I. (1998). DNA microsatellite analysis of Dolly. *Nature* **394**, 329.

3. Signer, E. N., Dubrova, Y. E., Jeffreys, A. J., Wilde, C., Finch, L. M. B., Wells, M. & Peaker, M. (1998). DNA fingerprinting Dolly. *Nature* **394**, 329–330.

4. Cibelli, J. B., Stice, S. L., Golueke, P. J., Kane, J. J., Jerry, J., Blackwell, C., Ponce de Leon, F. A. & Robl, J. M. (1998). Cloned transgenic calves produced from nonquiescent fetal fibroblasts. *Science* **280**, 1256–1258.

5. Wakayama, T., Perry, A. C. F., Zucotti, M., Johnson, K. R. & Yanagimachi, R. (1998). Full-term development of mice from enucleated oocytes injected with cumulus cell nuclei. *Nature* **394**, 369–374.

6. Wakayama, T. & Yanagimachi, R. (1999). Cloning of male mice from adult tail-tip cells. *Nat. Genet.* **22**, 127–128.

7. Kato, Y., Tani, Sotomaru, Y., Kurokawa, K., Kato, J., Doguchi, H., Yasue, H. & Tsunoda, Y. (1998). Eight calves cloned from somatic cells of a single adult. *Science* **282**, 2095–2098.

8. Baguisi, A., Behboodi, E., Melican, D. T., Pollock, J. S., Destrempes, M. M., Midura, P., Palacios, M. J., Ayres, S. L., Denniston, R. S., Hayes, M. L., Ziomek, C. A., Meade, H. M., Godke, R. A., Gavin, W. G., Overstroum, E. W. & Echelard, Y. (1999). Production of goats by somatic cell nuclear transfer. *Nat. Biotechol.* **17**, 456–461.

9. Polejaeva, I. A., Chen, S. H., Vaught, T. D., Page, R. L., Mullins, J., Ball, S., Dai, Y., Boone, J., Walker, S., Ayares, D., Colman, A. & Campbell, K. C. (2000). Cloned pigs produced by nuclear transfer from adult somatic cells. *Nature* **407**, 86–90.

10. Wakayama, T., Rodriguez, I., Perry, A. C. F., Yanagimachi, R. & Mombaerts, P. (1999). Mice cloned from embryonic stem cells. *Proc. Nat. Acad. Sci.* **96**, 14984–14989.

11. Rideout, W. M., Wakayama, T., Wutz, A., Eggan, K., Jackson-Grusby, L., Dausman, J., Yanagimachi, R. & Jaenisch, R. (2000). Generation of mice from wild-type and targeted ES cells by nuclear cloning. *Nat. Genet.* **24**, 109–110.

12. Shiels, P. G., Kind, A. J., Campbell, K. H. S., Waddington, D., Wilmut, I., Colman, A. & Schieke, A. E. (1999). Analysis of telomere lengths in cloned sheep. *Nature* **399**, 316–317.

13. Lanza, R. P., Cibelli, J. B., Blackwell, C., Cristofalo, V. J., Francis, M. K., Baerlocher, G. M., Mak, J., Schertzer, M., Chavez, E. A., Sawyer, N., Lansdorp, P. M. & West, M.D. (2000). Extension of cell life-span and telomere length in animals cloned from senescent somatic cells. *Science* **288**, 665–668.

14. Evans, M. J., Gurer, C., Loike, J. D., Wilmut, I., Schnieke, A. E. & Schon, E. A. (1999). Mitochondrial DNA genotypes in nuclear transfer-derived cloned sheep. *Nat. Genet.* **23**, 90–93.

15. McLaren, A. (1976). Mammalian chimaeras. Cambridge University Press: Cambridge.

5

Thinking Twice, or Thrice, about Cloning

Lee M. Silver

THOSE WONDERFUL CUMULUS CELLS

The patient is 40 years old. Her name is Jayne, and she is a divorced mother of an 8-year-old son. Jayne has come to the fertility clinic because she wants to get pregnant again—she wants to have a second child before her body says it's too late. Since her divorce, Jayne has dated various men, but none of these relationships was very serious. Indeed, none of the men she's slept with since she left her husband have met the very high standards she's set for the ideal father of her next child. Jayne has decided that she can no longer wait for Mr. Right to enter her life. Menopause can strike at any moment.

Jayne could have purchased sperm from a high-class Internet sperm bank to initiate her pregnancy. She could have chosen the characteristics of the sperm vendor from pull-down menus—height, skin tone, SAT scores, and more. But even with all this information, she would still be given her child unknown genes and, perhaps, hidden disease traits. Luckily, though, Jayne has a choice. She doesn't need to mix the genes of some anonymous man with those of her own to bring forth a child. With the new improved nuclear transfer technology, she can do it all by herself. And she asks herself, "Why not? Why give my child some stranger's genes when I don't have to?"

And so it has come to pass that Jayne now finds herself in the patient room of a bustling big city fertility center. Jayne is still quite fertile, and the hormones she has taken have caused 23 fluid-filled pimples to appear on her ovaries. One by one, each of those pimples is punctured, and the single ripe egg lying within is sucked out. The eggs are handed off to the highly skilled micromanipulation technician who releases them into a petri dish for viewing under the microscope.

Each egg looks like a peach surrounded by caviar. The caviar actually represents hundreds of tiny cumulus cells that normally provide a protective coating for eggs as they travel down the reproductive tract from Fallopian tube to uterus. But while the cumulus cells are small, they still have the entire complement of genes encoded within DNA that Jayne received from her mother and father before.

In the old days, before nuclear transfer became part of the fertility center's repertoire of techniques, the cumulus cells were simply a nuisance. They blocked the journey of sperm to the egg and had to be pushed aside to increase the chances of successful IVF. For the purposes of nuclear transfer, however, the cumulus cells are a gift made in heaven. How convenient to have donor and recipient cells micrometers apart from each other, to be recovered simultaneously from the woman desiring the procedure. How convenient that the cumulus cell arrive in the petri dish already in the dormant stage needed to act as high-efficiency donors.

What the technician does next is subtle and remarkable. First, he uses a tiny glass needle to remove the ball of DNA from the giant egg cell, so that it is now nothing more than a bag of cytoplasm. Next, he moves the needle over ever so slightly to one of the cumulus cells, and grabs hold of the small ball of genes present within. He moves the needle back again, and injects this ball of genes into the large bag of cytoplasm.

He repeats the same procedure with each caviar-coated peach until he has produced 23 new cells. He then places these cells into a specially prepared medium, and six turn into embryos. Two are chosen for reintroduction into Jayne's uterus that very same month. The rest are frozen away for future use, if needed.

The use of cumulus cells as *prêt-à-porter* donors was validated with their use to produce dozens of cloned mice in the summer of 1998 in Honolulu. Of course, as this book goes to press, the story just told is still fiction, but how long will it stay that way? In October 1998, it was announced that embryos produced by nuclear transfer between two unfertilized human eggs had already been introduced into the wombs of two women.[1] Although children who emerge from such a protocol will not be clones, the technology required to perform cloning is essentially the same—instead of donor nuclei from eggs, donor nuclei are obtained from adult cells.

I would be highly remiss if I didn't state here that I believe it would be unethical to attempt nuclear transfer with human cells before the safety of the procedure is demonstrated to be at the same level as normal IVF protocols, for both woman and child-to-be. But this hasn't held back other advances in reproductive technology, including ICSI (intracytoplasmic sperm injection) or the aforementioned nuclear transfer between egg cells.

At some point in the future, the story told here will come true. To believe otherwise is to misunderstand the power of the marketplace and the power of individual desires to reach very specific individual reproductive goals. These will be the driving forces behind the development and use of cloning technology. Yet there are many who believe that humanity will be destroyed when such a thing comes to pass. Why?

MISUNDERSTANDING CLONING

On Sunday morning, February 23, 1997, the world awoke to a technological advance that shook the foundations of biology and philosophy. On that day, we were introduced to Dolly, a 6-month-old lamb who had been cloned directly from a single cell taken from the breast tissue of an adult donor. Perhaps more astonished by this accomplishment than any of their neighbors were the scientists who actually worked in the field of mammalian genetics and embryology. Outside the lab where the cloning had actually taken place, most of us thought it could never happen. Oh, we would say that perhaps at some point in the distant future, cloning might become feasible through the use of sophisticated biotechnologies far beyond that available to us now. But what many of us really believed, deep in our hearts, was that this was one biological feat we could never master. New life—in the special sense of a conscious being—must have its origins in an embryo formed through the merger of gametes from a mother and father. It was impossible, we thought, for a cell from an adult mammal to become reprogrammed, to start all over again, to generate another entire animal or person in the image of the one born earlier.

How wrong we were.

Of course, it wasn't the cloning of a sheep that stirred the imaginations of hundreds of millions of people. It was the idea that humans could now be cloned as well, and many people were terrified by the prospect. Ninety percent of Americans polled within the first week after the story broke felt that human cloning should be outlawed[2] And while not unanimous, the opinions of many media pundits, ethicists, and policymakers seemed the same as that of the public at large. The idea that humans might be cloned was called "morally despicable," "repugnant," "totally inappropriate," as well as "ethically wrong, socially misguided and biologically mistaken."[3]

Scientists who work directly in the field of animal genetics and embryology were dismayed by all the attention that now bore down on their research. Most unhappy of all were those associated with the biotechnology industry, which has the most to gain in the short term from animal applications of the cloning technology[4] Their fears were not unfounded. In the aftermath of Dolly, polls found that two out of three Americans considered the cloning of *animals* to be morally unacceptable, while 56% said they would not eat meat from cloned animals.[5]

It should not be surprising, then, that scientists tried to play down the feasibility of human cloning. First they said that it might not be possible *at all* to transfer the technology to human cells.[6] And even if human cloning is possible in theory, they said, "it would take years of trial and error before it could be applied successfully," so that "cloning in humans is unlikely any time soon."[7] And even if it becomes possible to apply the technology successfully, they said, "there is no clinical reason why you would do this."[8] And even if a person wanted to clone themselves or someone else, they wouldn't be able to find trained medical professionals who would be willing to do it.

Really? That's not what science, history, or human nature suggest to me. The cloning of Dolly broke the technological barrier. There is no reason to expect that the technology couldn't be transferred to human cells. On the contrary, there is every reason to expect that it *can* be transferred. If nuclear transplantation works in every mammalian species in which it has been seriously tried, then nuclear transplantation *will* work with human cells as well. It requires only equipment and facilities that are already standard, or easy to obtain by biomedical laboratories and free-standing in vitro fertilization clinics across the world. Although the protocol itself demands the services of highly trained and skilled personnel, there are thousands of people with such skills in dozens of countries.

The initial horror elicited by the announcement of Dolly's birth was due in large part to a misunderstanding by the lay public and the media of what biological cloning is and is not. The science critic Jeremy Rifkin exclaimed: "It's a horrendous crime to make a Xerox (copy) of someone,"[9] and the Irvine, California, rabbi Bernard King was seriously frightened when he asked, "Can the cloning create a soul? Can scientists create the soul that would make a being ethical, moral, caring, loving, all the things we attribute humanity to?"[10] The Catholic priest William Saunders suggested that "cloning would only produce humanoids or androids—soulless replicas of human beings that could be used as slaves."[11] And the *New York Times* writer Brent Staples warned us that "synthetic humans would be easy prey for humanity's worst instincts."[12]

PROFESSIONAL OPINIONS

Anyone reading this volume already knows that real human clones will simply be later-born identical twins—nothing more and nothing less. Cloned children will be full-fledged human beings, indistinguishable in biological terms from all other members of the species. But even with this understanding, many ethicists, scholars, and scientists are still vehemently opposed to the use of cloning as means of human reproduction under any circumstances whatsoever. Why do they feel this way? Why does this new reproductive technology upset them so?

First, they say, it's a question of "safety." The cloning procedure has not been proven safe and, as a result, its application toward the generation of newborn children could produce deformities and other types of birth defects. Second, they say that even if physical defects can be avoided, there is the psychological well-being of the cloned child to consider. And third, above and beyond each individual child, they are worried about the horrible effect that cloning will have on society as a whole.

What I will argue here is that people who voice any one or more of these concerns are—either consciously or subconsciously—hiding the real reason they oppose cloning. They have latched on to arguments about safety, psychology, and society because they are simply unable to come up with an ethical argument that is not based on the religious notion that by cloning human beings, man will be playing God, and it is wrong to play God.

Let us take a look at the safety argument first. Throughout the twentieth century, medical scientists have sought to develop new protocols and drugs for treating disease and alleviating human suffering. The safety of all these new medical protocols was initially unknown. But through experimental testing on animals first, and then volunteer human subjects, safety could be ascertained and governmental agencies—such as the Food and Drug Administration in the United States—could make a decision as to whether the new protocol or drug should be approved for use in standard medical practice.

THE QUESTION OF SAFETY

It would be ludicrous to suggest that legislatures should pass laws banning the application of each newly imagined medical protocol before its safety has been determined. Professional ethics committees, Institutional Review Boards, and the individual ethics of each individual medical practitioner are relied upon to make sure that hundreds of new experimental protocols are tested and used in an appropriate manner each year. And yet the question of unknown safety alone was the single rationale used by the National Bioethics Advisory Commission (NBAC) to propose a ban on human cloning in the United States.

Opposition to cloning on the basis of safety alone is almost surely a losing proposition. Although the media have concocted fantasies of dozens of malformed monster lambs paving the way for the birth of Dolly, fantasy is all it was. Of the 277 fused cells created by Wilmut and his colleagues, only 29 developed into embryos. These 29 embryos were placed into 13 ewes, of which one became pregnant and gave birth to Dolly.[13] If safety is measured by the percentage of lambs born in good health, then the record, so far, is 100% for nuclear transplantation from an adult cell (albeit with a sample size of one). Indeed, there is no scientific basis for the belief that cloned children will be any more prone to genetic problems than naturally conceived children, but research will be necessary to confirm or reject this possibility.[14]

EFFECTS ON CHILDREN

Once safety has been eliminated as an objection to cloning, the next concern voiced is the psychological well-being of the child. Daniel Callahan, the former director of the Hastings Center [Editors note: and author of chapter 9 in the volume], argues that "engineering someone's entire genetic makeup would compromise his or her right to a unique identity."[15] But no such "right" has been granted by nature—identical twins are born every day as natural clones of each other. Dr. Callahan would have to concede this fact, but he might still argue that just because twins occur naturally does not mean we should create them on purpose.

Dr. Callahan might argue that a cloned child would be harmed by knowledge of her future condition. He might say that it's unfair to go through childhood knowing what you will look like as an adult, or being forced to consider future medical aliments that might befall you. But right now, there are children being born somewhere in the world who will mature into a spitting image of one parent or the other, just by chance. Other children will express a personality and behavior that is a replica of one parent, just by chance. And for a small number of children born every day, it will be both: a "chip off the old block," as the old saying goes. Indeed, there are sure to be people alive today, around the world, who are actually more similar in both looks ad personality to a parent than might be expected, on average, with a child who is a genetic clone! For this reason, observers will never know for sure (in the absence of DNA testing) whether a child is really a clone or just a parental look-alike.

Even in the absence of cloning, many children have some sense of the future possibilities encoded in the genes they got from their parents. Furthermore, genetic screening already provides people with the ability to learn about hundreds of disease predispositions. And as genetic knowledge and technology become more and more sophisticated, it will become possible for any human being to learn even more about their genetic future than a cloned child could learn from their progenitor's past.

It might also be argued that a cloned child will be harmed by having to live up to unrealistic expectations placed on her by her parents. But there is no reason to believe that her parents will be any more unreasonable than many other parents who expect their children to accomplish in their lives what they were unable to accomplish in their own. No one would argue that people with such tendencies should be prohibited from having children.

But let's grant that among the many cloned children brought into this world, some *will* feel bad about the fact that their genetic constitution is not unique. Is this alone a strong enough reason to ban the practice of cloning? Before answering this question, ask yourself another. Is a child having knowledge of an older twin worse off than a child born into poverty? If we ban the former, shouldn't we ban the latter? Why is it that so many politicians seem to care so much about cloning but so little about the welfare of children in general?

EFFECTS ON SOCIETY

Finally, there are those who argue against cloning based on the perception that it will harm society at large in some way. The *New York Times* columnist William Safire expresses the opinion of many others when he says that "cloning's identicality would restrict evolution."[16] This is bad, he argues, because "the continued interplay of genes . . . is central to humankind's progress." But Mr. Safire is wrong on both practical and theoretical grounds. On practical grounds, even if human cloning became efficient, legal, and popular among those in the moneyed classes (which is itself highly unlikely), it would still only account for a fraction of a percent of all the children born onto this earth. Furthermore, each of the children born by cloning to different families would be different from each other, so where does the identicality come from?

On theoretical grounds, Safire is wrong because humankind's progress has nothing to do with unfettered evolution, which is always unpredictable and not necessarily upward bound. H. G. Wells recognized this principle in his 1895 novel *The Time Machine*, which portrays the evolution of humankind into weak and dimwitted, but cuddly little creatures. And Kurt Vonnegut follows this same theme in "Galápagos," where he suggests that our "big brains" will be the cause of our downfall, and future humans with smaller brains and powerful flippers will be the only remnants of a once great species, a million years hence.

IS IT ALL RELIGION?

As is so often the case with new reproductive technologies, the real reason that people condemn cloning has nothing to do with technical feasibility, child psychology, societal well-being, or the preservation of the human species. The real reason derives from religious beliefs. It is the sense that cloning leaves God out of the process of human creation, and that man is venturing into places he does not belong. Of course, the "playing God" objection only makes sense in the context of one definition of God, as a supernatural being who plays a role in the birth of each new member of our species. And even if one holds this particular view of God, it does not necessarily follow that cloning is equivalent to playing God. Some who consider themselves to be religious have argued that if God didn't want man to clone, "he" wouldn't have made it possible.

Should public policy in a pluralistic society be based on a narrow religious point of view? Most people would say no, which is why those who hold this point of view are grasping for secular reasons to support their call for an unconditional ban on the cloning of human beings. When the dust clears from the cloning debate, however, the secular reasons will almost certainly have disappeared. And then, only religious objections will remain.

NOTES

1. Denise Grady, "Doctors Using Hybrid Egg to Tackle Infertility in Older Women," *New York Times*, October 10, 1998, p. 16.

2. Data extracted from a Time/CNN poll taken over February 26 and 27, 1997, and reported in *Time* on March 10, 1997; and an ABC Nightline poll taken over the same period, with results reported in the *Chicago Tribune* on March 2, 1997.

3. Quotes from the bioethicist Arthur Caplan in the *Denver Post*, February 24, 1997; the bioethicist Thomas Murray in the *New York Times*, March 6, 1997; Congressman Vernon Elders in the *New York Times*, March 6, 1997; and evolutionary biologist Francisco Ayala in the *Orange County Register*, February 25, 1997.

4. James A. Geraghty, president of Genzyme Transgenics Corporation (a Massachusetts biotech company) testified before a Senate Committee that "everyone in the biotechnology industry shares the unequivocal conviction that there is no place for the cloning of human beings in our society" (the *Washington Post*, March 13, 1997).

5. Data obtained from a Yankelovich poll of 1,005 adults reported in the *St. Louis Post-Dispatch* on March 9, 1997 and a Time/CNN poll reported in the *New York Times* on March 5, 1997.

6. Leonard Bell, president and chief executive of Alexion Pharmaceuticals, is quoted as saying, "There is a healthy skepticism whether you can accomplish this efficiently in another species" in the *New York Times*, March 3, 1997.

7. Interpretation of the judgments of scientists reported by Michael Specter and Gina Kolata in the *New York Times*, March 3, 1997, and by Wray Herbert, Jeffrey L. Sheler, and Traci Watson in *U.S. News and World Report*, March 10, 1997.

8. Quote from Ian Wilmut, the scientist who brought forth Dolly, in an article by Tim Friend for *USA Today*, February 24, 1997.

9. Quoted in an article by Jeffrey Kluger in *Time*, March 10, 1997.

10. Quoted in an article by Carol McGraw and Susan Kelleher for the *Orange County Register*, February 25, 1997.

11. Quoted in the on-line version of the *Arlington Catholic Herald* (http://www.catholicherald.com/bissues.htm) on May 16, 1997.

12. February 28, 1997, *New York Times* editorial.

13. L. Wilmut, A. E. Schnieke, J. McWhir, A. J. Kind, and K. H. S. Campbell (1997) "Viable offspring derived from fetal and adult mammalian cells," *Nature* 385: 810–813.

14. Silver, Lee M. *Remaking Eden: How Genetic Engineering and Cloning Will Transform the American Family.* New York: Avon Books, 1998.

15. Op-ed piece in the *New York Times*, February 26, 1997.

16. Column in the *New York Times*, February 27, 1997.

6

Would Cloned Humans Really Be Like Sheep?

Leon Eisenberg

The recent proof, by DNA-microsatellite analysis[1] and DNA-fingerprinting techniques,[2] that Dolly the sheep had indeed been cloned as Wilmut et al. claimed,[3] and the report by Wakayama et al.[4] of the successful cloning of more than 20 healthy female mice are likely to reactivate discussions of the ethics of cloning humans and to provoke more calls to ban experiments on mammalian cloning altogether. From the standpoint of biologic science, a ban on such laboratory experiments would be a severe setback to research in embryology.[5] From the standpoint of moral philosophy, the ethical debate has been so obscured by incorrect assumptions about the relation between a potential human clone and its adult progenitor that the scientific issues must be reexamined in order to clarify the relation between genotype and phenotype. There are powerful biologic objections to the use of cloning to alter the human species, objections that make speculations about the ethics of the process largely irrelevant.

EXPERIMENTS IN CLONING

A clone is the aggregate of the asexually produced progeny of an individual organism. Reproduction by cloning in horticulture involves the use of cut-

tings of a single plant to propagate desired botanical characteristics indefinitely. In microbiology, a colony of bacteria constitutes a clone if its members are the descendants of a single bacterium that has undergone repeated
asexual fission. The myriad bacteria in the clone each have precisely the
same genetic complement as that of the progenitor cell and are indistinguishable from one another. Success in cloning mammals demonstrates un-
equivocally that at least some of the nuclei in fully differentiated mammalian cells contain the full complement of potentially active genetic material
that is present in the zygote. What distinguishes differentiated cells is the
sets of genes that are turned "off" or "on." The cloning experiments in
animals suggest that similar techniques might make it possible to clone
humans. Such cloning would involve transferring a human ovum to a test
tube, removing its nucleus, replacing it with a somatic-cell nucleus from
the donor of the ovum or another person, allowing the ovum with its new
diploid nucleus to differentiate to the blastula stage, and then implanting it
in a "host" uterus. The resultant person, on attaining maturity, would be an
identical genetic twin of the adult nuclear donor. This hypothetical outcome, although remote, has given rise to speculation about the psychological, ethical, and social consequences of producing clones of human beings.

The futuristic scenarios evoked by the prospect of human cloning contain implicit assumptions about the mechanisms of human development.
Examination of these underlying premises highlights themes that can be
traced back to Greek antiquity, themes that recur in contemporary debates
about the sources of differences between groups with respect to such characteristics as intelligence and aggression.[6]

THEORIES OF DEVELOPMENT

The enigmas of human development have concerned philosophers and nat-
uralists since people first began to wonder how plants and animals emerged
from the products of fertilization.[7] Despite the fact that there is no resemblance between the physical appearance of the seed and the form of the
adult organism, the plant or animal to which it gives rise is an approximate
replica of its progenitors. The earliest Greek explanation was preformation—
that is, the seed contains all adult structures in miniature. This ancient
speculation, found in the Hippocratic corpus, was given poetic expression
by Seneca: "In the seed are enclosed all the parts of the body of the man
that shall be formed. The infant that is borne in his mother's wombe has
the rootes of the beard and hair that he shall weare one day."[8] The theory
of preformation was so powerful that 1600 years later, when the microscope
was invented, the first microscopists to examine a sperm were able to persuade themselves that they could see in its head a homunculus with all the
features of a tiny but complete man. Improvements in the microscope and
the establishment of embryology as an experimental science made the doc-
trine progressively more difficult to sustain in its original form. With better

microscopical resolution, the expected structures could not be seen, and experimental manipulation of embryos produced abnormal "monsters" that could not, by definition, have already been present in the seed.

The alternative view, that of epigenesis, was formulated by Aristotle. Having opened eggs at various stages of development, he observed that the individual organs did not all appear at the same time, as preformation theory demanded. He did not accept the argument that differences in the size of the organs could account for the failure to see them all at the same time. Others as well as Aristotle had noted that the heart is visible before the lungs, even though the lungs are ultimately much larger. Unlike his predecessors, Aristotle began with the observable data. He concluded that new parts were formed in succession and did not merely unfold from precursors already present:

> It is possible, then, that A should move B and B should move C, that, in fact, the case should be the same as with the automatic machines shown as curiosities. For the parts of such machines while at rest have a sort of potentiality of motion in them, and when any external force puts the first of them into motion, immediately the next is moved in actuality . . . in like manner also that from which the semen comes . . . sets up the movement in the embryo and makes the parts of it by having touched first something though not continuing to touch it. . . . While there is something which makes the parts, this does not exist as a definite object, nor does it exist in the semen at the first as a complete part.[9]

This is the first statement of the theory of epigenesis: successive stages of differentiation in the course of development give rise to new properties and new structures. The genetic code in the zygote determines the range of possible outcomes. Yet the genes that are active in the zygote serve only to initiate a sequence, the outcome of which is dependent on the moment-to-moment interactions between the products of successive stages in development. For example, the potential for differentiating into pancreatic tissue is limited to cells in a particular zone of the embryo. But these cells will produce prozymogen, the histologic marker of pancreatic tissue, only if they are in contact with neighboring mesenchymal cells; if they are separated from mesenchyme, their evolution is arrested, despite their genetic potential.[10] At the same time, the entire process is dependent on the adequacy of the uterine environment, defects in which lead to anomalous development and miscarriage.

OUTCOMES OF HUMAN CLONING

The methodologic barriers to successful human cloning are formidable. Nonetheless, even if the necessary virtuosity lies in a more distant future than science-fiction enthusiasts suggest, one can argue that a solution exists in principle and attempt to envisage the possible outcomes.

Restricting Genetic Diversity

One negative consequence of very wide-scale cloning is that it would lead to a marked restriction in the diversity of the human gene pool. Such a limitation would endanger the ability of our species to survive major environmental changes. Genetic homogeneity is compatible only with adaptation to a very narrow ecologic niche. Once that niche is perturbed (e.g., by the invasion of a new predator or a change in temperature or water supply), extinction may follow. For example, the "green revolution" in agriculture has led to the selective cultivation of grain seeds chosen for high yield under modern conditions of fertilization and pest control. Worldwide food production, as a result, is now highly vulnerable to new blights because of our reliance on a narrow range of genotypes.[11] Recognition of this threat has led to a call for the creation of seed banks containing representatives of "wild" species as protection against catastrophe from new blights or changed climatic conditions, to which the current high-yield grains prove particularly vulnerable.[12] Indeed, the loss of species (genetically distinct populations) is impoverishing global biodiversity as the result of shrinking habitats.[13]

Precisely the same threat would hold for humans, were we to replace sexual reproduction with cloning. The extraordinary biologic investment in sexual reproduction (as compared with asexual replication) provides a measure of its importance to the evolution of species. Courtship is expensive in its energy requirements, reproductive organs are elaborate, and there are extensive differences between male and female in secondary sexual characteristics. The benefit of sexual reproduction is the enhancement of diversity (by the crossover between homologous chromosomes during meiosis and by the combining of the haploid gametes of a male and a female). The new genetic combinations so produced enable the species to respond as a population to changing environmental conditions through the selective survival of adaptable genotypes.

Cloning Yesterday's People for Tomorrow's Problems

The choice of whom to clone could be made only on the basis of phenotypic characteristics manifested during the several decades when the persons being considered for cloning had come to maturity. Let us set aside the problem of assigning value to particular characteristics and assume that we agree on the traits to be valued, however unrealistic that assumption.

By definition, the genetic potential for these characteristics must have existed in the persons who now exhibit them. But the translation of that potential into the phenotype occurs in the particular environment in which development occurs. Even if we agree on the genotype we wish to preserve, we face a formidable barrier: we know so little of the environmental features necessary for the flowering of that genotype that we cannot specify in detail the environment we would have to provide, both before and after birth, to ensure a phenotypic outcome identical to the complex of traits we seek to perpetuate.

Let us make a further dubious assumption and suppose the day has arrived when we can specify the environment necessary for the flowering of the chosen phenotype. Nonetheless, the phenotype so admirably suited to the world in which it matured may not be adaptive to the world a generation hence. That is, the traits that lead a person to be creative or to exhibit leadership at one moment in history may not be appropriate at another. Not only is the environment not static, it is altered by our own extraordinary impact on our ecology. The proliferation of our species changes patterns of disease[14,15]; our methods of disease control, by altering population ratios, affect the physical environment itself.[16] Social evolution demands new types of men and women. Cloning would condemn us always to plan the future on the basis of the past (since the successful phenotype cannot be identified sooner than adulthood).

THE CONNECTION BETWEEN GENOTYPE AND PHENOTYPE

For the student of biology, cloning is a powerful and instructive method with great potential for deepening our understanding of the mechanisms of differentiation during development. The potential of a given genotype can only be estimated from the varied manifestations of the phenotype over as wide a range of environments as are compatible with its survival. The wider the range of environments, the greater the diversity observed in the phenotypic manifestations of the one genotype. Human populations possess an extraordinary range of latent variability. Dissimilar genotypes can produce remarkably similar phenotypes under the wide range of conditions that characterize the environments of the inhabitable portions of the globe.

The differences resulting from genotypic variability are manifested most clearly under extreme conditions, when severe stresses overwhelm the homeostatic mechanisms that ordinarily act as buffers against small perturbations.

Phenotypic identity requires identity between genotypes, which cloning can ensure, and identity between environmental interactions, which it cannot ensure. At the most trivial level, we can anticipate less similarity even in physical appearance between cell donor and cloned recipient than that which is observed between one-egg twins. Placental attachment and fetal-maternal circulation can vary substantially, even for uniovular twins housed in one uterus. Developmental circumstances will be more variable between donor and cloned recipient, who will have been carried by different women.

Postnatal Environmental Effects on the Human Brain

Let us force the argument one step further by assuming that the environmental conditions for the cloned infant have been identical to those of his or her progenitor, so that at birth the infant is a replica of the infant its "father" or "mother" was at birth. Under such circumstances (and within the limits of the precision of genetic specification), the immediate pattern of central nervous system connections and their responses to stimulation will be the same as those of the progenitor at birth.

However, even under these circumstances, the future is not predestined. The human species is notable for the proportion of brain development that occurs postnatally. Other primate brains increase in weight from birth to maturity by a factor of 2 to 2.5, but the human brain increases by a factor of 3.5 to 4. There is a fourfold increase in the neocortex, with a marked elaboration of the receiving areas for the teloreceptors, a disproportionate expansion of the motor area for the hand in relation to the representation of other parts, a representation of tongue and larynx many times greater, and a great increase in the "association" areas. The elaboration of pathways and interconnections is highly dependent on the quantity, quality, and timing of intellectual and emotional stimulation. The very structure of the brain, as well as the function of the mind, emerges from the interaction between maturation and experience.[17]

Nature and nurture jointly mold the structure of the brain. The basic plan of the central nervous system is laid down in the human genome, but the detailed pattern of connections results from competition between axons for common target neurons. Consider the steps in the formation of alternating ocular layers in the lateral geniculate bodies. Early in embryogenesis, axons from both eyes enter each of the geniculate nuclei and intermingle. How does the separation of layers for each eye, essential for vision, come about? It results from periodic waves of spontaneous electrical activity in retinal ganglion cells, because immature cell membranes are unstable. If these electrical outbursts are abolished experimentally, the layers simply do not become separated.[18] Competition between the two eyes, driven by spontaneous retinal activity, determines eye-specific lateral geniculate connections.[19] Neither the genes governing the retina nor the genes governing the geniculate specify the alternating ocular layers; it is the interaction between retina and geniculate during embryogenesis that brings it about. Furthermore, the precise targeting of projections from lateral geniculate to occipital cortex is dependent on electrical activity in the geniculate. Abolishing these action potentials with an infusion of tetrodotoxin results in projections to cortical areas that are normally bypassed and a marked reduction in projections to visual cortex.[20]

Postnatal stimulation is required to form the ocular dominance columns in the occipital cortex.[21] Both eyes of the newborn must receive precisely focused stimulation from the visual environment during the early months of postnatal life in order to fine-tune the structure of the cortex. If focused vision in one eye of a kitten or an infant monkey is interfered with, the normal eye "captures" most neurons in the occipital cortex in the absence of competition from the deprived eye. The change becomes irreversible if occlusion is maintained throughout the sensitive period. Amblyopia in humans, characterized by incongruent visual images from the two eyes, results in permanent loss of effective vision from the unused eye if the defect is not corrected within the first five years of life.

Thus, which of the overabundant neurons live and which die is determined by the amount and consistency of the stimulation they receive. Interaction between organism and environment leads to patterned neuronal activity that determines which synapses will persist.[22] Experience molds the brain in a process that continues throughout life. Myelination in a key relay zone in the hippocampal formation continues to increase from childhood through at least the sixth decade of life.[23] And recent research has provided evidence that neurons in the dentate gyrus of the hippocampus continue to divide in the adult brain.[24]

Changes in the Brain with Use

Techniques of functional brain mapping reveal marked variations in cortical representation that depend on prior experience. Manipulation of sensory inputs leads to reorganization of the cortex in monkeys[25] and humans.[26] The motor cortex in violinists displays a substantially larger representation of the fingers of the left hand (the one used to play the strings) than of the fingers of the right (or bowing) hand. Moreover, the area of the brain dedicated to finger representation is larger in musicians than in nonmusicians.[27] Sterr et al.[28] compared finger representation in the somatosensory cortex in blind persons who used three fingers on each hand to read Braille with that in Braille readers using only one finger on one hand and in sighted readers. They found a substantial enlargement of hand representation in the Braille readers who used two hands, with topographic changes on the postcentral gyrus.

If enlargement of cortical areas accompanies increases in activity, shrinkage follows loss. Within days after mastectomy, the amputation of an arm or leg, or the correction of syndactyly, the cortical sensory map changes. Intact areas have an enlarged representation at the expense of areas from which innervation has been moved.[29,30] What begins prenatally continues throughout life. Structure follows function.

BECOMING HUMAN

There is yet another level of complexity in the analysis of personality development. The human traits of interest to us are polygenic rather than monogenic; similar outcomes can result from the interaction between different genomes and different social environments. To produce another Wolfgang Amadeus Mozart, we would need not only Wolfgang's genome but his mother's uterus, his father's music lessons, his parents' friends and his own, the state of music in 18th-century Austria, Haydn's patronage, and on and on, in ever-widening circles. Without Mozart's set of genes, the rest would not suffice; there was, after all, only one Wolfgang Amadeus Mozart. But we have no right to the converse assumption: that his genome, cultivated in another world at another time, would result in the same musical genius. If a particular strain of wheat yields different harvests under different conditions of climate, soil, and cultivation, how can we assume that so much more complex a genome as that of a human being would yield its desired crop of operas, symphonies, and chamber music under different circumstances of nurture?

In sum, cloning would be a poor method indeed for improving on the human species. If widely adopted, it would have a devastating impact on the diversity of the human gene pool. Cloning would select for traits that have been successful in the past but that will not necessarily be adaptive to an unpredictable future. Whatever phenotypes might be produced would be extremely vulnerable to the uncontrollable vicissitudes of the environment.

Proposals for human cloning as a method for "improving" the species are biologic nonsense. To elevate the question to the level of an ethical issue is sheer casuistry. The problem lies not in the ethics of cloning a human but in the metaphysical cloud that surrounds this hypothetical cloned creature. Pseudobiology trivializes ethics and distracts our attention from real moral issues: the ways in which the genetic potential of humans born into impoverished environments today is stunted and thwarted. To improve our species, no biologic sleight of hand is needed. Had we the moral commitment to provide every child with what we desire for our own, what a flowering of humankind there would be.

REFERENCES

1. Ashworth D, Bishop M, Campbell K, et al. (1998). DNA microsatellite analysis of Dolly. *Nature* **394**:329.

2. Signer E N, Dubrova Y E, Jeffreys A J, et al. (1998). DNA fingerprinting Dolly. *Nature* **394**:329–30.

3. Wilmut I, Schnieke A E, McWhir J, Kind A J, Campbell K H. (1997). Viable offspring derived from fetal and adult mammalian cells. *Nature* **385**: 810–13. [Erratum, *Nature* 1997; 386: 200.]

4. Wakayama T, Perry A C, Zuccotti M, Johnson K R, Yanagimachi R. (1998). Full-term development of mice from enucleated oocytes injected with cumulus cell nuclei. (1998) *Nature* **394**:369–74.

5. Berg P, Singer M. (1998). Regulating human cloning. (1998). *Science* **282**:413.

6. Eisenberg L. (1972). The human nature of human nature. (1972). *Science* **176**:123–28.

7. Needham J. (1959). A history of embryology. (1959). New York: Abelard-Schuman.

8. Needham J. (1959). A history of embryology. New York: Abelard-Schuman, 66.

9. Needham J. (1959). A history of embryology. New York: Abelard-Schuman, 47–48.

10. Grobstein C. (1964). Cytodifferentiation and its controls. *Science* **143**:643–50.

11. Harlan J R. (1975). Our vanishing genetic resources. (1975). *Science* **188**:618–21.

12. National Research Council. (1972). Genetic vulnerability of major crops. (1972). Washington, D.C.: National Academy of Sciences.

13. Hughes J B, Daily G C, Ehrlich P R. (1997). Population diversity: its extent and extinction. (1997). *Science* **278**:689–92.

14. Black F L. (1975). Infectious diseases in primitive societies. (1975). *Science* **187**:515–18.

15. Black F L. (1992). Why did they die? (1992). *Science* 258:1739–40.

16. Ormerod W E. (1976). Ecological effect of control of African trypanosomiasis. *Science* **191**:815–21.

17. Eisenberg L. (1995). The social construction of the human brain. *Am J Psychiatry* **152**:1563–75.

18. Shatz C J, Stryker M P. (1988). Prenatal tetrodotoxin infusion blocks segregation of retinogeniculate afferents. *Science* **242**:87–89.

19. Penn A A, Riquelme P A, Feller M B, Shatz C J. (1998). Competition in retinogeniculate patterning driven by spontaneous activity. *Science* **279**: 2108–12.

20. Catalano S M, Shatz C J. (1998). Activity-dependent cortical target selection by thalamic axons. (1998). *Science* **281**:559–62.

21. Wiesel T N. (1982). The postnatal development of the visual cortex and the influence of environment (the 1981 Nobel Prize lecture). Stockholm, Sweden: Nobel Foundation.

22. Nelson C A, Bloom F E. (1997). Child development and neuroscience. *Child Dey* 68:970–87.

23. Benes F M, Turtle M, Khan Y, Farol P. (1994). Myelination of a key relay zone in the hippocampal formation occurs in the human brain during childhood, adolescence, and adulthood. (1994). *Arch Gen Psychiatry* 51: 447–84.

24. Eriksson P S, Perfilieva E, Bjork-Eriksson T, et al. (1998). Neurogenesis in the adult human hippocampus. *Nat Med* **4**:1313–17.

25. Wang X, Merzenich M M, Sameshima K, Jenkins W M. (1995). Remodelling of hand representation in adult cortex determined by timing of tactile stimulation. *Nature* **378**:71–75.

26. Hamdy S, Rothwell J C, Aziz Q, Singh K D, Thompson D G. (1998). Long-term reorganization of human motor cortex driven by short-term sensory stimulation. *Nat Neurosci* **1**:64–68.

27. Schlaug G, Jancke L, Huang Y, Steinmetz H. (1995). In vivo evidence of structural brain asymmetry in musicians. *Science* **267**:699–701.

28. Sterr A, Muller M M, Elbert T, Rockstroh B, Pantev C, Taub E. (1998). Perceptual correlates of changes in cortical representation of fingers in blind multifinger Braille readers. *J Neurosci* **18**:4417–23.

29. Yang T T, Gallen C C, Ramachandran V S, Cobb S, Schwartz B J, Bloom F E. (1994). Noninvasive detection of cerebral plasticity in adult human somatosensory cortex. *Neuroreport* **5**:701–4.

30. Mogilner A, Grossman J A, Ribary U, et al. (1993). Somatosensory cortical plasticity in adult humans revealed by magnetoencephalography. *Proc Natl Acad Sci USA* **90**:3593–97.

THE CONTEXT OF CLONING

7

Cloning in the Popular Imagination

Dorothy Nelkin & M. Susan Lindee

Dolly is a cloned sheep born in July 1996 at the Roslin Institute in Edinburgh by Ian Wilmut, a British embryologist. She was produced, after 276 failed attempts, from the genetic material of a six-year-old sheep. But Dolly is also a Rorschach test. The public response to the production of a lamb from an adult cell mirrors the futuristic fantasies and Frankenstein fears that have more broadly surrounded research in genetics, and especially genetic engineering. Dolly stands in for other monstrosities—both actual and fictional—that human knowledge and technique have produced. She provokes fear not so much because she is novel, but because she is such a familiar entity: a biological product of human design who appears to be a human surrogate. Dolly as "virtual" person is terrifying and seductive—despite her placid temperament.

Cloning was a term originally applied to a botanical technique of asexual reproduction. But following early experiments in the manipulation of hereditary and reproductive processes during the mid-1960s, the term became associated with human biological engineering. It also became a pervasive theme in horror films and science fiction fantasies. Appearing to promise both new control over nature and dehumanization, cloning attracted significant popular attention.

Underlying many depictions of cloning is the idea that human beings in all their complexity are simply readouts of a powerful molecular text. In *The DNA Mystique: The Gene as a Cultural Icon*, we called this idea *genetic*

83

essentialism, a deterministic tendency to reduce personality and behavior to the genes.[1] Exploring the popular appeal of genetic essentialism, we tracked its manifestations in the mass media—in television programs, advertising and marketing media, newspaper articles, films, child-care books, and popular magazines. We found repeated messages suggesting that the personal characteristics and identity of individuals are entirely encoded in a molecular text. We found references to genes for criminality, shyness, arson, directional ability, exhibitionism, tendencies to tease, social potency, sexual preferences, job success, divorce, religiosity, political leanings, traditionalism, zest for life, and preferred styles of dressing. We found pleasure-seeking genes, celebrity genes, couch-potato genes, genes for saving, and even genes for sinning. And we documented the public fears—or sometimes hopes—that geneticists will soon acquire the awesome power to manipulate the molecular text and thereby to determine the human future.

The responses to the Dolly phenomenon reflected these ideas. Dolly and cloning were immediately the subject of jokes on late-night talk shows and Internet web sites. Their humor depended largely on the assumption that human identity is contained entirely in the sequences of DNA in the human genome: Why not clone great athletes like Michael Jordan, or great scientists like Albert Einstein, or popular politicians like Tony Blair, or less popular politicians like Newt Gingrich, or wealthy entrepreneurs like Bill Gates? But there were also many anxious scenarios in the popular press, including futuristic stories about making new Frankenstein monsters, or creating Adolph Hitler clones, or producing "organ donors" only to harvest their (fully compatible) viscera.[2]

Dolly seems to lead to a future of highly managed, commercialized bodies, both animal and human. She is a manifestation of scientific rationality—a machine that can be tailored to human needs. And she is a symbol of human vulnerability—a sign that males may become obsolete or that commercial interests will dictate the human future. Speculations about Dolly reveal the patterns of current perceptions of science in the biotechnology age.

MORAL NARRATIVES

Cloning has long been a theme in novels and science fiction films. Most of these stories tend to be traditional narratives of divine retribution for violating the sanctity of human life. These days they employ the language of genetics, and they often dwell on the horrible consequences of genetic manipulation. A typical story appeared shortly after the 1976 controversy over recombinant DNA research. Stephen Donaldson's *Animal Lover* is about a famous geneticist named Avid Paracels who became the victim of "genetic riots" that took place when news spread about his efforts to create a superior human being. The public was morally outraged by his research. He had threatened the "sanctity of human life." The geneticist lost his grants and had to abandon his career. "I can't understand," he complained, "why the society won't bear biological improvements. . . . What's so sacred about biology?" Other novels, such as Robin Cook's *Mutation* and Michael Stewart's *Prodigy*, convey the same theme. "No man has the right to tamper with the building blocks of human life."

Real research projects associated with cloning have evoked a similar sense of horror and dismay. As early as 1938, a British magazine called *Titbits* reported on research taking place at the Srangeway laboratory and tissue archive in Cambridge, England, the first laboratory devoted to tissue cultivation. The writer predicted that "canned blood" would be used to create new lives, and he wondered: "What exactly will be created? Could you love a chemical baby? Will the sexless, soulless creatures of chemistry conquer the true human beings?"[3] Dreams of such creatures have been fueled by new biological technologies associated with agricultural and fertility research.

In 1993, scientists from George Washington University "twinned" a nonviable human embryo in an experiment intended to create embryos for in vitro fertilization. When they reported their work at the meeting of the American Fertility Society, newspapers, magazines, and television talk shows covered the experiment as if it involved a cloning technology for the mass production of human beings.[4] While the scientists viewed their research as a contribution to helping infertile patients, the media stories about the research envisioned selective breeding factories, cloning on consumer demand, the breeding of children as organ donors, a cloning industry for selling multiples of human beings, and even a freezer section of the "biomarket."[5] Journalists anticipated a "Brave New World of cookie cutter humans,"[6] and they asked if the GWU scientists were playing God. A *Time* magazine survey found that 75% of their respondents thought cloning was not a good thing, and 58% thought it was morally wrong. Thirty-seven percent wanted research on cloning to be banned; 40% called for a temporary halt to research.

Yet, public responses to the GWU experiment in 1993 and then to Wilmut's experiment four years later were not all so negative. For some, cloning held the promise of creating perfect cows, sheep, and chickens, or perhaps even perfect people. Reflecting deterministic assumptions of genetic essentialism, media stories have suggested that clones would surely be identical products of their genes.

Reproduction has often appeared in mass media stories as a commercial transaction where the goal is to produce good stock. Sperm banks are described as a place to shop for "Mr. Good Genes" where potential parents scan lists of desirable genetic traits.[7] Why not, in this context, use cloning to produce and reproduce perfect babies? They could, after all, be dependable reproductive products with proven performance.

Cloning has also been viewed as a way to assure a kind of immortality. Scientists have commonly constructed DNA as an immortal text. The Human Genome Diversity Project seeks to "immortalize" vanishing populations through saving their DNA. Molecular biologists have tried to extract "immortal" DNA from the remains of historical figures such as Abraham Lincoln and to reconstruct their health and personal characteristics long after they are dead. In his popular book *The Selfish Gene*, sociobiologist Richard Dawkins argues that DNA is immortalized through the reproductive process, for we are blindly programmed to preserve and pass on our genes. And, of course, in *Jurassic Park*, the DNA lives on forever in fossilized form and contains the complete instruction code of the living organism. In Michael Crichton's story, if you want a dinosaur, all you need is dinosaur DNA.

Post-Dolly narratives build on these assumptions. Again and again media stories have predicted that cloning will allow the resurrection of the dead (bereaved parents, for example, might clone a beloved deceased child). Or the technology could provide life everlasting for the deserving (narcissists could arrange to have themselves cloned). Dawkins confessed his own desire to be cloned: "I think it would be mind-bogglingly fascinating to watch a younger edition of myself growing up in the twenty-first century instead of the 1940s."[8] Indeed, psychiatrist Robert Coles, in a *New York Times* interview, suggested that the very idea of cloning "tempts our narcissism enormously because it gives a physical dimension to a fantasy that one can keep going on through the reproduction of oneself."[9]

Not surprisingly, in the United States, where demands and desires are frequently framed in terms of rights, cloning too has been defined as a "right." Infertile women and their physicians have been among the most ardent advocates of cloning as a right; for if a single embryo could be used to create identical embryos for later fertilization, this could avoid the hormonal overload and painful procedures that women undergo for in vitro fertilization. The technology of cloning thus spawned not only Dolly but an association called "Cloning Rights United Front." Its members insisted that cloning is part of the reproductive rights of every human being, and, in tune with the political sentiments of the 1990s, they wanted "the government to keep out."[10]

Dolly also spawned an amazing range of humor—some silly, some funny—about the implications of cloning. Poems, cartoons, one-liners and puns about cloning appeared almost immediately on the Internet and in mainstream publications. Jokes can reveal cultural fault lines and social tensions; for their humor often plays on the contradictions and ironies of familiar contexts, events, or situations. Dolly jokes were no exception.

While cloning could theoretically make both sexes irrelevant to reproduction, it was suggested that the technology could be a threat to the male of the species—men will no longer be necessary! Writer Wendy Wasserman wondered what you would say to your shrink if you are your own mother.[11] An Internet inquirer wondered: "If I have sex with my clone, will I go blind?" A cartoonist in the London *Guardian* depicted a women comforting a cab driver who had just run over her husband: "That's alright, I have another one upstairs." Even the issue of scientific fraud became a source of cloning humor. What if the cloning experiment was in fact a fraud? "Well, they really would have pulled the wool over our eyes."

Meanwhile, a journalist predicted a new action movie called *Speed Sheep* in which thousands of cloned sheep clogged Interstate 95. Headlines of cloning stories reveled in puns: "An Udder Way of Making Lambs," "Send in the Clones," "Little Lamb, Who Made Thee?" "Will There Ever Be Another Ewe?" and "Getting Stranger in the Manger." And inevitably there was the anticipation of "Double Trouble."

Many of the cloning quips were comments on social, political, and professional tensions.[12] A divorce lawyer predicted the doubling of his business. Historians wondered if the Founding Fathers could be cloned for display in a "living history" exhibit in a theme park: they suggested that the park might be called "Clonial Williamsburg." Some cynical policy commentators announced that cloning experiments could be developed to solve social problems: The race problem could be resolved by manipulating the balance between melanin and IQ genes. The age-old nature-nurture dispute could be definitively settled by creating clones and raising them systematically in different environments.

Religious ethicists and theologians had a lot to say about the cloning experiment.[13] One writer quipped that cloning offered a "second chance for the soul." If you sin the first time, try again. But a theologian, Rabbi Mosher Tendler, a professor of medical ethics at Yeshiva University in New York, warned that "whenever man has shown mastery over man, it has always meant the enslavement of man." Other theologians, long concerned about the implications of genetic engineering, worried that the scientists who experimented with cloning were "playing God" and "tampering with God's creation."[14] However, a less reverent wag wondered about the implications of cloning the Pope: Would they both be infallible? And what if they disagree?

ECONOMIC NARRATIVES

In his scientific paper itself, Wilmut fussed over the problem of whether "a differentiated adult nucleus can be fully reprogrammed." He called the lamb in question 6LL3 rather than Dolly, and made it clear, in diagrams and illustrations of gels, that there is some question about the precise genetic relationship between Dolly and the "donor."[15] Somatic DNA, which was the source of Dolly's genes, is constantly mutating. Dolly, in fact, may not be genetically identical in every way to her "mother," a point that is of some importance for the possible agricultural uses of cloning techniques.

For writers in the popular press, however, such technical details were less important than symbolic associations. The cloning of a lamb was immediately set in a context of other fears about genetics and genetic manipulation, and especially about rapid and sometimes startling advances in reproductive technology. The technological changes allowed by the possibility of freezing sperm and embryos and by the improvements in techniques of in vitro fertilization (IVF) have been remarkable. But they have also been controversial. They have included, for example, a controversial proposal, put forth by a fertility specialist, to harvest eggs from the ovaries of aborted fetuses and then mature the fetal cells and fertilize them in a petri dish for use in research and implantation. And they have included a plan for creating embryos through parthenogenesis.[16] The debates over such reproductive techniques set the stage for the response to cloning.

So too, responses to Dolly reflected public debates about other uses and abuses of science and technology. One journalist compared cloning to weapons development. Another worried that the shortage of organs for transplantation would be resolved by cloning anencephalic babies (who are born without a brain but are otherwise normal), so that their organs could be harvested for patients in need. Many news stories have reflected mistrust of scientists, and the fear that the outrageous possibilities suggested by cloning a sheep will eventually, perhaps inevitably, be realized in human beings. News headlines frequently suggest that science cannot be controlled: "Science Fiction Has Become a Social Reality." "Whatever's Next?" And, of course, "Pandora's Box."

Many news and popular culture accounts have expressed the growing tensions over commercial trends in genetics and biotechnology and their implications for the commodification of the body. A series of legal developments in the 1980s set the stage for commercial developments in biotechnology. They encouraged collaborations between university researchers and biotechnology companies and allowed the patenting of products of nature, including human genes.[17] In this context, business interests welcomed Dolly; for cloning has huge potential economic implications especially for agricultural and pharmaceutical applications. As predicted by a *Business Week* cover article in March 1997, called "The Biotech Century," "cloning animals is just the beginning." Such advances "will define progress in the 21st century. It's all happening faster than anyone expected."[18]

But there is a downside of these commercial trends that also helped to shape responses to the cloning experiments. Critics have documented the growing conflicts of interest in science, the increased secrecy, and the reluctance to share data.[19] Reporters noted that Wilmut held back the announcement of Dolly's birth until he registered a patent. And other observers speculated about the implications of patenting clones for perceptions of the person. Is the body to become little more than a commodity, a commercial entity that can be simply constructed as a product?[20]

Just as the GWU experiment evoked images of a cloning industry and breeding factories, so Dolly evoked cynical references to "test-tube capitalists," and sardonic queries about a market for genetic "factory seconds" and "irregulars." A World Wide Web site called Dreamtech satirized the issues by advertising a commercial service to create either "custom clones" or "designer clones." The "company" would clone various celebrities for a range of licensing fees, depending on the anticipated value of the product. The advertisement also offered a personal extraction kit, surrogate services, rapid delivery, and a backup embryo.[21]

In the 1990s, the bar-code has become a popular image, to be seen, for example, as a common body tattoo. It is also a symbol of protest. In London, protesters organized a demonstration against the granting of a patent for the processing of umbilical cord blood, a useful source of stem cells. The demonstration featured a pregnant woman with a bar code on her belly. And in a casual but revealing conversation after a television interview on cloning with one of the authors, a camera technician who was cleaning up the gear quipped, only half in jest: "I used to be a person, then I became a social security number. Now I am just a bar code—just a commodity like the cloned sheep."[22]

The commercialization of fertility procedures through the growth of an IVF industry have compounded concerns about commodifying the body. The full-page advertisements for private fertility clinics, the calls for female egg donors as well as sperm donors, the incidents of embryo theft, have tainted this thriving enterprise. There is a sense, especially in feminist writings, that the human body is being devalued as reproduction has become a commercial enterprise.[23] And there is a fear that private clinics would not be constrained at all by moral or ethical reservations about cloning.

NARRATIVES OF CONTROL

The messages evoked by Dolly have ranged from promises of progress to portents of peril, from images of miracles to visions of apocalypse. There were many calls for regulation and for a moratorium on cloning experiments. Just a week after the cloning of Dolly, President Clinton issued a directive banning the use of federal funds to support research on the cloning of human beings. So too, the president of France, the president of the European Commission, the director-general of Unesco, and the Vatican all called for a moratorium on research on cloning, which had clearly become politically unacceptable.

As political and social pressures grew, scientists responded, defending the importance of the work. Media images were "selling science short." The calls for regulations and restrictions, they argued, ignored the medical benefits that could follow from cloning experiments and their potential contribution to the development of life-saving treatments, skin grafts for burn victims, treatments for infertile couples, and a means of testing new drugs.[24] We are not interested in playing God, said James Geraghty, president of the biotechnology firm Genzyme, but in "playing doctor."[25] Mammalian cloning could help to generate tissue for organ transplantation and encourage transgenics experimentation. And certainly research using cloning would enhance scientific knowledge about cell differentiation. The politicians who sought a ban on cloning research, said scientists, were "shooting from the hip."[26] Science fiction, they insisted, should not be the guide to science policy.

Responding to the growing threat of regulation, a group of prominent scholars from the International Academy of Humanism signed a "Defense of Cloning and the Integrity of Research." This academy, a group of self-identified "secular humanists," have, since the debates over teaching creation theory in the schools during the 1970s, been inclined to interpret every critique of science and every discussion about regulation as a manifestation of antiscience sentiment. There were, they claimed, no particularly profound moral issues related to cloning, but only a "Luddite rejection" of cloning by "advocates of supernatural and spiritual agendas." They included in this Luddite group the President's National Bioethics Advisory Commission, which was convened in 1997 to consider moral issues and to recommend government policy.

This 18-member commission had focused on potential physical and psychological risks as well as the moral acceptability of cloning. After three months of intensive deliberation, it concluded that the government should continue its moratorium on federal funding for cloning research. Perhaps most interesting, the commission members worried that private IVF clinics were likely to break the moratorium and to clone babies in response to their private patients' demands. Thus, the commission recommended legislation that would ban all research on the cloning of people. However, the group was reluctant to permanently fetter research and proposed that legislation be crafted as temporary until there was time for further deliberation over the coming years. And no prohibitions at all were placed on the cloning of individual cells or animals for research purposes. Whether the legislature has the constitutional authority to regulate scientific procedures, and whether federal laws have authority over IVF clinics operating within state boundaries and normally regulated by state laws is a matter of some disagreement within the legal community and remains to be tested.

President Clinton accepted the report and sent legislative recommendations to Congress. In a speech in the Rose Garden, he said: "Cloning has the potential to threaten the sacred family bonds at the core of our ideals and our society . . . to make our children objects rather than cherished individuals." But the very same day, a Switzerland-based group, supported by a group of investors, launched an international company called Valiant Ventures Ltd. It claims to provide a "Clonaid" service for wealthy parents worldwide who want to have a child cloned. The cost would be just $200,000.[27] In addition, the company would provide safe storage of the tissue from any "beloved person" so that it could be cloned at a later date in case of death. This company also offered to support the efforts of Dr. Richard Seed, the physicist who said, in 1998, that he intended to open a commercial clinic to clone people.

RORSCHACH READINGS

Dolly, after all, is only a sheep, and she is depicted again and again as cuddly and cute. But as a symbolic site for the exploration of identity, heredity, destiny, and the social meaning of science, she is a spectacular beast. She is a compelling actor in contemporary dreams about science—evoking for some euphoric fantasies; for others horrible nightmares and the fear of science out of control. She offers up the possibility of hyperrationality in the management of bodies and of complete genetic control of cows, sheep, and humans as well. She offers the specter of technical decisions that will turn all bodies (human and animal) into intentional products, manufactured and designed "on purpose." She evokes a way of thinking about bodies as little more than efficient mechanisms for the production of "value"—be it milk, or meat, or creative imagination. But she is also a focus of popular mistrust of research that is tied to commercial interests.

Dolly can thus be regarded less as a sheep than as a microcosm of the history of science—a symbol of the rich interconnections between animals and human beings, of the struggles between technological changes and moral tenets, of the tensions between the advance of scientific knowledge and demands for political expediency in the face of public concerns.

Popular speculations about science and its terrors have often been dismissed as based on journalistic ignorance of science, sensationalism, or willful misinterpretation for the sake of making news. But media messages matter, and often reflect legitimate concerns. Widely disseminated images and narratives have real effects, regardless of their relationship to the technical details of the scientific work. They shape the way people think about new technologies, assess their impacts, and develop ways to control them.

The popular responses to Dolly are especially important because they convey meanings that extend well beyond the single experiment. Dolly has become far more than a biological entity; she is a cultural icon, a symbol, a way to define the meaning of personhood and to express concerns about the forces shaping our lives. She provides a window on popular beliefs about human nature and the social order, on public fears of science and its power in society, and on concerns about the human future in the biotechnology age. She is a stunning image in the popular imagination.

NOTES

1. Dorothy Nelkin and M. Susan Lindee, *The DNA Mystique: The Gene as a Cultural Icon*, New York: W. H. Freeman, 1995.

2. Robert Langreth, "Cloning Has Fascinating, Disturbing Potential," *The Wall Street Journal*, February 24, 1997.

3. Strangeway Archives, Spears Archive, Box 5, Wellcome Trust Library, London.

4. See, for example, the report on the George Washington University experiment in Philip Elmer Dewitt, "Cloning: Where Do We Draw the Line?" *Time*, November 8; 1993, pp. 64–70.

5. Some of the scenarios appeared in a review of the media coverage of human embryo cloning in the *Newsletter* of the Center for Biotechnology Policy and Ethics of Texas A & M, January 1, 1994.

6. Dewitt, "Cloning: Where Do We Draw the Line?"

7. See a review of the media reports on reproductive technologies in Nelkin and Lindee, *The DNA Mystique*, p. 97.

8. Richard Dawkins, quoted in *Nature* **386**, March 6, 1997. Dawkins called himself a "closet clone" in an interview with the *Evening Standard* (London), February 25, 1997.

9. Robert Coles, quoted in Gustav Niebuhr, "Suddenly Religious Ethicists Face a Quandary on Cloning," *New York Times*, March 1, 1997.

10. Reported in the *New York Times Magazine*, May 25, 1997, p. 18.

11. Quoted in *New York Times*, February 27, 1997.

12. Some of these quips come from the Oracle Service List on the Internet (oracle-list@synapse.net, April 30, 1997). Others come from other Internet communications and from talk shows. Still others are jokes we have heard in casual conversations. We selected them from the many jokes and stories we found in popular sources because they suggest the range of concerns and issues raised in popular culture.

13. For a summary of religious views about cloning, see Gustav Niebuhr, "Suddenly, Religious Ethicists Face a Quandary on Cloning," *New York Times*, March 1, 1997.

14. Articles in evangelical magazines such as *Christianity Today* have regular articles opposing genetic engineering as "tampering with genes." See, for example, *The Plain Truth* **55**, pp. 3–8. Pope John Paul II has taken a position on genetic manipulation, arguing that: "All interference in the genome must be done in a way that absolutely respects the specific nature of the human species, the transcendental vocation of every being and his incomparable dignity." Address to the Pontifical Academy of Sciences, October 8, 1994, quoted in *Family Resource Center News* winter 1996, p. 1.

15. I. Wilmut et al., "Viable Offspring Derived from Fetal and Adult Mammalian Cells," *Nature* **35**, February 27, 1997. pp. 810–812.

16. See discussion in National Institutes of Health, *Report of the Human Embryo Research Panel*, Washington, DC, September 1994, volume I.

17. Among the most important federal laws was 15 U.S.C.S. 3701 et seq. 1995, which provided tax incentives to companies investing in academic research. At the same time, a landmark Supreme Court case, Diamond v. Chakrabarty, 147 US 303 (1980), granted a patent on a life form (a bacterium), setting the stage for the patenting of human genes.

18. Cover Story of *Business Week*, March 10, 1997.

19. For example, see David Blumenthal et al., "Participation of Life Science Faculty in Research Relations with Industry," *New England Journal of Medicine* **335** (1996): pp. 1734–1739; and Sheldon Krimsky et al., "Financial Interests of Authors in Scientific Journals," *Science and Energy Ethics* **21** (1996): pp. 395–410.

20. Dorothy Nelkin and Lori Andrews, "Homo Economicas: Commercialization of Body Tissue in the Age of Biotechnology," *Hastings Center Report* 28, September 1998, pp. 30–39, and Andrews and Nelkin, *Body Bazaar*, New York: Crown Books, 2001.

21. Dreamtech, http//www.D-B.Net/DT1/Intro2.HTML.

22. Personal communication to Nelkin, May 1997.

23. Gena Corea, "Current Developments and Issues: A Summary," *Reproductive and Genetic Engineering* **2**, (1989): p. 3.

24. Medical News and Perspectives, "Threatened Bans on Human Cloning Research Could Hamper Advances," *JAMA* 277, April 2, 1997, p. 1023. Also see statement from the International Academy of Humanism called, "Defense of Cloning and the Integrity of Scientific Research," quoted in *Science* **276**, May 30, 1997, p. 1341.

25. James Geraghty, quoted in *Genetic Engineering News*, April 1, 1997, p. 10.

26. Brigid Hogan, professor in cell biology, quoted in Meredith Wadman, "Politicians Accused of 'Shooting from the Hip' on Human Cloning," *Nature* **386**, March 13, 1997, p. 97.

27. It advertises on the Internet at http://www.clonaid.com.

8

The Two-Edged Sword

Biotechnology and Mythology

Kenneth M. Boyd

The public debate provoked by Dolly the sheep has repeatedly returned to the following question. Cloning large animals by nuclear transfer is now possible. Such techniques *could* eventually be used to clone not just human tissues or organs, but also individual human beings. Why should this *not* be done? Behind this question is the suspicion, even the conviction, that eventually it *will* be done. Scientific curiosity, or the prospect of profits to be made, will be too compelling to resist.

Why should nuclear transfer technology *not* be used to clone individual human beings? There is a strong, perhaps compelling, argument against doing this in the foreseeable future. But this argument does not necessarily rule out the long-term possibility of cloning human beings. Nor, in the foreseeable future, does it exclude related scientific work in animals or in human cells. The moral considerations surrounding human cloning and cloning humans, moreover, raise wider questions about the ethics of biotechnology. These wider questions are the main concern of this paper. But let me begin with the immediate ethical objection to cloning human beings.

CLONING HUMAN BEINGS

The immediate ethical objection to cloning human beings is not about creating a so-called "carbon copy" of an existing person. A person born as a result of such an experiment, if it was successful would be less similar to the person cloned than two identical twins are to one another. Someone born as a result of successful cloning by nuclear transfer, in other words, would be a unique individual in his or her own right. He or she might have special psychological problems to overcome—just as people born as a result of sperm, egg, or embryo donation, for example, may (but do not necessarily) have special problems. This possibility certainly has to be taken seriously. But it is not the most immediate ethical objection to cloning humans.

The most immediate ethical objection, rather, is that the risks of such an experiment with a human being are incalculable until the experiment is done; and while society may be willing to accept unsuccessful experiments in sheep, it is unlikely to accept such unsuccessful experiments in humans. Or at least—because the question is a little more complicated than that—society is unlikely to accept the risk of unsuccessful experiments, unless the potential benefits of cloning humans can be shown to be so great as to make the risk worth taking.

But are such great potential benefits in prospect? Few if any have been suggested. Perhaps making a clone from a husband who cannot produce sperm might be more acceptable to some couples than using sperm from an anonymous donor. That *might* be justifiable—provided that nuclear transfer technology had become sufficiently successful in animals to make the risk of an unsuccessful human experiment reasonably unlikely. But in order to sanction such an experiment, society and its regulatory bodies would need to be pretty convinced that the experimental risks were minimal; and so of course would scientists and their backers—since people, unlike sheep, could sue them for large sums if things went badly wrong.

It might be wiser, then, to wait and see whether scientific work in animals or preimplantation human embryos is successful. Perhaps further research may lead to ways of generating sperm-making or egg-making capacity in people without it. A similar case might even be made, for example, for not running the risks of transplanting genetically modified animal organs into humans, in particular the risk of transmitting new viruses. Further research might eventually discover ways of regenerating damaged organs, or even of growing new ones. Here, of course, the argument is weaker, because the stakes are so much higher. The number of human organs available for transplantation falls far short of the number needed—people whose lives could be saved will die, unless or until suitable animal organs or some yet-to-be-developed alternative become available. But perhaps this only serves to make the point. It is difficult to see any such compelling reason for cloning people.

Yet who knows what the more distant biotechnological future holds? Many benefits to humanity that advocates of current scientific or technological projects currently hold out either may never materialize, or may come about by quite different and at present unforeseen scientific routes. Many of these routes, moreover, may turn out to be moral minefields. With foreknowledge no one would go down them. Still, one should never say "never" about scientific possibilities. The potential benefits as well as the potential risks of new scientific pathways are unpredictable.

In the light—or darkness—of our ignorance of the future, is there any way in which we can gain sufficient understanding of what is actually going on in biotechnology, to save ourselves and our descendants from the worst potential risks, while not sacrificing at least some of the most promising potential benefits? Not with any certainty. But an effort to understanding what is going on in biotechnology, and to relate that to what we value about our humanity, is worth making. It may not give us answers as clear as we might like to questions about which scientific routes should be pursued and which should be off-limits. But it may make us more discriminating and less easily taken in, either by scientific hype or by antiscientific propaganda.

The best advice on this perhaps, comes from the great scientist himself, Francis Bacon. In his essay *Of Innovations,* he argues that we should not reject novelties, but have a healthy scepticism about them, and that in discriminating between them we should "make a stand upon the ancient way, and then look about us, and discover what is the straight and right way, and so walk in it."[1]

With that advice in mind then, let me now turn to the wider moral implications raised not just by cloning, but by biotechnology generally. I shall try to say something first about *biotechnology* as a *technology*, then about *biotechnology* and *mythology*, and finally about *the moral status* of animals and science and the market.

BIOTECHNOLOGY AS A TECHNOLOGY

Biotechnology, plant and microbial as well as animal, is potentially an enormous force for good. If its potential is fully realized, our dependence on many nonrenewable natural resources will be overcome; a growing world population will be fed from crops that are pest- or disease-resistant or can be grown in hostile environments; and hitherto fatal or disabling human and animal diseases will be treated and cured, either with new medicines or by organ and tissue regeneration. How much of this potential will be realized is as yet unknown. But the signs are encouraging.

This utopian vision of the biotechnological future, however, depends upon biotechnology's *anticipated beneficial* consequences not being outweighed by its *unanticipated harmful* consequences. The word *technology* suggests a process over which we have control. But the dead metaphor buried in the word should make us wary of that. *Technē*, art, skill, or craft, comes from *tiktō,* to bring into the world, beget or bear; and *teknon,* that which is borne or born, is, like the Scots *bairn,* a child. What may have to be borne about children, as countless parents have discovered, is that they do not turn out to be what was hoped for or expected. That could also be true of biotechnology.

What makes biotechnology particularly problematic in this respect is what it works with, the organism. Every good artist, craftsman, engineer and technologist realizes the importance of working *with* their materials. But for the biotechnologist, this is important in a special sense, because the material with which he works is neither passive nor predictable in its response to what the technologist does to it. As Hans Jonas once put it:

> In hardware engineering, the number of "unknowns" is practically nil, and the engineer can accurately predict the properties of his product. For the biological engineer, who has to take over, "sight-unseen," the untold complexity of the given determinants with their self-functioning dynamics, the number of unknowns in the design is immense.[2]

The number of unknowns in the organism's active complexity is so immense, Jonas argued, that "the simple ethics of the case are enough to rule out the direct tampering with human genotypes (which cannot be other than amateurish) from the very beginning of the road."[3] *Human* biotechnology should be ruled out even in its first experimental stages, he believed, because for such a research programme to achieve its objectives, it would be necessary, somewhere along the road, to conduct "sight unseen" and irreversible experiments on human subjects. This is a point similar to the objection to cloning humans I mentioned earlier.

Jonas made this point however, almost a quarter of a century ago. In the meantime, there have been significant advances in identifying genetic factors contributing to many rare and some more common human diseases and disabilities. This suggests that Jonas's argument for ruling out human biotechnology "from the very beginning of the road" may have been mistaken. On the other hand it is still unclear whether these advances will lead to effective gene therapy; and even if they do, it will be difficult to decide how far to go down a road that leads, with no apparent break, from preventing the most serious diseases and disabilities related to genes of high penetrance to enhancing human genetic capabilities. The problem here is not only that of deciding what such enhanced or better adapted people would be like—"Better adapted to what?", as Jonas puts it. It is also, again, that of the peculiar unpredictability of irreversible biotechnological experimentation on human subjects.

This peculiar unpredictability of biotechnological experimentation may prove to be the crucial factor in society's moral cost-benefit analysis, when deciding how far to proceed down the road of human genetic modification. Governments, regulatory bodies, companies, and the courts will need to be convinced that the condition to be treated is serious enough, the genetic influence clear enough, and the techniques proposed safe and effective enough, before they will be prepared to countenance researchers enrolling human subjects into clinical trials. This gate may be sufficiently narrow to allow gene therapy for serious conditions to get through, while keeping out a Gadarene rush down the slippery slope to enhancement and eugenics.

Or again it may not. The devil, as they say, is in the detail; and the details of biotechnology and of the relations between fundamental and applied research are nothing if not complex. The unanticipated harmful consequences in the end *may* outweigh its anticipated benefits. Or, again, they may not. At this stage, we simply do not know what the final balance sheet will show.

That conclusion applies not only to human genetics, but also to plant and animal biotechnology. They have more benefits to show, of course, having been at it longer—indeed for millennia, if plant and animal biotechnology is regarded as simply plant cultivation and animal breeding by more scientific means. But the risks of "evolution in the fast lane," as Bernard Rollin puts it,[4] are much more unpredictable, and an increasing number of voices worldwide now are asking if it really is wise to proceed further with this risky new business of biotechnology.

It is easier to ask that question than to answer it, however. Science cannot answer it: the scientific unpredictability of possible consequences is precisely the point. Nor, perhaps, can society, if by that we mean deciding science policy by consulting opinion polls. Such decisions cannot responsibly be delegated to public opinion or public sentiment alone—most people's awareness of the issues is too occasional, or too vulnerable to the spin put upon the issues by interested parties or sensation-hungry media. For decisions about the future of biotechnology to be responsible decisions, they need to be not only scientifically well informed, but also based on principles we can respect and appeal to.

BIOTECHNOLOGY AND MYTHOLOGY

But where are such principles to be found? They cannot be plucked from the air or handed down by government diktat. One place where they may be found and, surprisingly, seem to remain accessible today, is in mythical thinking. I say "surprisingly" because for many years we have tended to regard mythical thinking as something modern people have outgrown. But this appears not to be so; and a case in point is that when discussion of biotechnology becomes heated, someone is sure to bring up the issue of *hubris*, or "playing God."

The notion of *hubris* comes from ancient Greece, where it meant "the madness of human pride, arrogantly setting out to defy the gods," and it is in roughly the same sense that talk about hubris or "playing God," or even the odd erudite reference to Promethean fire seems to surface when people nowadays worry about biotechnology "going too far." These terms, incidentally, do not appear to be used especially by people with recognizable religious beliefs; and why they should appear so regularly in a modern secular context seems at first sight puzzling. But it may seem less puzzling if, putting on one side the prejudice that mythical thinking is something we (should) have outgrown, we try to entertain the idea that myths represent, in pictorial or symbolic terms, the sedimented experience of generations of human beings. For if that is even part of what mythical thinking is, we may have something to learn from it, including something about the principles and values we can respect and appeal to, in making responsible decisions about the future of biotechnology.

In trying to learn something from mythical thinking however, we also need to think critically. Some myths may be more reliable and relevant than others. The notion of hubris is a good example, since there is a risk of our taking over this idea uncritically from a mythology which may not represent our reflective experience of the world we live in. The mythological context from which the notion of hubris derives is that of a world ruled by gods who, in their lusts, jealousies, and touchiness are utterly capricious. In order not to offend them, the best thing is always to keep your head below the parapet, or in the loose translation of hubris beloved by parents, priests, and other paternalists from time immemorial, not to "get above yourself." The sedimented experience which this kind of mythological thinking represents, in other words, is experienced shaped by the assumption that you live in a world not governed by ultimately intelligible laws. But this view of the world is incompatible with that of modern science and of its origins in the cultural synthesis of classical philosophy and Judeo-Christian religion. Science and religion both assume that the world *is* ultimately intelligible, however little we may yet understand it. But if the world *is* ultimately intelligible, it is difficult to understand why fear of hubris or of playing God, should play a significant, let alone a decisive part, in making responsible decisions about the future of biotechnology. We may, of course, still be trying to say something important when we use these terms. But is there is some better way of saying it?

To suggest one possible way, let me briefly mention two important moments in one of the founding myths of the culture within which modern scientific inquiry has grown up. In the second chapter of Genesis, the newly created animals are brought to Adam "to see what he would call them; and whatsoever Adam called every every living creature, that was the name thereof." In this mythological context, naming something, deciding what it is, or how to act toward it, is the first step either toward real power over it, or, if it is more powerful than you, toward coming to terms with it in your own mind. The story of Adam naming the animals can be read as a mythical representation of the fact that the human animal, by possessing language, can and does gain knowledge about, and exercise power over, the other animals. Moreover, the fact that what Adam calls each animal *is* its name, implies in mythological terms, that human knowledge of the world is reliable. The myth does not suggest, of course, that reliable knowledge of the world is immediately and intuitively accessible. What myth represents is not a moment in time, but the human condition as such—in this case the whole process of working painstakingly toward knowledge. But if we work painstakingly enough, the myth suggest, reliable knowledge can be arrived at.

In the light of this mythological understanding, modern biotechnology can be seen as the late flowering of a seminal idea in the ancient mythic imagination, brought to mature growth by millennia of animal breeding and centuries of scientific refinement. This does not mean that everything done in the name of modern biotechnology is morally justified. But it does suggest that biotechnology is not an unnatural activity for humans to engage in. As Bernard Rollin again has observed, what we sometimes call "interfering" with nature is in fact what humans, by their very nature, do.[5]

But there is also the second important moment in the founding myth. In the third chapter of Genesis, Adam and Eve eat the forbidden fruit of the tree of the knowledge of good and evil, and lose their primal innocence or harmlessness. This again is a mythical representation of a familiar fact about the human animal, namely that while our actions can have beneficial or harmful consequences, it is often difficult to predict, not only what the consequences of our actions will be, but also whether even the consequences we can envisage will be truly good or evil for us, for others, and in the larger scheme of things. The fruit of the tree of the knowledge of good and evil, Genesis implies, was food not for humans but for gods, because to distinguish clearly between what is good and what is evil, you also need to be both omniscient and undistracted by self-interest—both of which are divine rather than human attributes. When humans eat the fruit, they become confused, at least at first. Their only hope of escaping from moral confusion is by achieving as close an approximation to these divine attributes as is possible for mortals. In practice this means working to become both less ignorant and also less distracted by short-term self-interest—goals which historically have been those, respectively, of science and philosophy and of religion and morality.

Let me briefly sum up what I think these two moments in the Genesis myth suggest. Seeking both to understand the world and to change it for the better are natural aspirations of the human animal. Understanding the world is possible, but only by a long, painstaking process. Changing the world for the better also is possible. But to achieve this, an equally long and painstaking process is required, to overcome not only ignorance, but also short-term self-interest, which prevents us from understanding what good and evil really are. Our best hope of understanding the world and changing it for the better, in other words, is by becoming not only better scientists, but also better people—more skillful and impartial interpreters of our own and other people's motives, limitations, and human potential.

What principles can be derived from this mythical understanding? In the European tradition, mature reflection replaces the superstitious fear of hubris with the philosophical and religious principle of *prudence*. Prudence is not, as sometimes supposed, the same thing as caution. It is, rather, a positive virtue, which seeks through practical wisdom, based on mature experience, to steer a middle course between rashness and timidity, and to determine the right means to the best ends. Steering this middle course, in relation to an enterprise such as biotechnology, requires a willingness to take calculated risks—but only when our calculations are based on a disinterested attempt to judge what is truly good or evil not only for ourselves, but also for others and in the larger scheme of things as far as we can understand it. Our capacity to understand that larger scheme of things however, grows in proportion not just to our knowledge but also to the sort of people we are.

THE MORAL STATUS OF ANIMALS

Let me now turn from biotechnology in general to animal biotechnology in particular, and beginning from the mythological context I have mentioned, briefly discuss the question of the moral status of animals.

The Judaic founding myths appear to endorse the view that animals exist primarily for human use. There is just a hint in Genesis that vegetarianism might be God's preferred option for human, and later there are some laws against cruelty to domestic animals. But otherwise man has dominion. For the experience sedimented in mythical thinking, that is just the way things are. But as and when humans secured time to think a little more deeply about their condition, there are signs that people may have begun to question whether this was the way things *ought* to be.

The main sign perhaps, is the enormous effort people have always devoted to justifying human dominion by arguing that humans are radically different from other animals. Being "animal" or "beastly" is precisely what, it is claimed, humans ought not to be. This strenuous denial of our kinship with other animals has received some severe blows in recent times however, above all from Darwin. Man continues to have dominion, but with a less easy conscience. In ancient Greek, the word *bios* or life was distinguished from *zōē* or animal life and meant a course of life, a manner of living, or a life in the sense of biography. A troubling question about animal biotechnology for the uneasy modern conscience is whether, and if so when, we are treating as *zōē* what we ought to be treating as *bios*.

One does not need to be a full-blown antivisectionist to ask such questions. Government regulations in Britain, for example, do not allow animal work on great apes or domestic strays, and have included the invertebrate octopus as well as vertebrates under the provisions of the Animals (Scientific Procedures) Act. Such decisions make sense only in the light of scientists' descriptions of being *watched* by octopuses in their laboratory tanks, of what Dian Fossey *felt* about her gorillas, and of what the man in the street *knows* about his dog. Each of them, in their different ways, are including their animals among the "others" from whose point of view, the philosopher Kant tells us[6] we need to try to think if we are to reach moral maturity.

Attempts to see things from the animal's point of view, of course, often have been dismissed on the grounds that we have no way of knowing what it is like to be a bat, for example; and it certainly is true that we have no way of knowing other animals' experience from the inside. But the same, strictly speaking, holds of other people. I can't get inside your head, nor you inside mine. Yet that does not stop each of us exercising our imagination about what the other might be feeling. Extending this courtesy to the members of other species, I think, is a factor that will need increasingly to be included in our moral cost-benefit analysis if animal biotechnology is to continue to receive public support.

This point has not been lost, I think, on the owners of Dolly the sheep, who frequently emphasize what a natural and well-husbanded existence their cloned or transgenic sheep enjoy, safely grazing in their elysian field. But things may be more difficult for researchers proposing to create transgenic whole animals models of some of the nastier human diseases. They may find themselves in a situation increasingly similar to that of researchers proposing to use human subjects—the condition to be investigated will have to be serious and, probably, widespread enough, and the scientific evidence and techniques robust enough, for such work to be allowed to go ahead in certain animal species.

Such ethical considerations are complicated, of course, by the possibility that the precision of transgenic technology, after an initial increase in the number of experimental transgenic animals, might lead in due course to the use of far fewer animals in research than at present. Or again they might be complicated in a different way, if it became possible to create decerebrate but otherwise viable transgenic animal models—a prospect that some people regard as the ideal solution but others as a moral nightmare. Ethical cost-benefit analysis in relation to animal biotechnology, in other words, is unlikely to become easier.

SCIENCE AND THE MARKET

The final issue I want to mention briefly concerns the financial profits that can be made from biotechnology. "Scientists at work," Lewis Thomas once wrote,

> have the look of creatures following genetic instruction; they seem to be under the influence of a deeply placed human instinct. They are, despite their efforts at dignity, rather like young animals engaged in savage play. When they are near to an answer their hair stands on end, they sweat, they are awash in their own adrenalin. To grab the answer, and grab it first, is for them a more powerful drive than feeding or breeding or protecting themselves against the elements.[7]

Where else, except in the laboratory, can you observe such a combination of intellectual acumen and visceral excitement? In the tents of traders and the temples of commerce, where there is all to play for. Bring scientists and businessmen together and don't be surprised if prudence flies out the window.

There is no simple remedy for this, because the instincts of merchants are just as deeply placed in human nature as those of scientists, markets are just as necessary to society as hospitals, and there may be benefits in mixing them up, as we do nowadays, with boutiques in hospital concourses and surgeries in supermarkets. But there may be a downside too. Some commentators believe that biotechnology, unlike many earlier technologies, now tends to be disproportionately driven by what can be sold to the rich, rather than by seeking more effective ways of meeting the basic needs of the poor. This tendency, they warn, has already begun to provoke a backlash of antiscientific sentiment in public opinion worldwide; and if this becomes sufficiently widespread, the practical results will ultimately impoverish all of us.[8]

Again there is no easy remedy for this. The only remedy is the difficult one. In order to handle the beautiful but dangerous two-edged sword of biotechnology with the skill it requires, we need to learn to become more reflective in our public debates and more discriminating in our political and market choices. Whether we shall learn that in time, or before some biotechnological Chernobyl erupts, is an open question.

NOTES

1. Bacon F. "Of Innovations," in *Essays* (1579). London: Dent, 1955: 74–75.

2. Jonas H. *Philosophical Essays.* Chicago: University of Chicago Press, 1974: 143.

3. Ibid., 167.

4. Rollin B. *The Frankenstein Syndrome.* Cambridge: Cambridge University Press, 1995.

5. *Ibid.*

6. Kant I. *The Critique of Judgement* (1790). Oxford: Oxford University Press, 1952: 152.

7. Thomas L. *The Lives of a Cell.* London: Futura, 1976: 118.

8. Dyson F. "Can Science be Ethical?" *New York Review of Books*, XLIV 1997, 6: 46–49.

9

Cloning, Then and Now

Daniel Callahan

The possibility of human cloning first surfaced in the 1960s, stimulated by the report that a salamander had been cloned. James D. Watson and Joshua Lederberg, distinguished Nobel laureates, speculated that the cloning of human beings might one day be within reach; it was only a matter of time. Bioethics was still at that point in its infancy—indeed, the term "bioethics" was not even widely used then—and cloning immediately caught the eye of a number of those beginning to write in the field. They included Paul Ramsey, Hans Jonas, and Leon Kass. Cloning became one of the symbolic issues of what was, at that time, called "the new biology," a biology that would be dominated by molecular genetics. Over a period of five years or so in the early 1970s a number of articles and book chapters on the ethical issues appeared, discussing cloning in its own right and cloning as a token of the radical genetic possibilities.

While here and there a supportive voice could be found for the prospect of human cloning, the overwhelming reaction, professional and lay, was negative. Although there was comparatively little public discussion, my guess is that there would have been as great a sense of repugnance then as there has been recently. And if there had been some kind of government commission to study the subject, it would almost certainly have recommended a ban on any efforts to clone a human being.

Now if my speculation about the situation 20 to 25 years ago is correct, one might easily conclude that nothing much has changed. Is not the present debate simply a rerun of the earlier debate, with nothing very new added? In essence that is true. No arguments have been advanced this time that were not anticipated and discussed in the 1970s. As had happened with other problems in bioethics (and with genetic engineering most notably), the speculative discussions prior to important scientific breakthroughs were remarkably prescient. The actuality of biological progress often adds little to what can be imagined in advance.

Yet if it is true that no substantially new arguments have appeared over the past two decades, there are, I believe, some subtle differences this time. Three of them are worth some comment. In bioethics, there is by far a more favorable response to scientific and technological developments than was then the case. Permissive, quasi-libertarian attitudes toward reproductive rights that were barely noticeable earlier now have far more substance and support. And imagined or projected research benefits have a stronger prima facie claim now, particularly for the relief of infertility.

1. *The response to scientific and technological developments.* Bioethics came to life in the mid-to late 1960s, at a time not only of great technological advances in medicine but also of great social upheaval in many areas of American cultural life. Almost forgotten now as part of the "sixties" phenomenon was a strong antitechnology strain. A common phrase, "the greening of America," caught well some of that spirit, and a number of writers were as prepared to indict technology for America's failings as they were to indict sexism, racism, and militarism.

While it would be a mistake to see Ramsey, Jonas, and Kass as characteristic sixties thinkers—they would have been appalled at such a label— their thinking about biological and genetic technology was surely compatible with the general suspicion of technology that was then current. In strongly opposing the idea of human cloning, they were not regarded as Luddites or radicals, nor were they swimming against the tide. In mainline intellectual circles it was acceptable enough to be wary of technology, even to assault it. It is probably no accident that Hans Jonas, who wrote so compellingly on technology and its potentially deleterious effects, was lauded in Germany well into the 1990s, that same contemporary Germany that has seen the most radical "green" movement and the most open, enduring hostility to genetic technology.

There has been considerable change since the 1960s and 1970s. Biomedical research and technological innovation now encounter little intellectual resistance. Enthusiasm and support are more likely. There is no serious "green" movement in biotechnology here as in Germany. Save possibly for Jeremy Rifkin, there are no regular, much less celebrated, critics of biotechnology. Technology bashing has gone out of style. The National Institutes of Health, and *particularly* its Human Genome Project, receive constant budget increases, and that at a time of budget cutting of government programs. The genome project, moreover, has no notable opponents in bioethics—and it would probably have support *even* if it did not lavish so much money on bioethics.

Cloning, in a word, now has behind it a culture far more supportive of biotechnological innovation than was the case in the 1960s and 1970s. Even if human cloning itself has been, for the moment, rejected, animal cloning will go forward. If some *clear* potential benefits can be envisioned for human cloning, the research will find a background culture likely to be welcoming rather than hostile. And if money can be made off of such a development, its chances will be greatly enhanced.

2. *Reproductive rights.* The right to procreate, as a claimed human right, is primarily of post–World War II vintage. It took hold first in the United States with the acceptance of artificial insemination (AID) and was strengthened by a series of court decisions upholding contraception and abortion. The emergence of in vitro fertilization in 1978, widespread surrogate motherhood in the 1980s, and a continuous stream of other technological developments over the past three decades have provided a wide range of techniques to pursue reproductive choice. It is not clear what, if any, limits remain any longer to an exercise of those rights. Consider the progression of a claimed right: from a right to have or not have children as one chooses, to a right to have them any way one can, and then to a right to have the kind of child one wants.

While some have contended that there is no natural right to knowingly procreate a defective or severely handicapped child, there have been no serious moves to legally or otherwise limit such procreation. The right to procreation has, then, slowly become almost a moral absolute. But that was not the case in the early 1970s, when the reproductive rights movement was just getting off the ground. It was the 1973 *Roe* v. *Wade* abortion decision that greatly accelerated it.

While the National Bioethics Advisory Commission ultimately rejected a reproductive rights claim for human cloning, it is important to note that it felt the need to give that viewpoint ample exposure. Moreover, when the commission called for a five-year ban followed by a sunset provision—to allow time for more scientific information to develop and for public discussion to go forward—it surely left the door open for another round of reproductive rights advocacy. For that matter, if the proposed five-year ban is eventually to be lifted because of a change in public attitudes, then it is likely that putative reproductive rights will be a principal reason for that happening. Together with the possibility of more effective relief from infertility (to which I will next turn) it is the most powerful viewpoint waiting in the wings to be successfully deployed. If procreation is, as claimed, purely a private matter, and if it is thought wrong to morally judge the means people choose to have children, or their reasons for having them, then it is hard to see how cloning can long be resisted.

3. *Infertility relief and research possibilities.* The potential benefits of scientific research have long been recognized in the United States, going

back to the enthusiasm of Thomas Jefferson in the early years of American history. Biomedical research has in recent years had a particularly privileged status, commanding constant increases in government support even in the face of budget restrictions and cuts. Meanwhile, lay groups supportive of research on one undesirable medical condition or another have proliferated. Together they constitute a powerful advocacy force. The fact that the private sector profits enormously from the fruits of research adds still another potent factor supportive of research.

A practical outcome of all these factors working together is that in the face of ethical objections to some biotechnological aspirations there is no more powerful antidote than the claim of potential scientific and clinical benefits. Whether it be the basic biological knowledge that research can bring, or the direct improvements to health, it is a claim difficult to resist. What seems notably different now from two decades ago is the extent of the imaginative projections of research and clinical benefits from cloning. This is most striking in the area of infertility relief. It is estimated that one in seven people desiring to procreate are infertile for one reason or another. Among the important social causes of infertility are late procreation and the effects of sexually transmitted diseases. The relief of infertility has thus emerged as a major growth area in medicine. And, save for the now traditional claims that some new line of research may lead to a cure for cancer, no claim seems so powerful as the possibility of curing infertility or otherwise dealing with complex procreation issues.

In its report, the bioethics commission envisioned, through three hypothetical cases, some reasons why people would turn to cloning: to help a couple both of whom are carriers of a lethal recessive gene; to procreate a child with the cells of a deceased husband; and to save the life of a child who needs a bone marrow transplant. What is striking about the offering of them, however, is that it now seems to be considered plausible to take seriously rare cases, as if—because they show how human cloning could benefit some few individuals—that creates reasons to accept it. The commission did not give in to such claims, but it treated them with a seriousness that I doubt would have been present in the 1970s.

Hardly anyone, so far as I can recall, came forward earlier with comparable idiosyncratic scenarios and offered them as serious reasons to support human cloning. But it was also the case in those days that the relief of infertility, and complex procreative problems, simply did not command the kind of attention or have the kind of political and advocacy support now present. It is as if infertility, once accepted as a fact of life, even if a sad one, is now thought to be some enormous menace to personal happiness, to be eradicated by every means possible. It is an odd turn in a world not suffering from underpopulation and in a society where a large number of couples deliberately choose not to have children.

WHAT OF THE FUTURE?

In citing what I take to be three subtle but important shifts in the cultural and medical climate since the 1960s and 1970s, I believe the way has now been opened just enough to increase the likelihood that human cloning will be hard to resist in the future. It is that change also, I suggest, that is responsible for the sunset clause proposed by the bioethics commission. That clause makes no particular sense unless there was on the part of the commission some intuition that both the scientific community and the general public could change their minds in the relatively near future—and that the idea of such a change would not be preposterous, much less unthinkable.

In pointing to the changes in the cultural climate since the 1970s, I do not want to imply any approval. The new romance with technology, the seemingly unlimited aims of the reproductive rights movement, and the obsession with scientific progress generally, and the relief of infertility particularly, are nothing to be proud of. I would like to say it is time to turn back the clock. But since it is the very nature of a progress-driven culture to find such a desire reprehensible, I will suggest instead that we turn the clock forward, skipping the present era, and moving on to one that is more sensible and balanced. It may not be too late to do that.

10

On Re-Doing Man

Kurt Hirschhorn

[When Dan Callahan was the editor of *Commonweal* he organized a symposium in 1967 to address many aspects, cultural and scientific, of the rapid changes going on at the time in human biology and behavior. He asked me to address the genetics aspects, which I did in an article entitled, "On Re-Doing Man," originally published in *Commonweal*.[1] This was later reprinted in Annals of the New York Academy of Sciences.[2] I have been asked to comment on my predictions and concerns in that paper and perhaps look somewhat into the future.

One thing is certain: the speed and the amount of activity designed to study and influence human biology, especially human genetics, has increased exponentially and continues to accelerate. The completion of the mapping and sequencing of the entire human genome, will lead to yet another quantum leap in these activities.

Many of the predictions in the paper have come to pass, but many of the concerns remain. Let me begin with the topic of cloning, defined as producing an identical copy of an individual by nuclear transfer techniques. These techniques were suggested in the 1960s by Lederberg and, of course, have recently been successful in sheep, mice, cows and pigs. I predicted at the time that such cloning would be possible and, therefore, would eventually be tried because, as a general rule, what can be done will be done. This prediction was agreed to by Waddington in his comments on my paper in *Annals of the New York Academy of Sciences*.[3]

I pointed out the danger of doing this on a large scale since in the long run it would increase homozygosity (identical genes in both chromosomes of the pair transmitted by father and mother) for a number of genes if such individuals eventually breed with each other. Such homozygosity would lead to a decrease in the adaptation of the human species to changes in environment which are also happening at an increasing rate. While we know of no attempt to clone a human being by this method, the success in large mammals implies that this should not be too difficult. It is important, however, to point out some of the misconceptions regarding the nature of cloned individuals. It is assumed that they will be identical in genotype (the actual genes in an individual's coded DNA) and to a great extent in phenotype (expressed characteristics, appearance, and traits determined by the genotype) to the individual whose somatic nucleus is used to put into the nucleated egg.

There are several errors in this assumption. First if the egg did not come from the same person as the nucleus, it would bring with it a host of mitochondrial genes that will have an impact on the cloned individual, making him or her different from the donor of the nucleus. The cloned embryo will develop in a different uterine environment than did the donor of the nucleus, bringing numerous phenotypic and adaptive changes produced by the intrauterine environment. Perhaps most important is that the development of the brain and its myriad connections is guided by the experiences of the young child and not simply by the genes in the neurons. Therefore, the character and thought processes of the cloned individual will be considerably different from those of the donor. In addition, the clone will have to function in a world which is significantly different from that in which the donor functioned; thereby changing the thoughts, activities, and endeavors of the new individual. This obviously does away with all the predictions of hordes of either Einsteins or fascist armies.

I believe that cloning will happen probably first to help some desperate couples suffering from infertility, and such plans are already being discussed. More positively, embryonic cloning could lead to development of replacement organs for the donor of the nucleus. This technique would never lead to a completely cloned human being but only to some stem cells which probably would be able to be induced to specific types of cellular differentiation (see chapter 27).

I also predicted that gene therapy would become possible and perhaps practical. Early experiments have in fact been successful to a greater or less extent. Most people in the field advocate restricting gene therapy to somatic gene therapy in order to avoid introducing foreign genes into the germ line to be passed to the offspring of the treated individual. The recent suggestions for intrauterine gene therapy may result in the inadvertent introduction of the therapeutic gene into the germ line of the fetus. In fact, with other new techniques now becoming practical, germ line gene therapy, even as a concept, becomes unnecessary. Preimplantation diagnosis in cells taken from an 8-cell embryo permits the implantation into the mother's uterus of those embryos that do not carry the genetic conditions for which they are at risk. Early work has even shown that by genetic examination of the polar body, one can guarantee that the egg used in in vitro fertilization does not carry the detrimental gene or the abnormal chromosome. (When the germ cell divides to produce the egg containing one chromosome of each pair, the other chromosome of each pair ends up in a cell called the polar body.) I believe that the future will bring efficient sperm selection in that unaffected sperm can be used for natural fertilization.

Many laboratories and industries are gearing up to expand current screening and testing by molecular techniques. Such screening for breast cancer genes and a variety of inborn errors is already available. The future will bring technology putting DNA arrays on microscopic slides (chips) which will be able to predict with greater or lesser certainty what common illnesses individuals are susceptible to, when they might get ill and perhaps even when they might die. Such knowledge would clearly be of great interest for insurance companies and perhaps employers. Society must force legislators to protect genetic (and medical) confidentiality from potential discriminatory practices.

While one could continue to speculate on novel techniques, reality will be more exciting and surprising than anything additional that I could predict at this time. However, one principle which I stressed in the paper remains important to remember. Large attempts at positive eugenics would lead to increased genetic uniformity of the human species. As I said then, and I report now, even a small shift in this uniformity could seriously impact the survival of the human species, which is dependent upon maximum heterogeneity for maximum adaptability to changing environments. It, therefore, remains a major duty of science and medicine to address adaptation of the environment to allow optimal realization of full genetic potential for all individuals, however naturally endowed, as far as possible.]

The past 20 years and, more particularly, the past five years have seen an exponential growth of scientific technology. The chemical structure of the hereditary material and its language have essentially been resolved. Cells can be routinely grown in test tubes by tissue-culture techniques. The exact biochemical mechanisms of many hereditary disorders have been clarified. Computer programs for genetic analysis are in common use. All these advances and many others have inevitably led to discussions and suggestions for the modification of human heredity, both in individuals and in populations: genetic engineering.

One of the principal concerns of the pioneers in the field is the problem of the human genetic load, that is, the frequency of disadvantageous genes in the population. Each of us carries between three and eight genes that, if present in double dose in the offspring of two carriers of identical genes, would lead to severe genetic abnormality, or even to death of the affected individual before or after birth. In view of the rapid medical advances in the treatment of such diseases, it is likely that affected individuals will be able to reproduce more frequently than in the past. Therefore, instead of a loss of genes due to death or sterility of the abnormal, the mutant gene will be transmitted to future generations in a slowly but steadily increasing frequency. This is leading the pessimists to predict that we will become a race of genetic cripples requiring a host of therapeutic crutches. The optimists, on the other hand, have a great faith that the forces of natural evolution will continue to select favorably those individuals who are best adapted to the then current environment. It is important to remember in this context that the "natural" environment necessarily includes man-made medical, technical, and social factors.

Because it appears that at least some of the aspects of evolution and a great deal of genetic planning will be in human and, specifically, scientific hands, it is crucial at this relatively early stage to consider the ethical implications of these proposed maneuvers. Few scientists today doubt the feasibility of genetic engineering, and there is considerable danger that common use of this practice will be upon us before its ethical applications are defined.

A number of different methods have been proposed for the control and modification of human hereditary material. Some of these methods are meant to work on the population level, some on the family level, and others directly on the affected individual. Interest in the alteration of the genetic pool of human populations originated shortly after the time of Mendel and Darwin, in the latter part of the ninth century. The leaders were the English group of eugenicists headed by Galton. Eugenics is nothing more than planned breeding. This technique, of course, has been successfully used in the development of hybrid breeds of cattle, corn, and other food products.

Human eugenics can be positive or negative. Positive eugenics is the preferential breeding of so-called superior individuals in order to improve the genetic stock of the human race. The most famous of the many proponents of positive eugenics was the late Nobel Prize winner Herman J. Muller. He suggested that sperm banks be established for a relatively small number of donors, chosen by some appropriate panel, and that this frozen sperm remain in storage until some future panel had decided that the chosen donors truly represented desirable genetic studs. If the decision is favorable, a relatively large number of women would be inseminated with these samples of sperm; proponents of this method hope that a better world would result. The qualifications for such a donor would include high intellectual achievement and a socially desirable personality, qualities assumed to be affected by the genetic make-up of the individual, as well as an absence of obvious genetically determined physical anomalies.

A much more common effort is in the application of negative eugenics. This is defined as the discouragement or the legal prohibition of reproduction by individuals carrying genes leading to disease or disability. This can be achieved by genetic counseling or by sterilization, either voluntary or enforced. There are, however, quite divergent opinions as to which genetic traits are to be considered sufficiently disadvantageous to warrant the application of negative eugenics.

A diametrically opposite solution is that of euthenics, which is a modification of the environment in such a way as to allow the genetically abnormal individual to develop normally and to live a relatively normal life. Euthenics can be applied both medically and socially. The prescription of glasses for nearsighted individuals is an example of medical euthenics. Special schools for the deaf, a great proportion of whom are genetically abnormal, is an example of social euthenics. The humanitarianism of such efforts is obvious, but it is exactly these types of activities that have led to the concern of the pessimists, who assume that evolution has selected for the best of possible variations in man and that further accumulations of genes considered abnormal can only lead to decline.

One of the most talked-about advances for the future is the possibility of altering an individual's genetic complement. Since we are well on the way to understanding the genetic code, as well as to deciphering it, it is suggested that we can alter it. This code is written in a language of 64 letters, each one determined by a special arrangement of three out of four possible nucleotide bases. A chain of these bases is called deoxyribonucleic acid, or DNA, and makes up the genetic material of the chromosomes. If the altered letter responsible for an abnormal gene can be located and the appropriate nucleotide base substituted, the corrected message would again produce its normal product, which would be either a structurally or enzymologically functional protein.

Another method of providing a proper gene, or code word, to an individual having a defect has been suggested from an analysis of viral behavior in bacteria. It has long been known that certain types of viruses can carry genetic information from one bacterium to another or instruct a bacterium carrying it to produce what is essentially a viral product. Viruses are functional only when they live in a host cell. They use the host's genetic machinery to translate their own genetic codes. Viruses living parasitically in human cells can cause such diseases as poliomyelitis and have been implicated in the causation of tumors. Other viruses have been shown to live in cells and to be reproduced along with the cells without causing damage either to the cell or to the organism. If such a harmless virus either produces a protein that will serve the function of one lacking in an affected individual, or if it can be made to carry the genetic material required for such functions into the cells of the affected individual, it could permanently cure the disease without additional therapy. If carried on to the next generation, it could even prevent the inheritance of the disease.

TRANSPLANTING NUCLEI

An even more radical approach has been outlined by Lederberg. It has become possible to transplant whole nuclei, the structures that carry the DNA, from one cell to another. It has become easy to grow cells from various tissues of any individual in tissue culture. Such tissue cultures can be examined for a variety of genetic markers and thereby screened for evidence of new mutations. Lederberg suggests that it would be possible to use nuclei from such cells derived from known human individuals, again with favorable genetic traits, for the asexual human reproduction of replicas of the individuals whose nuclei are being used. For example, a nucleus from a cell of the chosen individual could be transplanted into a human egg whose own nucleus has been removed. This egg, implanted in a womb, could then divide just like a normal fertilized egg, to produce an individual genetically identical to the one whose nucleus was used. One of the proposed advantages of such a method would be that, as in positive eugenics, one could choose the traits that appear to be favorable, and do so with greater efficiency by eliminating the somewhat randomly chosen female parent necessary for the sperm bank approach. Another advantage is that one can mimic what has developed in plants as a system for the preservation of genetic stability over limited periods of time. Many plants reproduce intermittently by such vegetative or parthenogenetic mechanisms, always followed by periods of sexual reproduction for the purpose of elimination of disadvantageous mutants and increase in variability.

Another possibility derives from two other technological advances. Tissue typing, similar to blood typing, has made it possible to transplant cells, tissues, and organs from one individual to another with reasonably long-term success. During the past few years, scientists have also succeeded in producing cell hybrids containing some of the genetic material from each of two cell types, either from two different species or from two different individuals of the same species. Very recently, Weiss and Green at New York University have succeeded in hybridizing normal human culture cells with cells from a long-established mouse tissue-culture line. Different products from such fusions contain varying numbers of human chromosomes and, therefore, varying amounts of human genes. If such hybrids can be produced which carry primarily that genetic information which is lacking or abnormal in an affected individual, transplantation of these cultured cells into the individual may produce a correction of his defect.

These are the proposed methods. It is now fair to consider the question of feasibility. Feasibility must be considered not only from a technical point of view; of equal importance is the effect of each of these methods on the evolution of the human population and the effect of evolution on the efficacy of the method. In general, it can be stated that most of the proposed methods either now or in the not too distant future will be technically possible. We are, therefore, not dealing with hypothesis in science fiction but with scientific reality. Let us consider each of the propositions independently.

Positive eugenics by means of artificial insemination from sperm banks has been practiced successfully in cattle for years. Artificial insemination in man is an everyday occurrence. But what are some of its effects? There is now ample evidence in many species, including man, of the advantages for the population of individual genetic variation, mainly in terms of flexibility of adaptation to a changing environment. Changes in environment can produce drastic effects on some individuals, but a population that contains many genetic variations of that set of genes affected by the particular environmental change will contain numerous individuals who can adapt. There is also good evidence that individuals who carry two different forms of the same gene, that is, are heterozygous, appear to have an advantage. This is true even if the gene in double dose—that is, in the homozygous state—produces a severe disease. For example, individuals homozygous for the gene coding for sickle cell hemoglobin invariably develop sickle cell anemia, which is generally fatal before the reproductive years. Heterozygotes for the gene are, however, protected more than normals from the effects of the most malignant form of malaria. It has been shown that women who carry the gene in single does have a higher fertility in malarial areas than do normals. This effect is well known to agricultural geneticists and is referred to as hybrid vigor. Fertilization of many women by sperm from few men will have an adverse effect on both of these advantages of genetic variability because the population will tend to be more and more alike in its genetic characteristics. Also, selection for a few genetically advantageous factors will carry with it selection for a host of other genes present in the same individuals, genes whose effects are unknown when present in high numbers in the population. Therefore, the interaction between positive eugenics and evolution makes this method not feasible on its own.

ABNORMAL OFFSPRING

Negative eugenics is, of course, currently practiced by most human geneticists. It is possible to detect carriers of many genes that, when inherited from both parents, will produce abnormal offspring. Parents, both of whom carry such a gene, can be told that they have a one-in-four chance of producing an abnormal child. Individuals who carry chromosomal translocations are informed that they have a high risk of producing offspring with congenital malformations and mental retardation. But how far can one carry out such a program? Some states have laws prescribing the sterilization of individuals mentally retarded to a certain degree. These laws are frequently based on false information regarding the heredity of the conditions. The marriage of people with reduced intelligence is forbidden in some localities, again without adequate genetic information. While the effects of negative eugenics may be quite desirable in individual families with a high risk of known hereditary diseases, it is important to examine its effects on the general population.

These effects must be looked at individually for conditions determined by genes that express themselves in a single dose (dominant) or in double dose (recessive) and those which are due to an interaction of many genes (polygenic inheritance). With a few exceptions, dominant diseases are rare and interfere severely with reproductive ability. They are generally maintained in the population by new mutations. Therefore, there is either no need or essentially no need for discouraging these individuals from reproduction. Any discouragement, if applicable, will be useful only within that family but not have any significance for the general population. One possible exception is the severe neurological disorder, Huntington's chorea, which does not express itself until most of the patient's children are already born. In such a situation it may be useful to advise the child of an affected individual that he has a 50% chance of developing the disease and a 25% chance of any of his children's being affected. Negative eugenics in such a case would at least keep the gene frequency at the level usually maintained by new mutations.

The story is quite different for recessive conditions. Although detection of the clinically normal carriers of these genes is currently possible only for a few diseases, the techniques are rapidly developing whereby many of these conditions can be diagnosed even if the gene is present only in single dose and will not cause the disease. Again, with any particular married couple it would be possible to advise them that they are both carriers of the gene and that any child of theirs would have a 25% chance of being affected. However, any attempt to decrease the gene frequency of these common genetic disorders in the population by prevention of fertility of all carriers would be doomed to failure. First, we all carry between three and eight of these genes in single doses. Second, for many of these conditions, the frequency in the population of carriers is about one in 50 or even greater. Prevention of fertility for even one of these disorders would stop a sizable proportion of the population from reproducing and for all of these disorders would prevent the entire human population from having any children. Reduction in fertility of a sizable proportion of the population would also prevent the passing on to future generations of a great number of favorable genes and would, therefore, interfere with the selective aspects of evolution, which can function only to improve the population within a changing environment by selecting from a gene pool containing enormous variability. It has now been shown that in fact no two individuals, with the exception of identical twins, are likely to be genetically and biochemically identical, thereby allowing the greatest possible adaptation to changing environment and the most efficient selection of the fittest.

The most complex problem is that of negative eugenics for traits determined by polygenic inheritance. Characteristics inherited in this manner include many measurements that are distributed over a wide range throughout the population, such as height, birth weight, and intelligence. The last of these can serve as a good example of the problems encountered. Severe mental retardation in a child is not frequently associated with perfectly normal intelligence or in some cases even superior intelligence in the parents. These cases can, a priori, be assumed to be due to the homozygous state in the child of a gene leading to mental retardation, the parents rep-

resenting heterozygous carriers. On the other hand, borderline mental retardation shows a high association with subnormal intelligence in other
family members. This type of deficiency can be assumed to be due to polygenic factors, more of the pertinent genes in these families being of the
variety that tends to lower intelligence. However, among the offspring of
these families there is also a high proportion of individuals with normal
intelligence and a sprinkling of individuals with superior intelligence.

All of these comments are made with the realization that our current
measurements of intelligence are very crude and cannot be compared between different population groups. It is estimated that, on the whole, people
with superior intelligence have fewer offspring than do those of average or
somewhat below average intelligence. If people of normal intelligence were
restricted to producing only two offspring and people of reduced intelligence were by negative eugenics prevented from having any offspring at all,
the result, as has been calculated by the British geneticist Lionel Penrose,
would be a gradual shift downward in the mean intelligence level of the
population. This is due to the lack of replacement of intellectually superior
individuals from offspring of the majority of the population, that is, those
not superior in intellect.

CURRENT POSSIBILITIES

It can be seen, therefore, that neither positive nor negative eugenics can
ever significantly improve the gene pool of the population and simultaneously allow for adequate evolutionary improvement of the human race. The
only useful aspect of negative eugenics is in individual counseling of specific families in order to prevent some of the births of abnormal individuals.
One recent advance in this sphere has important implications from both a
genetic and a social point of view. It is now possible to diagnose genetic
and chromosomal abnormalities in an unborn child by obtaining cells from
the amniotic fluid in which the child lives in the mother. Although the
future may bring further advances, allowing one to start treatment on the
unborn child and to produce a functionally normal infant, the only currently possible solution is restricted to termination of particular pregnancies
by therapeutic abortion. This is, of course, applied negative eugenics in its
most extreme form.

Euthenics, the alteration of the environment to allow aberrant individuals to develop normally and to lead a normal life, is currently being employed. Medical examples include special diets for children with a variety
of inborn errors of metabolism who would, in the absence of such diets,
either die or grow up mentally retarded. Such action, of course, requires
very early diagnosis of these diseases, and programs are currently in effect
to routinely examine newborns for such defects. Other examples include
the treatment of diabetics with insulin and the provision of special devices
for children with skeletal deformities. Social measures are of extreme im-

portance in this regard. As has many times been pointed out by Dobzhansky, it is useless to plan for any type of genetic improvement if we do not provide an environment within which an individual can best use his strong qualities and obtain support for his weak qualities. One need only mention the availability of an environment conducive to artistic endeavor for Toulouse-Lautree, who was deformed by an inherited disease.

The feasibility of alteration of an individual's genes by direct chemical change of his DNA is technically an enormously difficult task. Even if it became possible to do this, the chance of error would be enormous. Such an error, of course, would have the diametrically opposite effect of that desired; in other words, the individual would become even more abnormal. The introduction of corrective genetic material by viruses or transplantation or appropriately hybridized cells is technically predictable and, since it would be performed only in a single affected individual, would have no direct effect on the population. If it became widespread, it could, like euthenics, increase the frequency in the population of so-called abnormal genes, but if this treatment became a routine phenomenon, it would not develop into an evolutionarily disadvantageous situation. It must also be constantly kept in mind that medical advances are occurring at a much more rapid rate than any conceivable deterioration of the genetic endowment of man. It is, therefore, very likely that such corrective procedures will become commonplace long before there is a noticeable increase in the load of disadvantageous genes in the population.

The growing of human beings from cultured cells, while again possibly feasible, would interfere with the action of evolutionary forces. There would be an increase, just as with positive eugenics, of a number of individuals who would be alike in their genetic complement, with no opportunity for the high degree of genetic recombination that occurs during the formation of sperm and eggs and which is evident in the resultant progeny. This would diminish the adaptability of the population to changes in the environment and, if these genetic replicas were later permitted to return to sexual reproduction, would lead to a marked increase in homozygosity for a number of genes with the disadvantages pointed out before.

WHO WILL BE THE JUDGES?

We see, therefore, that many of the proposed techniques are feasible although not necessarily practical in producing their desired results. We may now ask the question, which of these are ethical from a humanistic point of view? Both positive and negative eugenics as applied to populations presume a judgment of what is genetically good and what is bad. Who will be the judges, and where will the line be between good and bad? We have had at least one example of a sad experience with eugenics in Nazi Germany. This alone can serve as a lesson on the impossibility of separating science and politics. The most difficult decisions will come in defining the borderline cases. Will we breed against tallness because space requirements become more critical? Will we breed against nearsightedness because people with glasses may not make good astronauts? Will we forbid intellectually inferior individuals from procreating despite their ability to produce a number of superior individuals? Or should we, rather, provide an adequate environment for the offspring of such individuals to realize their full genetic potential?

C. C. Li, in his presidential address to the American Society of Human Genetics in 1960, pointed out the real fallacy in eugenic arguments. Man has continuously improved his environment to allow so-called inferior individuals to survive and reproduce. The movement into the cave and the putting on of clothes protected the individual unable to survive the stress of the elements. Should we then consider that we have reached the peak of man's progress, largely determined by environmental improvements designed to increase fertility and longevity, and that any future improvements designed to permit anyone to live a normal life will only lead to deterioration? Nineteenth-century scientists, including such eminent biologists as Galton, firmly believed that this peak was reached in their time. This obviously fallacious reasoning must not allow a lapse in ethical considerations on the part of the individual and by humanity as a whole, just to placate the genetic pessimists.

The tired axiom of democracy that all men are created equal must not be considered from the geneticist's point of view, since genetically all men are created unequal. Equality must be defined purely and simply as equality of opportunity to do what one is best equipped to do. When we achieve this, the forces of natural evolution will choose those individuals best adapted to this egalitarian environment. No matter how we change the genetic make-up of individuals, we cannot do away with natural selection. We must always remember that natural selection is determined by a combination of truly natural events and the artificial modifications that we are introducing into our environment in an exponentially increasing number.

With these points in mind, we can try to decide what, in all of these methods, is both feasible and ethical. I believe that the only logical conclusion is that all maneuvers of genetic engineering must be judged for each individual and in each case must take into primary consideration the rights of the individual. This is by definition impossible in any attempt at positive eugenics. Negative eugenics in the form of intelligent genetic counseling is the answer for some. Our currently unreasonable attitude toward practicing negative eugenics by means of intelligent selection for therapeutic abortion must be changed. Basic to such a change is a more accurate definition of a living human being. Such restricted uses of negative eugenics will prevent individual tragedies. Correction of unprevented genetic disease, or that due to new mutation, by introduction of new genetic material may be one answer for the future; but until such a new world becomes universally feasible, we must on the whole restrict ourselves to environmental manipulations from the points of view both of allowing affected individuals to live normally and of permitting each individual to realize his full genetic potential. There is no question that genetic engineering will come about. But both the scientists directly involved and, perhaps more important, the political and social leaders of our civilization must exercise utmost caution in order to prevent genetic, evolutionary, and social tragedies.

NOTES

Reprinted from *Annals of the New York Academy of Sciences* (1971) 184: 103–112. Based on a paper originally given at a symposium jointly sponsored by Marymount College, N.Y., and *Commonweal* magazine. Reprinted by permission from *Commonweal* 88: 257–261 (1968), 232 Madison Ave., New York, N.Y.

1. Hirschhorn, K. On Re-Doing Man. *Commonweal* (1968) 88:257–261.
2. Hirschhorn, K. On Re-Doing Man. *Annals of the New York Academy of Sciences* (1971) 184:103–110.
3. Waddington, C. H. On Re-Doing Man. *Annals of the New York Academy of Sciences* (1971) 184:110–112.

II

A Report from America

The Debate about Dolly

Arlene Judith Klotzko

THE SCIENTIFIC BACKGROUND

On February 23, 1997, a British newspaper, the *Observer*, broke a story
that truly shocked the world. Dr. Ian Wilmut and his team at the Roslin
Institute, near Edinburgh, Scotland, had succeeded in a task many had believed to be impossible. They had cloned a mammal—a sheep, named
Dolly—from an adult mammary cell. (In the same experiment, the scientists also cloned sheep from cell lines composed of fetal and embryo
cells.)[1]

In order to create Dolly, the scientists performed nuclear transfer. Two
hundred and seventy seven times. Again and again, they removed a nucleus
from an egg cell and replaced that nucleus with one that had been taken
out of a mammary cell from an adult sheep. They applied an electric current, which caused the egg and its new nucleus to fuse and develop into
an embryo. Any embryos that resulted were implanted into surrogates.
Dolly was the only lamb to be born.

She is virtually a genetically identical copy of the adult sheep whose
mammary cell was used in the experiment—hence, a clone. Dolly can be

understood as a kind of later-born identical twin but, unlike those twins that occur naturally, there is some genetic difference between Dolly and her twin. Although the egg with which the nucleus of the adult sheep cell was fused had been stripped of its nucleus, it was not empty; it contained its own DNA—mitochondrial DNA.

Before this experiment, conventional wisdom held that a mammal could not be cloned from an adult cell. Unlike early embryo cells, which are totipotent—capable of becoming any and every cell in the body—adult cells are differentiated. In such cells—those that form the skin, muscle, and brain, for example—genes not needed to perform the required specialized function are switched off. In contrast, a totipotent cell can give rise to any cell in the body because it is capable of activating any gene on any chromosome.

In essence, the Roslin team had to trick the adult cell's DNA into reverting to its undifferentiated past by placing the mammary cells in a culture and starving them of nutrients for several days. In this quiescent state, few if any genes remained switched on. When the nuclei were removed from the adult cells, placed next to the enucleated egg cells, and fused by electricity, the eggs were able to reprogram the donor nuclei into behaving as if they had come from undifferentiated cells. The precise mechanism of this reprogramming is still not completely understood. [There is now great controversy over the question of quiescence. See chapter 1.]

Cloning of sheep and cows had been done successfully for some time, but only from early embryonic cells, and only in relatively small numbers at each attempt. In March 1996—almost exactly one year before the world got news of Dolly—the Roslin team announced a stunning breakthrough: the birth of Megan and Morag, two sheep cloned from mature embryo cells.[2]

The technique that they used—nuclear transfer performed with cells that had grown in a culture and then been rendered quiescent—was the same technique used to clone Dolly. Five identical lambs were born; two died at birth, and a third soon after. From one fertilized egg, Wilmut and his colleagues were able to grow a collection of cells in whatever quantities they wished, and thereby enable the production of very large numbers of genetically identical animals—even entire flocks.

The ultimate purpose of these experiments is to produce—efficiently and predictably—large numbers of transgenic animals (animals with human genes) that can secrete pharmaceutical compounds in their milk. In Britain, the creation of Megan and Morag was greeted with frenzied media speculation about creating legions of identical copies of animals. And humans. Concern was somewhat muted, however, because they were clones of embryos—not existing adults. Perhaps, for that reason, the American media did not take note.

THE MEDIA RESPONSE

In contrast, the cloning of Dolly produced an explosion of coverage in the American print and television media. Cloning stories appeared on the covers of *Time* and *Newsweek*. Most newspapers ran front page stories, and the coverage persisted for weeks. Many stories appeared accompanied by diagrams of nuclear transfer. Public consternation, already at fever pitch, only grew when, on March 2, it was announced that two monkeys had been cloned from embryo cells at the Primate Research Center in Beaverton, Oregon. Although this experiment marked the first time that live primates had been produced through cloning, embryo cells were used—not adult cells. The public inferred—erroneously—that we were now a step closer to cloning humans. As more than one observer dryly noted, last week it was a sheep, this week a monkey, so next week it will probably be a human.

There has been excellent reporting on the scientific accomplishments of Ian Wilmut's team—notably in the *New York Times* and the *Washington Post*. But there has been a lot of irresponsible reporting as well. *Newsweek*, for example, wondered whether society could "stuff the cloning genie back into the bottle," and compared Wilmut's discovery with nuclear bombs (Hiroshima and Nagasaki) and chemical weapons (the attack on the Tokyo subway).[3]

Americans were treated to descriptions of Mary Shelley's *Frankenstein*, armies of drones, and clone farms to produce spare parts. One particularly imaginative journalist painted an image of grandiose dictators cloning generations of themselves instead of building monuments,[4] evoking thoughts of another Shelley—Percy Bysshe. His poem "Ozymandias" describes a statue of a once powerful ruler, reduced by the passage of time to a pedestal, a shattered visage, and two trunkless legs. Cloning could have provided the story of Ozymandias with a happy ending. Or, perhaps, no ending.

The sensational character of U.S. press coverage was matched in the United Kingdom, where the most egregious example came from the *Daily Mail*. The headline of its February 24 issue asked, "Could We Now Raise the Dead?"—obviously confusing replication with resurrection. George Annas has written that "novels such as *Frankenstein* and *Brave New World* and films such as *Jurassic Park* and *Blade Runner* have prepared the public to discuss deep ethical issues in human cloning."[5] I emphatically disagree.

THEOLOGICAL VOICES

The American media have been full of bioethics and bioethicists on talk shows, in television interviews, and in newspaper and magazine articles. While American bioethics—fortunately or unfortunately, depending upon one's point of view—often has a consequentialist cast, the cloning discussion has had a pronounced deontological flavor. To be sure, opponents of human cloning made secular arguments—consequentialist and deontological—but many of the concerns articulated had a religious character that is particularly noteworthy.

Dr. Stanley Hauerwas, a divinity professor at Duke University, told the *New York Times* that he saw " 'a kind of drive behind this for us to be our own creators.' " In the same article, Dr. Kevin FitzGerald, a Jesuit priest and geneticist, said that a human clone would "have to have a different soul."[6] In a commentary in the *Washington Post*, theologian Nancy Duff stated her view that "if there were ever a Tower of Babel—which originally was an attempt to elevate ourselves through human accomplishments to the level of God—surely this is it."[7]

According to Daniel Callahan, American bioethics began with theologians—most notably, Joseph Fletcher and Paul Ramsey—but they were soon joined by lawyers, philosophers, and social scientists, and the field was secularized.[8] The ethical discussion of cloning, however, seems to have taken us back in time. And the customary public and media excitement over the latest advances in medical technology was eclipsed by talk of moral repugnance, evil, wrongness, playing God, and impermissible interventions. In this respect, there was a striking difference between American and British media coverage. In the United Kingdom, the overall character of the reporting was secular; religious arguments did appear, but they were few and far between.

Ronald Dworkin has noted the profound ambivalence of Americans on the subject of religion. Despite its constitutional separation of church and state, the United States is among the most religious of modern Western countries. And in the tone of its most powerful religious groups, by far the most fundamentalist.[9] In the opposition to abortion and to euthanasia, the Catholic Church has played an important role, as have fundamentalist Christians—particularly in regard to abortion, fetal tissue research, and embryo research.

But there is no anticloning lobby; religious objections to human cloning have been much more broadly based than in those other contexts. And they have been influential. The National Bioethics Advisory Commission (NBAC) devoted much of its first public hearing on human cloning to taking testimony from four Protestant theologians, and their Catholic, Jewish, and Muslim counterparts.[10]

The consequentialists might have been outshone, but they were not invisible. As Daniel Callahan observed, "The argument about Dolly saw two camps instantly formed—one was alarmed by the development and opposed to any further movement toward cloning humans; the other (seemingly much smaller) touted a potential gain in health and more reproductive choice if cloning went forward."[11] And, of course, consequentialist arguments can cut both ways. They can also point to the dire consequences of human cloning. And they did.

THE GOVERNMENT RESPONSE

Despite the enormous impact American bioethics has had in the United States—and worldwide—Prof. Alexander Capron, a member of NBAC, has noted that "the mere mention of the term 'bioethics' stirs up controversies in some quarters—particularly among right-to-life advocates."[12] In the late 1980s, the agenda of the Biomedical Ethics Advisory Committee—created by the Congress to study issues raised by the new genetics—ran afoul of the politics of abortion. The committee expired in 1990 after holding only two meetings.

Subsequently, there was a move to set up a national bioethics commission to be chartered by Congress. When Congress failed to enact legislation, the president's Office of Science and Technology Policy proposed to charter a bioethics commission as a subcommittee of the National Science and Technology Council, which advised the assistant to the president for science and technology. President Clinton issued an executive order creating NBAC in October 1995.[13] The members were not named until July 1996. Unless its charter is renewed, the commission is scheduled to go out of business in October 1997—one year after holding its first meeting.

In the words of Prof. Arthur Caplan, NBAC was given a "very low priority. . . . All of a sudden, cloning explodes and the president looks desperately. . . . for help and advice. The only group he can go to is the National Bioethics Advisory Commission."[14] On February 24, 1997, the president asked NBAC for a report on the legal and ethical implications of the cloning of Dolly—a report playfully described by one British commentator as "a quick-roasted ethical attitude without mint sauce within 90 days."[15]

On March 4, 1997, President Clinton issued an executive order banning the use of federal funding for human cloning research. Such a ban is a mechanism that has been seen before. A ban on the use of federal funds for research on human fetal tissue was instituted in 1988 and lifted only in 1993. A ban on federal funding for human embryo research has been in effect since 1994. The president also asked privately funded scientists to halt any human cloning research until NBAC issued its report.

But legislators did not wait for the report. On February 27, 1997, legislation (S368) was introduced in the Senate by Sen. Christopher Bond (R-Mo.):

> (a) IN GENERAL—No Federal funds may be used for research with respect to the cloning of a human individual.
> (b) DEFINITION—For purposes of this section, the term "cloning" means the replication of a human individual by the taking of a cell with genetic material and the cultivation of the cell through the egg, embryo, fetal, and newborn stages into a new human individual.

On March 5, 1997, legislation (HR 923) was introduced in the House of Representatives by Rep. Vernon Ehlers (R-Mioh.). His Human Cloning

Prohibition Act is broader than the Senate bill. It not only denies federal funds, it makes using "a human somatic cell for the process of producing a human clone" a civil wrong, and assesses a civil penalty of up to $5,000. [These bills did not become law, nor did a bill proposed later by President Clinton. For an updated discussion of anti-cloning legislation at the state level, see Chapter 23.]

By the fall 1997, legislators in New York, California, Illinois, Alabama, Florida, Maryland, Missouri, New Jersey, North Carolina, Oregon, South Carolina, and West Virginia had introduced bills that would do one or more of the following: prohibit human cloning research, prohibit use of state funds for such research, urge the president and Congress to do the same with federal funds, create a panel to advise the state legislature, or make human cloning a criminal offense.

In the Senate, a remarkable effort to foster an informed public (and legislative) discussion was quickly put together. On March 12, 1997, a hearing was held by the Senate Labor Committee's Subcommittee on Public Health and Safety. Dr. Ian Wilmut was one of 10 witnesses invited to testify. The chair of the subcommittee, Sen. William Frist, a former heart transplant surgeon, stated that he hoped to use the hearing to begin a public discussion about cloning and that the precondition for such a discussion was a thorough understanding of the science and of the underlying facts. The entire hearing was broadcast several times on television, and Dr. Wilmut's remarks were reported in virtually every newspaper and television news program.

Before the session, Dr. Wilmut met with Senator Frist and Senator Kennedy, the subcommittee's ranking Democratic member. In an interview, he characterized that meeting as follows: "It was clear that their aim was to allow careful thought before legislation. They were concerned that there was a knee-jerk response that said, 'We must stop that,' which could inadvertently prohibit uses which society would accept. I was very impressed with the senators and their staff assistants. And I hope that they achieve their objective, because one difficulty with this subject is that people use words carelessly. There may be uses of nuclear transfer with human cells that do not involve producing a new person. It is very important to make sure that in considering the prohibition of the production of new people, you don't inadvertently prohibit acceptable uses."[16]

Dr. Wilmut did not participate in the public hearings held by the National Bioethics Advisory Commission in Washington on March 13 and 14. When asked about his absence from these meetings, he said that he had not been invited to attend. "I am surprised that my colleagues and I have not been asked to present any information to the National Bioethics Advisory Commission either in person or in writing."[17]

In an April 22 speech before the National Press Club, Dr. Neal Lane, director of the National Science Foundation, said that the newfound ability to clone a mature sheep demands extensive public discussion and debate. But he expressed concern at the low level of scientific literacy in the United States. Senator Bill Frist addressed this deficit by asking the two scientist witnesses—Dr. Harold Vasmus, director of the National Institutes of Health, and Dr. Wilmut—to explain the science of cloning and its implications in a way that the public could understand. And they did—brilliantly.

But, after the saturation coverage of the Frist hearing, the media seemed to lose interest. The two-day NBAC hearing that began the next day received meager attention, and its subsequent meetings even less. When the ban on federal funding of human cloning research expires, there will be—one hopes—a public discussion about what should be done in the face of the unusual, if not unique, American legal situation concerning assisted reproduction.

THE LEGAL LANDSCAPE

In the United States there is no comprehensive regulatory scheme for IVF and embryo research. Human IVF has been developed and advanced in the private sector. More than 400 clinics operate with no federal money and little federal oversight. Guidelines have been developed by the industry, but compliance is voluntary.[18] There is also a set of ethical considerations, published by the American Society for Reproductive Medicine in 1994.[19]

According to biologist Prof. Lee Silver, "There are hundreds of private IVF clinics in America, where doctors and technicians are capable of. . . . artificially inseminating donated human eggs. Cloning would be no problem to such people."[20] There is no existing federal prohibition involving the cloning of humans or human embryos other than the president's temporary one. And no federal laws regulate embryo research. The subject is such a contentious one that Congress has simply been unable to address it.

In 1993, following a report that scientists at George Washington University, in Washington D.C., had cloned human embryos by embryo splitting,[21] there was a blast of media coverage and public concern somewhat similar to recent events. But the American scientists had not employed nuclear transfer; they had used a less advanced technique, called blastomere separation, by which the totipotent cells in two to eight cell embryos are made to separate, and then form smaller than normal embryos. And they used private—not federal—research funds.[22]

In February 1994, a Human Embryo Research Panel was appointed by the director of the National Institutes of Health. Later that year, the panel recommended that federal funding be allowed for certain types of embryo research. Included within this category was research on those embryos remaining after IVF or preimplantation diagnosis, up to the appearance of the primitive streak (day 14), and on embryos purposely created when a "compelling case' " can be made for the scientific and therapeutic value of the research. Among the categories of research determined to be unacceptable for federal funding was research involving cloning, followed by transfer.[23]

When the report was released, the political uproar was immediate and fierce. In December 1994, President Clinton banned the use of federal funds to create embryos for research purposes. Congress went even further—extending the ban to research on so-called spare embryos. In amendments to the appropriations bills funding the National Institutes of Health in 1996 and 1997, Congress prohibited the NIH from using federal funds to finance any research involving the destruction of embryos.

At the Senate cloning hearing, Drs. Varmus and Wilmut argued strongly for the preservation of a distinction between research to achieve the cloning of human beings and research on human cloning at the cellular level [more recently termed, "therapeutic cloning".]

> One potential use for [the latter] . . . would involve taking differentiated cells, such as skin cells, from a human patient suffering from a genetic disease. These cells inevitably have a limited potential to do other things. One could take these cells and—using the technique that we have developed—get them back to the beginning of their lives by nuclear transfer into an oocyte to produce a new embryo.
>
> From that new embryo, you would be able to obtain relatively simple, undifferentiated cells, which would retain the ability to colonize the tissues of the patient. Once these cells were in the laboratory, there would be the ability to make genetic changes or even add a gene.[24] Thus, for human cloning at the cellular level to achieve its medical promise, it will become necessary to do research on very early embryos, created specifically for this purpose. If the ban on federal funding for embryo research remains, such research could be crippled.

POLITICS AND ETHICS

Logic would dictate that a long overdue comprehensive scheme to regulate IVF and embryo research would be designed, and cloning addressed within that framework. But, unfortunately, when the discussion turns to reproductive questions in the United States the imperatives are more often political than logical. And, of course, there are issues, like the status of the embryo, that provoke profound moral disagreement that remains impervious to someone else's logic. In Britain, assisted conception and embryo research are regulated by a law that is essentially without norms. The United Kingdom's Human Fertilisation and Embryology Act is a legislative *tour de force* in that it creates a legal framework for embryo research in the absence of moral consensus.[25] [In the UK, embryos may legally be created for research purposes and it is this factor that makes therapeutic cloning at least an option.]

But it must be remembered that questions of abortion (and embryo research) are far less contentious in the United Kingdom than in the United States, where, as Daniel Callahan has noted, bioethical questions often get "overlaid with ideology and swept up in the culture wars."[26] Particularly when they are related to abortion. NBAC "has explicitly not been asked to

address issues of embryo research and abortion, ethical problems that are
. . . so contentious that any committee venturing into those waters would
likely sink without a bubble."[27] [NBAC was later asked to address the ethics
of embryo stem cell research.] Because human cloning at the cellular level
would involve the deliberate creation of embryos not destined to be implanted, government federally funded research in this area has a dim future
in the United States.

Those who try to navigate waters that are both morally troubled and
highly political may find the experience enormously frustrating. Prof. Alta
Charo, a member of NBAC, who also served on the Human Embryo Research Panel, told the *New York Times* that the panel's work had taught her
a valuable lesson.

That group, she said, relied on logic to make its case that research with
early human embryos was ethically acceptable. "The logic was airtight, but
it did not change anybody's mind and there was a lot of resentment," Ms.
Charo said. Now, she said, she realizes that "logical arguments are only
rationalizations for gut feelings or religious viewpoints." And, she said, that
is where the group ought to start in analyzing what it wants to say about
the cloning of humans.[28]

On January 22, 1993, when President Clinton announced the lifting of
the ban on federal funding for fetal tissue research, he said that "we must
free science and medicine from the grasp of politics." Surely, we must, but
we all too often don't. Let us hope that bioethics—and, with it, the promise
of human cloning at the cellular level—will not suffer a similar a fate.

A longer and now slightly out of date version of this chapter was written
in the late spring of 1997. If expressed the hope—shrouded in pessimism—
that the stunning scientific advances embodied in the Roslin breakthrough
would be allowed to yield their promised benefit. That research involving
human cloning at the cellular level—a practice now referred to as "therapeutic cloning"[29]—would not be crippled by hasty and ill-advised legislation. Fortunately, cool heads prevailed. Research is going forward, and
the prospect of using cloning techniques to create tailor-made and immu-
nocompatible cells, tissues, and perhaps, one day, organs is on the horizon.
Although it never could or would receive federal funding in the United
States.

Another hope voiced in the paper with even more pessimism—that the
demonstrable need to legislate in order to prevent or at least control reproductive cloning would lead to comprehensive regulation of assisted reproduction generally—was a faint one at best. And here the outcome was not
nearly so felicitous. Cloning exceptionalism remains the order of the day.
Proposed legislation has been directed solely at controlling or banning the
cloning of human beings. And it is naive in the extreme to believe that
passage of such laws will make us safe from the actual and potential abuses
that occur in the anything goes, Wild West atmosphere of assisted reproduction.

Meanwhile, enormous strides have been made in the realm of animal cloning. Mice, cows, and then pigs have been cloned from adult cells. Cloning may soon enable us to realize the dream of xenotransplantation, by making possible the production of herds of genetically modified pigs that could provide hearts, livers, and kidneys to human patients who would otherwise die waiting for human organ transplants. What could be more dramatic than these developments?

Certainly not their grip on the public imagination. Cloning has come to be seen by all too many people as an old and tired and largely irrelevant story. Last year's news. Or the year before's. In June 1997, when the National Bioethics Advisory Commission issued its report to the president[30] just four months after publication of the *Nature* paper, public interest was already on the wane. The commission recommended that Congress enact federal legislation that would ban, for three to five years, the creation of a child through cloning. At the end of that time, the question could be reconsidered.

Meanwhile, the commission suggested that President Clinton's ban on federal funding of research involving the creation of a child by somatic cell nuclear transfer cloning be extended, and they requested that private organizations not receiving federal funds voluntarily comply with the ban. The president responded by sending a bill to Congress that would outlaw for five years all somatic cell nuclear transfer cloning to produce a child. No action was taken.

In their testimony to a Senate subcommittee, Drs. Varmus and Wilmut had argued eloquently for the preservation of the distinction between human cloning to produce a person and human cloning to produce cells and tissues. NBAC's recommendations preserved this distinction, and the private sector was left free to conduct research in the area of therapeutic cloning.

As summer turned into autumn, the cloning story seemed virtually dead. Then in December 1997, media attention and public apprehension were reawakened when a real-life Dr. Frankenstein actually appeared—in the person of a physicist named Richard Seed. Seed was identified as "a Chicago-born Harvard graduate who has no medical license and has said that cloning will make man immortal and closer to God."[31] The time frame for this miracle was equally bizarre. "My target," Seed told mesmerized journalists, "is to have a two-month pregnancy in a year and a half."[32]

And who was to be his first creation? A genetic replica of himself! As a gift to humankind, we were to be presented with yet another Richard Seed. This one a clone to be carried to term by a surrogate mother—his wife. Who can forget the sight of Seed, eyes glittering, telling a bemused Ted Koppel, the U.S. television journalist, that he could be cloned without his consent, or even knowledge, from just one strand of his hair! The public's reaction to all of this was unadulterated horror. And the politicians

shared their concern. The president renewed his call for a ban on reproductive human cloning for five years.

On February 3, 1998, the Human Cloning Prohibition Act (S. 1601),
sponsored by Bill Frist and Christopher Bond, was introduced in the Senate.
The next day, another bill was introduced, S. 1611: Prohibition on Cloning
of Human Beings Act of 1998, sponsored by Ted Kennedy and Diane Fein-
stein. The difference in title says it all: The first bill sought to ban all human
cloning, while the second carved out an exception for therapeutic cloning—
cloning of cells, tissues, and organs—in short, all human cloning that did
not result in the birth of a baby.

Scientists and groups representing the disabled and chronically ill criticized the Bond-Frist bill as too broad and likely to cripple promising lines
of research. Some conservative Republicans—such as 95-year-old Strom
Thurmond—deserted their ideological soul mates, and voted against taking
up their bill. On the other side, the Kennedy-Feinstein bill failed because
it implicitly allowed embryo research, prohibiting only cloning that created
an embryo that was implanted in a woman's uterus. There was a legislative
standoff. There is still no law against human reproductive cloning as this
volume goes to press, more than three years after the Dolly announcement.

Meanwhile, Richard Seed, the cause of all the uproar, was being
roundly discredited as unqualified and even unstable.[33] Back in the realm
of real science, rumblings about the validity of the Roslin experiment were
increasing. Dolly was still the lone clone derived from an adult cell. Dark
suspicions were advanced of some mistake,and even outright fraud. The
loudest and most influential voices in the chorus of doubters belonged to
Dr. Norton Zinder and his colleague Vittorio Sgaramella, who questioned
Dolly's status in a letter to the journal *Science*.[34]

If the Roslin scientists did what they said they did, why had no one
else been able to do it too? Replicating the Roslin results would have been
a great coup. But all was silence. Until July 1998, when scientists at the
University of Hawaii announced that they had used a somewhat similar
method to clone three generations of mice from adult cells, including clones
of clones. Newspaper coverage tended to be technical, a bit dry, and not at
all sensational—unlike the Dolly coverage. We were told that this accomplishment overturned biological wisdom. No one had believed that mice
could be cloned because cell differentiation in mice begins so quickly; there
was simply no time for the egg to reprogram the nucleus. According to *Time*
magazine, "cloning has, with a speed no one anticipated, been transformed
from an astonishing tour de force into what seems almost a mundane laboratory procedure."[35]

And "mundane" was the operative word in the public mind. Most people seemed to snooze right through this chapter of the cloning saga, perhaps
because three generations of frisky mice was a less sensational and much
more domestic kind of nightmare vision. Besides, mice weren't really like
us, so what could it mean? Quite a lot, actually. Because of the physiological

closeness between man and mouse and because the Hawaiian technique was stunningly efficient compared to the method that produced Dolly.

Then in December came another crucial step: Japanese scientists cloned eight cows from a single adult cow. This was important in two respects: another species could be cloned and in a highly efficient manner. Cloning was here to stay.

But it was the news of the previous month that provided perhaps the strangest chapter in the cloning saga. Advanced Cell Technology (ACT), a small biotechnology company, announced that in 1995 and 1996 one of its scientists had fused a human skin cell to a cow's egg and thereby created an embryo of uncertain biological and moral status. There were no live births. Was it a human embryo? How much cow DNA must there be before a human embryo ceases to be human? The work of ACT has never been published in a scientific journal; rather, it was reported in the *New York Times*. But the president took note, and NBAC was asked to examine the ethical issues related to the work of ACT and also the question of embryo stem cells generally.

This *New York Times* story followed the publication of two stunning pieces of research. Biologist Dr. James Thomson and his group at the University of Wisconsin, Madison isolated stem cells from human embryos left over to be ultimately discarded after IVF treatment and grew them into five immortal cell lines. Pluripotent stem cells were derived from fetal tissue obtained from terminated pregnancies by Dr. John Gearhart and his team at Johns Hopkins University. In 1999, there was so much new and exciting work in the area that the journal *Science* hailed stem cell research as the most significant scientific breakthrough of the year.

The hope of going forward with such research—particularly when federal funding is involved (which it was not for Gearhart and Thomson)—is clearly fraught with the same ethical and policy problems that have long plagued embryo research. The already acrimonious debate is certain to get worse. The Patient's Coalition for Urgent Research, an umbrella group composed of dozens of patient organizations, has been lobbying Congress to allow federal funding for embryonic stem cell research, but not for the derivation of such cells. This is in keeping with an opinion issuing from the General Counsel of the Department of Health and Human Services that drew a distinction between the two. Many professional organizations have also argued the merits of this research. But powerful forces are arrayed on the other side, including the National Council of Catholic Bishops. What has all this got to do with cloning? Well, quite a lot, but not in the United States, where no one is seriously proposing that federal government funding be used to create embryos for the purpose of deriving compatible stem cells for very ill patients. It is in the United Kingdom that the stem cell and cloning questions have been linked—in the discussion and debate about therapeutic cloning. (See chapter 27.)

On the human reproductive cloning front the last big news story appeared just before Christmas 1998. Scientists at South Korea's Kyunghee

University announced that on December 16 they used the Hawaiian cloning technique to create an embryo that was the genetic replica of a 30-year-old fertility patient at their clinic. They said that they allowed it to reach the four-cell stage before they destroyed it. Most scientists did not believe that the Koreans accomplished what they said they did. But, clearly, cloning from adult cells had already been shown to work in sheep, mice, and cows. There was no reason to think that it could not work in humans.

Indeed, we may soon find out. A cult that believes in UFOs has recently announced its plan to clone a child to replace a child who had suddenly died. (See chapter 15.) While this claim might well be a hoax, there is simply no way to know whether somewhere, someone is attempting to clone a human being, not to an early embryo stage to harvest embryonic stem cells, but to produce a live baby. If and when this happens, the media frenzy that greeted Dolly the sheep will seem quite tame in comparison.

NOTES

I am grateful to the following persons for their varying contributions to this paper: Dr. Kenneth Boyd, Mrs. Patricia Boyd, Dr. Tony Hope, Ms. Rachel McWilliams, Ms. Laurie Petrick, Dr. Zbigniew Szawarski, Dr. Ian Wilmut, the members of his research team, and his personal assistant, Ms. Jaki Young. And my special thanks to Ms. Elizabeth Graham, senior information officer at the Wellcome Trust Information Service, for her indefatigable research support and unfailing good humor.

1. Wilmut, I., A. E. Schnieke, J. McWhir, A. J. Kind, and K. H. S. Campbell, "Viable Offspring Derived from Fetal and Adult Mammalian Cells," *Nature* **385**, 810–813 (February 27, 1997).

2. Campbell, K. H. S., J. McWhir, W. A. Ritchie, and I. Wilmut, "Sheep Cloned by Nuclear Transfer from a Cultured Cell Line," *Nature* **380** (March 7, 1996), 64–66.

3. Begley, S., "Little Lamb, Who Made Thee?" *Newsweek* (March 10, 1997), 56.

4. Kluger, J., "Will We Follow the Sheep?" *Time* (March 10, 1997), 71.

5. Annas, G., "Human Cloning: Should the United States Legislate against It?" *ABA Journal* **80** (May 1997).

6. Kolata, G., "With the Cloning of a Sheep, the Ethical Ground Shifts," *New York Times* (February 24, 1997), A1.

7. Duff, N., "Clone with Caution: Don't Take Playing God Lightly," *Washington Post* (March 2, 1997), C1.

8. Callahan, D., "Bioethics and the Culture Wars," *The Nation* (April 14, 1997), 24.

9. Dworkin, R., *Life's Dominion: An Argument about Abortion, Euthanasia and Individual Freedom*. New York: Knopf, 1993, p. 6.

10. Witham, L., "Christians Oppose Human Clones," *Washington Times* (March 14, 1997), 3.

11. Callahan, "Bioethics and the Culture Wars," 23.

12. Capron, A. M., "An Egg Takes Flight: The Once and Future Life of the National Bioethics Advisory Commission," *Kennedy Institute of Ethics Journal* **7**, 1, (1997), 67.

13. *Ibid.*, 65.

14. Kolata, G., "Little-Known Panel Challenged To Make Quick Cloning Study," *New York Times* (March 18, 1997), C9.

15. Radford, T., "Well, Hello Dolly . . . ," *Guardian* (February 27, 1997), 5.

16. See chapter 1.

17. *Ibid.*

18. Congressional Research Service, *CRS Report for Congress* 97–335 SPR (March 11, 1997).

19. The Ethics Committee of the American Fertility Society, "Ethical Considerations of Assisted Reproductive Technologies," 62 *Fertility and Sterility Suppl.* (November 1994).

20. McKie, R., "But Will There Ever Be Another You?" *The Observer* (March 2, 1997), 18.

21. Hall, J. L., D. Engel, P. R. Gindoff et al., "Experimental Cloning of Human Polyploid Embryos Using An Artificial Zona Pellucida," The American Fertility Society conjointly with the Canadian Fertility and Andrology Society, Program Supplement, Abstracts of the Scientific Oral and Poster Sessions, Abstract 0–001, S1 (1993).

22. See note 18.

23. "What Research? Which Embryos?" *Hastings Center Report* **25**, 1 (1995), 36.

24. See chapter 1.

25. Klotzko, A. J., "The Regulation of Embryo Research under the Human Fertilisation and Embryology Act of 1990," in *Conceiving the Embryo*, ed. D. Evans (The Hague: Martinus Nijhoff Publishers, 1996), 303–314.

26. Callahan, "Bioethics and the Culture Wars," 23.

27. Palca, J., "A New National Bioethics, Commission—Maybe," *Hastings Center Report* **26** 1 (January–February 1996), 5.

28. See note 14.

29. See chapter 27.

30. National Bioethics Advisory Commission, *Cloning Human Beings* (Rockville, Md.: NBAC, June 1997).

31. "US Scientist Setting up Cloning Lab in Japan," *Milwaukee Journal Sentinel* (December 6, 1998), 17.

32. "Support Grows for Ban on Human Cloning," *Atlanta Journal and Constitution* (January 12, 1998), 1A.

33. "Image of Human Cloning Proponent: Odd and Mercurial," *Chicago Tribune* (January 12, 1998), 6.

34. *Science* **279**, 635.

35. "Dolly, You're History," *Time* **152**, 5 (August 3, 1998).

12

Power without Responsibility

Media Portrayals of British Science

Tom Wilkie & Elizabeth Graham

The majority of adults in Britain cite the mass media as their main source of information about developments in science and technology.[1] This alone makes it worth studying how the press covered the story of Dolly the cloned sheep. But the media's reporting of Dolly is interesting for reasons quite apart from its bearing on public information and, potentially, public attitudes toward science. Rather the way in which science and the processes of science stood revealed through the prism of mass media coverage suggested that there may be serious difficulties in the relationship between science and society. Although there were failures of journalistic accuracy and balance, these should not be allowed to obscure such deeper issues. The coverage provided ample vindication of the contention put forward by Dorothy Nelkin and Susan Lindee in their book.[2] *The DNA Mystique*, that the "gene has a cultural meaning independent of its precise biological properties." They continue, "The gene of popular culture is not a biological entity. . . . its symbolic meaning is independent of biological definitions." [Editors note: For a discussion by Nelkin and Lindee of the cultural meaning of cloning, see chapter 7, this volume.] Time and again as Dolly's story unfolded, one would see a tension between the scientists' desire to keep their discourse to the scientific context where they were figures of authority,[3] and the desire of the press to discuss the cultural context of cloning.

In this article, we will analyse principally the British broadsheet newspaper coverage of the Dolly story. We also look at some of the corresponding U.S. newspaper coverage and find striking contrasts, relating not just to journalistic practices but also to the public status and position of science in the two countries. For reasons of space and difficulty of obtaining archive material, we do not address the role of the broadcast media.

A word of caution: the focus of attention here is the press coverage, and although newspapers profess to reflect the concerns of their readers one cannot necessarily extrapolate from a short burst of media interest to infer the long-term public reaction to these events. Several measures of public attitudes and opinion have been taken since the story first broke, and they reveal a richly textured public response. One opinion poll survey, conducted for the public relations company for the institution which cloned Dolly,[4] found that there was a very high level of public knowledge and awareness of cloning, Dolly, and related issues—even 18 months after the event. Qualitative research undertaken by the Wellcome Trust to probe the wellsprings of public values, found comparatively little direct reference to factual media news reports.[5] Instead, members of the public appear to ground their moral belief and attitudes in narratives drawn from their personal lives and from popular culture: films, TV soaps, and the lives of film stars and other highly public figures.

However, if one looks to the media not as a proxy for the expression of public opinion but as a mirror reflecting aspects of national life (in this case the relation between science and society), the following striking characteristics are apparent in media portrayals of Dolly.

- Traditional methods of scientific communication are inadequate in modern, technologically sophisticated societies.
- The scientific community in Britain failed to address the cultural significance of Dolly's birth.
- There is a strange dislocation between science and British society, illustrated by the scarcity of scientists quoted as experts in British newspaper stories, in marked contrast to reports in the United States.

Some of the reservations expressed above against taking newspaper coverage as a proxy measure of public opinion can be illuminated by the Coverage-Attitude Hypothesis,[6] put forward in 1981 by Allan Mazur. He argued that "the rise in reaction against a scientific technology appears to coincide with a rise in quantity of media coverage, suggesting that media attention tends to elicit a conservative public bias." Thus it is not necessarily the content but the quantity of media coverage which is important in eliciting a response. Nonetheless, a conservative bias was evident in the content of the press coverage of, and in public and political reaction to, Dolly.

There are at least two strategies for minimizing the extent to which the media might elicit this conservative bias. One is to intervene directly in the reporting process and this was the approach adopted by the team at Roslin. A second approach is more indirect and involves recruiting "allies" from other sectors of society to form constituencies in favor of the new development.[7] We shall argue in this paper that the first strategy is necessary but insufficient and at one point was counterproductive; and that the second path might have been more fruitful, but that scientific institutions in Britain seem reluctant to engage in this type of action.

THE PUBLIC COMMUNICATION OF SCIENCE

It is now a commonplace for the major scientific journals to seek publicity for themselves by sending out press releases in advance of the publication date of the journal itself. The press releases are embargoed—that is to say, their contents are not to be made public until the day of publication of the journal—but contain a simplified synopsis of some of the scientific papers, together with contact telephone numbers for the scientists who carried out the work and for other researchers who might be able to comment on its (scientific) significance.

Among the most important journals to do this are *Science, Nature*, and, in the United Kingdom the *BMJ* and the *Lancet*. The system works to everyone's satisfaction. It eases the workload of the journalists reporting on the "news event"—publication in a scientific journal—because they have privileged access to the information before everyone else and thus have several days in which to construct their stories for publication. This is a consideration which is particularly pertinent in Britain where economic recession has made fast turnover of stories a necessity of print journalism.[8] It suits the journals because, by making it easy for the journalists, they increase the chances of getting free publicity for their journal. It also serves the interests of the scientific community as a whole, since it also increases the likelihood that stories in the mass media about science will be better informed and more accurate than would be the case if journalists had to turn copy around on the day of publication without access to phone numbers and other aids.

But there have been problems with the system. *Science* has already suffered one notorious episode of embargo breaking,[9] where advance knowledge of a paper on the genetics of human obesity stemming from the press release led to fluctuations in share prices on the U.S. stock exchanges prior to publication of the paper itself.

The Dolly story similarly began with an embargoed press release sent out by *Nature. Nature* is published every Thursday but makes its press release available by fax or electronically on the preceding Friday. *Nature* also operates a "fax-back" service by which bona fide journalists can obtain faxed copies of the scientific papers themselves by quoting numbers listed on the press release. According to the rules governing the use of embargoed press releases, although journalists are free to contact scientists in advance for comment, they must refrain from publishing those comments until the Thursday morning—*Nature*'s publication day.

How the Story Broke

However, this procedure was not followed for Dolly. It appears that, in addition to *Nature*'s publicity efforts, the scientists at Roslin had attempted to condition the way in which the story broke by hiring a PR company and cooperating with a television documentary intended for broadcast after the *Nature* publication.

There are several examples in biomedical science where attempts to manage the way in which the news breaks have failed—most notably the announcement that new variant Creutzfeld Jakob disease might be connected with the outbreak of madcow disease in British cattle.[10] A common feature is that giving one media outlet privileged access tends to be counterproductive. Robin McKie, the immensely experienced science editor of the *Observer* got wind of the Dolly story because of the TV program rather than the *Nature* press release.[11] The arcane rules of embargoes are such that, having obtained the significant information—that a cloned sheep had been born—from a source other than the *Nature* press release, the *Observer* no longer had to respect the vow of *omerta* which bound all the other print journalists. It ran the story on its front page.[12] The story won for Robin McKie the top award for science journalism: a Glaxo Wellcome/Association of British Science Writers prize for science writing.

Significantly, the story made no mention of publication of a paper in *Nature* but did credit the television company by name,[13] saying that the Roslin team's work "will feature in a forthcoming edition of Carlton TV's *Network First*." This is the journalistic equivalent of the references at the end of an academic paper and an implicit indicator of the original source of the information. However, some information contained in the *Observer* story was not in the TV documentary,[14] indicating that a further source must have been available. In addition, the story was illustrated with a photograph credited to a photographer who had taken advance pictures of Dolly for a weekly journal with permission of *Nature*.

Balance of the Coverage

The *Observer* story is generally upbeat in its description of Dolly's cloning describing it in the second paragraph as "a landmark in biological research—and a triumph for UK science, one that should lead to breakthroughs in work on ageing, genetics, and medicines." The third paragraph, however, moves from the biomedical to the wider cultural context of this research: "But cloning is also likely to cause alarm. The technique could be used on humans, drawing parallels with Huxley's *Brave New World* and the film *The Boys from Brazil*, in which clones of Hitler are made." This point is tempered with the report that human cloning would be illegal in Britain and the assertion, attributed to unnamed scientists, that no responsible biologist would support work on human cloning. The rest of the 13-paragraph story discusses the biomedical background: sketching how the work was done, the history of Roslin's work, and the beneficial consequences of this cloning experiment.

Only in the last paragraph does the discourse of concern[15] reappear: "it is the prospect of cloning people, creating armies of dictators, that will attract most attention." The story notes that "a sheep is a complex mammal, after all, so cloning one raises concerns" [about cloning humans, from the context]. In an attempt at reassurance it concludes with the assertion, "Whether anyone would wish to clone a human is a different matter."

Because the information in the *Nature* press release was now in the public domain with the *Observer* story, no one was any longer bound by the *Nature* embargo. It is customary for each British national newspaper to obtain copies of the first edition of its rivals and to incorporate any rival's exclusive stories into later editions. This practice is facilitated by the heavily metropolitan character of British national papers—all are published out of London.[16] Thus both the *Sunday Telegraph* and the *Sunday Times*[17] managed to run stories about Dolly, on the front page and page 2, respectively, on the same day as the *Observer*. These were much shorter, contained less scientific detail, and the newspapers clearly did not have prior access to a photograph—all of which pointed to the writers not knowing about the story in advance.

Significantly, both the *Sunday Telegraph* and the *Sunday Times* mentioned the publication in *Nature* but did not mention the TV documentary. Significantly also, their articles contained rather more "discourse of concern" than of promise, compared with the *Observer*. The *Nature* press release, which, together with the *Observer* article, is all they would have had to work with at first, is comparatively scant on technical detail. Moreover, it contains *no information at all* on the practical benefits of the work. Journalistically, if there is little in the way of factual information, then it is relatively easy to fill space with cautionary remarks. The *Sunday Telegraph* reported that "the scientists have raised the spectre of a race of 'perfect' humans by a process once dismissed as nothing more than science fiction." It also produced quotes from two socioethical commentators expressing concern and worry about the wider implications. The *Sunday Times* headline set the tone for its coverage: "Sheep Clone Raises Alarm over Humans."

International Pickup

The *Observer* story was picked up by national and international news agencies—the wire services as they are traditionally and now inaccurately known. Agence France Press, the Associated Press, and Reuters were among those who ran stories, which would have been automatically available to any of their subscribers. This, together with the time difference between Britain and America, allowed North American newspapers to come out with the story on the same date as the *Observer*. Thus among those carrying the story in their Sunday editions were the *New York Times, Toronto Star*, and the *Ottawa Citizen*.

The *New York Times* article[18] was long and full of scientific detail. At 34 lengthy paragraphs, it was about three times the length of the *Observer* story that had broken the news. The *Times* story included concerns in its very first sentence: "In a feat that may be the one bit of genetic engineering that has been anticipated and dreaded more than any other, researchers in Britain are reporting that they have cloned an adult mammal for the first time." However, most of the text was devoted to the scientific technique, and the ethical concerns were articulated only in the last nine paragraphs of the story.

The story of Dolly really took off with the publication of daily newspapers on the Monday. Having run the story on the front page of its Sunday edition, the *New York Times* ran a follow-up story (also on the front page).[19] Continued inside, the *Times* articles covered almost an full page. Its 31-paragraph main story went over the scientific ground once again in some detail, but this time the wider implications of the work were explored more fully and were the main focus of both the early and the late paragraphs—with the science sandwiched in the middle. Two lengthy sidebars covered the commercial prospects of the technology and profiled the chief scientist, Dr. Ian Wilmut. The *Washington Post* covered the story in about 25 paragraphs in its Monday edition.[20] It too focused largely on the scientific detail, but pointed out that there was a "regulatory vacuum" in the United States such that human embryo research was prohibited in federal laboratories but not in private ones.

Subsequent Development of the Story

In contrast, the British press reaction was comparatively muted that Monday. The London *Times*[21] published the longest story, and the only one to be run on the front page. The *Times* covered the issue in 27 paragraphs—almost all of them given over to the wider implications of the research. Very little scientific detail was given. Significantly, the story was written by the paper's health correspondent rather than its science correspondent. The paper's confusion on the issue can be gauged from the fact that the news story cross-referenced an editorial comment that was concerned entirely with labeling genetically engineered food and not with cloning.[22]

On Monday, the *Daily Telegraph* ran half a page on Dolly on page 5; the *Independent* ran 16 paragraphs on page 3; the *Guardian* just 11 paragraphs on page 7, together with a jokey question-and-answer guide in its humorous "Pass Notes" section. If anything, even less was published in the

British press on Tuesday 25, with the exception of a lengthy comment and opinion piece by the philosopher Mary Midgely in the *Guardian*.

However, the weight of press coverage in the United States fed back to British newspapers.[23] When President Clinton announced an inquiry into the ethical issues, this transformed the British newspaper coverage. The story had shifted from the abstruse, unfamiliar ground of developmental biology onto something very familiar indeed, international politics.

In addition, the Roslin team met the British press for the first time.[24] On Wednesday, February 26, all the main British newspapers profiled the research team—two days later than the *New York Times* had. They also started to explore more precisely the legal situation governing human cloning in Britain—again a couple of days later than the U.S. press had examined the legal position in their country. However, the interviews with Dr. Wilmut did not lead to greater scientific detail appearing in the British press; rather the story had moved on to the wider implications of cloning people and even of bringing people back from the dead.

By now, the story was still running on the news pages, but increasingly bylines were being shared between science and foreign correspondents. Space was also made available inside the newspaper on the comment and opinion pages, illustrating once more how this had ceased to be a science story but was now the purview of those whose profession was to have an opinion, no matter what the subject matter. This is a routine dynamic of major news stories, and a similar pattern was discernible in the British press coverage of the possible connection between new variant Creutzfeld Jakob disease and the epidemic of mad cow disease in British cattle.[25]

British newspaper coverage continued at great length throughout March even though very little in the way of fresh information was forthcoming. Again, it was economic and political developments that provided the peg on which to hang cloning news stories. Thus pronouncements by Jacques Santer, president of the European Commission, and by Jacques Chirac, the French president, both revived the story on the news pages. Other scientific claims provoked news coverage. The Oregon Regional Primate Research Center announced that it had cloned rhesus monkeys. This was confused in some accounts[26] with the nuclear substitution technique of Roslin.

For the most part, however, the coverage consisted of comment and discussion in the form of lengthy features, rather than news stories, about the implications for human cloning. Whereas Ian Wilmut had stated a year earlier (on the occasion of the birth of twin sheep cloned by a different technique) that "I cannot see why anybody would want to do such a thing,"[27] by the end of Dolly's first week the newspapers had produced feature-length interviews with individuals who asked to be cloned.[28] An eloquently simple letter in the *Sunday Times* on March 9, 1997 expressed a father's wish to clone his elder son who had been killed in a road accident so that "our family would be complete again."[29] By April, supermodel Claudia Schiffer had joined the list of those who had applied to be cloned.[30]

The true legal position in Britain became apparent only after the Human Fertilization and Embryology Authority was called to give evidence to the House of Commons Science and Technology Committee. The committee's hearings also provided the first public arena in which claims for the beneficial consequences of the technology could be made and assessed. After nearly a fortnight in which human cloning had been held out as a terrible prospect against which international legislation was urgently needed, stories appeared for the first time stressing that cloning could have beneficial consequences and that legislators should not rush to ban the technology wholesale.[31] This discourse of promise was reported absolutely straight, with no editorializing. However, the upbeat effect was rather undone the following day when Dr. Ian Wilmut incautiously admitted to the Select Committee that human cloning might be possible in a couple of years.[32]

British press coverage of Dolly became more sporadic in April as the general election campaign overshadowed interest in the story. Among the month's events which did spark an interest were the pregnancy of Megan and Morag (the cloned sheep born a year earlier). However, a striking exception to the metropolitan nature of the British press is the vigor of the Scottish newspapers. Roslin is situated just outside Edinburgh, so Dolly was in many ways a local story for the Scottish newspapers, and they continued to cover aspects even when the London-based dailies had lost interest.

Although interest in Dolly specifically may have declined, the concept of cloning had entered general usage: there were 259 references to cloning in British papers in April, only a handful of them referring to Dolly. Judging by the number of references, interest continued at about the same rate in May. There was considerable press interest when Dolly was first sheared and her fleece auctioned for charity. The story took off again in June with coverage triggered by the report of the U.S. National Bioethics Commission. This once again exemplified how the British press agenda is dominated by U.S. concerns.

Comparisons between U.S. and U.K. Coverage

One of the striking contrasts between British and American press coverage is the copious scientific detail provided in U.S. broadsheets, whereas British papers provided parsimonious scientific reporting. This contrast has been noted before in a comparison of U.K. and U.S. newspaper coverage of the "gay gene" story.[33]

Several factors could be at work here. One is clearly manpower: there are more science writers on the *New York Times* alone than on all British broadsheet newspapers put together. It is therefore impossible for individual British science writers to maintain expertise across the whole field that they have to cover. Stories which draw upon their own background knowledge will be better informed than others. As it happens, those science reporters on British newspapers with science degrees studied in the physical sciences (mainly chemistry, in fact) rather than developmental biology. They therefore came to the Dolly story with a built-in disadvantage.

The narrow base of expertise among British science journalists thus means that they are dependent on assistance from the scientific community. Yet very few scientific sources were cited in British newspapers other than the scientists who carried out the work. The major focus of press attention was the implications for human beings, but the scientific institutions in Britain which fund human biomedical research—the Medical Research Council, for example—were conspicuous by their silence. The head of the Biotechnology and Biological Sciences Research Council (which partly finances Roslin) was quoted in support of the research[35] but not until February 28 five days after the story broke, and his comments were tempered with the accompanying sentence, "He did so [supported Roslin] as the Vatican and the European Commission followed President Clinton and called for an inquiry into the ethical implications and a German Euro-MP said the Roslin Institute . . . should turn to fighting AIDS or cancer instead." The antithesis perfectly illustrates the conservative bias elicited by a science portrayed as being without allies of international stature.

Science in Its Public and Social Context?

A news event such as Dolly does not happen in a vacuum. There is already a social cultural, and scientific context which will color the way in which the news is reported and interpreted. In a sense this is a trivial observation, for if such a context did not exist the news event would be literally incomprehensible. We have searched the FT Profile computerized database of British newspaper clippings for 1996—the year before Dolly and discovered that the words "clone," "cloned," "clones," or "cloning" appeared in 1,440 articles. The term appeared with "sheep" just 101 times and was linked with "fear," "peril," "danger," or "warning" just 47 times. For reference, the terms appeared 1,820 times in the first six months of 1997, whereas items relating to nuclear power appeared 2,280 times in 1996 and 886 times in the first half of 1997. The cloning issue does not compare in the British press with mad cow disease, which appeared in some 10,117 items in 1996 and 2,970 items between January and June 1997. There was clearly a pre-existing level of interest and public consciousness about cloning. We have discovered several unusual appropriations of the term—a Gaelic football team is named after a village "Clones" (pronounced Cló-nass) in county Monaghan, Ireland, and the term is frequently employed in discussions of mobile phone fraud.

Despite this general background, very little in the way of guidance for the perplexed was available when the story of Dolly broke. According to De Facto, the PR company engaged by Roslin, *Nature* "provided no media handling guidance to the scientists and were not prepared to list direct line telephone numbers until persuaded to use De Facto's as the co-ordinating contact."[36] By Sunday, the company had press releases and briefing notes available by fax and from its web site (but accessible only via a password). However, it is unclear whether, in the messy circumstances in which the story broke prematurely—somewhat akin to "the fog of war"—many journalists actually had that information to hand. The primary source for this story was the highly technical scientific paper due for publication in *Nature*—even an accompanying commentary was couched in forbidding language. But given the focus of the press coverage, at least in the United Kingdom, assistance with technical accuracy, while necessary, would not have been sufficient: no easy guide was available as to *why* this research had been done and what benefits it could bring. The overriding need was to address the questions of benefit and rationale. Although the *Nature* press release was available, it was sketchy in content and really addressed only the technical aspects. Roslin released its own press information sheet on February 24, but it too tended to focus on technical issues when the media focus was elsewhere.

As American press coverage showed, none of these problems was necessarily fatal to accurate and informed reporting, because the U.S. scientific community was willing to articulate the issues freely. In Britain however, there is little tradition of scientists speaking direct to the media about the work of other researchers. British coverage was inevitably affected by the way in which journalists were trying to understand the technical details of what *had* been done at the same time as they were trying to explore the implications of what *could* be done. It is difficult to assimilate and understand information presented in highly technical language such as *Nature*, and there can be a lengthy time lag between *reporting* the event and *understanding* what is going on, and thus being able to put it in its proper context. Had the scientists at Roslin made public their research aims before they embarked on cloning Dolly there would doubtless have been a vigorous public and media debate but it would have been different in character.

Instead, Dolly was initially treated as an event within the scientific community, to be handled in the traditional fashion of reporting the technical details of the experiment in a learned journal, with a PR company hired to provide extra resources to manage the media angle. When the inadequacy of the traditional means of scientific communication was revealed, it became evident also that few other channels of communication exist between science and the rest of society. Once in the public eye, British science found itself with few institutional allies from other areas of society. The possibility of preparing the public ground in advance was raised when the Roslin scientists were called to give evidence[37] to the House of Commons Select Committee on Science and Technology on March 6, 1997. The response was that Roslin scientists could not go public until they knew that they had actually achieved a clone. Once they knew that, they were bound by the publication rules of the journal *Nature*—no prepublication disclosure.

One MP expressed surprise that the committee had been given no hint of the work which led to Dolly when the committee had visited Roslin in the course of its investigations into human genetics two years earlier. Since Roslin had had to apply three years earlier for a Home Office license to conduct the experiments (the nuclear substitution procedure falls within the purview of the Animals (Scientific Procedures) Act 1986, which is administered by the Home Office), government officials certainly knew of the work and its aims. Yet there seems to have been no realization on the part either of the scientists or officials that the work would affect a much wider constituency and that other organizations (such as the Commons Committee or the HFEA) should be prepared well in advance to deal with the fallout from the announcement of Dolly's birth.

From the press coverage, few other institutions joined the HFEA in trying to reassure the public that Parliament's intention in passing the Human Fertilization and Embryology Act in 1990 had been clearly to ban human cloning. There was a deafening silence from the scientific and policy world. No major funders of medical research stepped forward instantly to state that they would not fund human cloning research. No one reiterated the legal point being made by the HFEA. Because of the human implications, Dolly would have been a sensation whatever happened, but an authoritative statement, from the Medical Research Council for example, on the first day of the crisis might have helped dampen some of the extreme comments. In the end, only when Sir Colin Campbell, chairman of the newly constituted Human Genetics Advisory Commission, announced publicly that his commission would investigate the matter did a public sense grow that the issue was under control.

The dynamic of news stories in the media is that such claims (even if true) have to be more than just "one-source" stories. Much reporting in British newspapers is taken up with party politics, and the habits and underlying assumptions of political journalism spill over into other areas. Thus the media almost unconsciously tried to assess the size of the "constituencies" in the cloning debate. Rather than recruiting institutional allies before the event, however, Roslin had attempted to condition the reception of their news by direct intervention in the news-gathering process—by the engagement of a PR company and cooperation with a TV documentary.

Had the provision of information to other relevant bodies been conducted much earlier in the process, then Ministry of Agriculture Food and Fisheries, Biotechnology and Biological Sciences Research Council, and others might more easily have entered the public debate as allies of Roslin. This point does not reflect upon the efforts or competence of the PR company, since that provision of information and recruitment of allies is a matter that goes far beyond PR purposes. Equally in our view, this does not reflect personally on the scientists at Roslin, but arises from structural problems facing British science as a whole.

In reporting the Dolly story, the press portrayed British science as isolated from the mainstream of British society and cut off from Parliament and other public bodies. Our view is that in this respect the newspaper coverage provided a reasonably nondistorting mirror.

British press coverage conveys a strange impression of the isolation of science. Scientists appear as figures possessed of great power—in this case, the ability to create life—but also as remote from the public at large and from the familiar social institutions by which power is diluted and distributed through society. As portrayed in the press, it is science rather than the media that appears to enjoy "power without responsibility."[38]

POSTSCRIPT: 18 MONTHS LATER

As we have indicated, the press coverage of Dolly continued after the initial excitement, albeit somewhat abated. It was revivified in January 1998 when the American Dr. Richard Seed announced plans to finance human cloning and, later in the month, when the Human Fertilization and Embryology Authority (HFEA) together with the Human Genetics Advisory Commission (HGAC) announced a public consultation on human cloning.

At first sight, the Seed affair appears to highlight some of the themes developed earlier in this article. The British media picked up upon a story from the United States, one that had been stoked by presidential statements of concern, and flagged it in the traditional discourse of concern. However, this time some things were different. In particular, some elements of the British media examined Dr. Seed's scientific credentials very closely and found them lacking.[39] Just one quotation will suffice to give the flavor: "Take one slow news day, a failed physicist and a measure of millennial paranoia over things scientific and, voila, a hash of an argument over cloning." Interestingly, as this particular story indicates, some elements of the media reported Dr. Seed in a self-reflexive manner: it was not just Dr. Seed nor his scientific credentials (or lack thereof) which formed the story; part of the media story was the way in which the media had turned it into a story.

This slightly more skeptical tone in terms of reporting potential dangers of cloning was reinforced later in the month with the publication of the HFEA/HGAC consultation paper on human cloning.[40] The consultation paper was the fulfillment of the announcement, nearly a year earlier, by Sir Colin Campbell, chairman of the HGAC, that there would be an official investigation into cloning. His original announcement had been prompted by a perception that some statement had to be made to show that the issue was under control; by a coincidence of timing, the publication of the consultation statement met a second need to demonstrate that the issue was under control—a need stimulated earlier in the month by Dr. Seed's suggestions.

The consultation paper drew a distinction between two purposes for which cloned embryos might be created: "reproductive cloning," in which a cloned baby might be born, and "therapeutic cloning," in which cells might be taken from a cloned embryo before the fourteenth day of development in order to grow donor-compatible tissues or for other purposes that would not lead to the implantation or growth of a human fetus. The force of this distinction was readily accepted in the press, even if there was some uncertainty about the specifics of therapeutic cloning. In fact, the possibility of a distinction had been trailed several times earlier, so the groundwork had already been laid in public. The paper left the door to reproductive cloning slightly ajar, as a possibility for those with mitochrondrial diseases. Press accounts varied. The *Financial Times* and the London *Times* took the view that this was not likely to come to pass, and so reproductive cloning would effectively remain illegal; whereas the *Independent* was still concerned that the door had not been shut completely.[41]

The consultation paper did not spell the end for Dolly and cloning as items on news agendas. All manner of subsequent events were reported, from allegations that Dolly might not be a true clone of an adult cell, through the birth of Dolly's lamb, to the cloning of mice. However, the publication of the HFEA/HGAC paper had the effect of demonstrating that the national machinery for dealing with difficult issues in reproduction, science, and ethics had finally swung into action. The institutions of society had finally caught up with the science. Paradoxically, therefore, the announcement of a call for public views resulted in a press coverage tending to the effect that cloning was less of a public issue and could now be subcontracted to the "experts."

NOTES

This article represents the authors' personal views and should not be construed in any way as reflecting the policy or attitudes of the Wellcome Trust.

1. Ciba Foundation Conference, *Communicating Science to the Public.* John Wiley, 1987.

2. Dorothy Nelkin and Susan Lindee, *The DNA Mystique: The Gene as Cultural Icon.* W. H. Freeman, 1995, pages 2 and 16.

3. The recombinant DNA controversy, culminating in the Asilomar conference of 1975, is now interpreted by many scholars as a strategy whereby scientists asserted their authority over the social agendas, by ensuring that discussions were framed in a technical discourse. See Susan Wright "Molecular Politics in a Global Economy," *Politics and the Life Sciences* **15**, 2 (1996): 249–263 and references therein.

4. Speech by Susan Charles, chief executive of HCC De Facto, to the 3rd European Biotechnology Symposium, Glasgow, September 16, 1998.

5. "Public Perspectives on Human Cloning: A Social Research Study," The Wellcome Trust, Medicine in Society Programme, London, December 1998.

6. Allan Mazur, "Media Coverage and Public Opinion in Scientific Controversies," *Journal of Communication*, winter 1981, pages 106–115.

7. The concept of recruiting allies is fundamental to Bruno Latour's anthropological description of scientific research (*Science in Action*, Harvard University Press, 1988). He deploys the notion to argue that even the construction of factual knowledge in science is a political process of negotiation and that those who win—those who are credited with having made a scientific discovery—are those who have forged the best alliances. We use the concept in a more conventional sense relating to the process by which public acceptance is secured that a given piece of scientific work is an inherently valuable addition to the stock of knowledge and that it will have beneficial consequences. This process has been fully explored by Michael Mulkay in *The Embryo research Debate: Science and the Politics of Reproduction* Cambridge University Press 1997). However, there is an echo of Latour's usage in the argument, made toward the end of this article, that social and legal facts will not readily be accepted as such unless constituencies or alliances exist to defend those facts in the public domain.

8. Tom Wilkie, "From Labs to Hacks: Are Scientific Journals Doing Their Job?" Paper given at seminar on *Scientific Journals and the Public*, University College London, April 18, 1997.

9. Wade Roush, "Fat Hormone Poses Hefty Problem for Journal Embargo," *Science* **269**, August 4, 1995, page 627.

10. The British government hired an advertising company to develop a PR campaign to persuade consumers that British beef was safe. The story broke out of control of the government because the advertising industry's journal *Campaign* got wind of the story, and this was picked up by the media correspondent of the *Daily Mirror*, on the morning of the day the government made its announcement. Another example was the story of "Boxgrove Man"—an important archaeological discovery of human remains in Britain.

11. In fact, what appears to be a case of embargo breaking did occur with the publication of a two-paragraph story credited to the news agency ANSA, on page 16 of the Italian newspaper *La Stampa* on Saturday February 22. But the size and positioning of this story indicate that the paper did not understand its full significance, and it appears to have had no impact on the English-language media.

12. Robin McKie "Scientists Clone Adult Sheep," *The Observer*, February 23 1997, page 1.

13. In the London editions. It appears that this reference was not included in the Scottish edition read by the Roslin scientists and led them to believe the information had come from a different source.

14. Dr. Ian Wilmut, private communication, July 22, 1997.

15. Durant and colleagues have distinguished two discourses in which the new genetics are often described: the discourse of great promise and the discourse of concern. John Durant, Anders Hansen, and Martin Bauer, "Public Understanding of the New Genetics." In *The Troubled Helix*, Ed. Theresa Marteau et al. Cambridge University Press 1996. See also Bauer et al., *Science and Technology in the British Press 1946–1990*, Science Museum, London, 1995.

16. Tom Wilkie "Sources in Science: Who Can We Trust?" *The Lancet* **347** (1996): 1308–1311.

17. Robert Matthews and Jacqui Thornton, "Scientists Create an Adult Sheep," *Sunday Telegraph*, February 23, 1997, page 1, and Cherry Norton, "Sheep Clone Raises Alarm over Humans," *Sunday Times*. February 23, 1997, page 2.

18. Gina Kolata, "Scientist Reports First Cloning Ever of Adult Mammal," *New York Times*, February 23, 1997, page 1.

19. Gina Kolata, "With Cloning of a Sheep, the Ethical Ground Shifts," *New York Times*, February 24, 1997, page A1.

20. Rick Weiss, "Scottish Scientists Clone Adult Sheep; Technique's Use with Humans is Feared," *Washington Post* February 24, 1997, page A1.

21. Jeremy Laurance and Michael Hornsby, "Warning on 'Human Clones': Fears Follow Production of Sheep from Single Cell," London *Times*, February 24, 1997, page 1.

22. Jon Turney, of University College London, has suggested that "equally plausibly it could be read as an astute recognition that the public do not parcel up these objects of concern into discrete bundles but regard the biotechnological enterprise as interconnected" (private communication).

23. David Felton, news editor, *The Independent*, private communication.

24. According to Roslin's site on the World Wide Web, in the first week after the story broke, it fielded 2,000 telephone calls, talked to 100 reporters and arranged for 16 film crews and 50 photographers to photograph Dolly.

25. Tom Wilkie, "Prions, Politicians and the Press." Talk given to the annual meeting of the British Association for the Advancement of Science, Birmingham, September 10, 1996.

26. See for example Quentin Letts, "Scientists Make Monkeys by Cloning Technology," London *Times*, March 4, 1997, page 12.

27. Dr. Ian Wilmut, as quoted by Nigel Hawkes, "Cloning Breakthrough Sounds Ethical Alarm," London *Times*, March 8 1996, page 9.

28. Sarah Boseley and Ed Vulliamy, "Fearful Symmetry: Nobody would want to clone a human being? Meet David Pizer, the man who wants to be cloned and so escape mortality," *The Guardian*, March 1, 1997, page 1 (*The Week* Section).

29. Harry Harris, "Cloning Can Bring Back My Dead Son," *Sunday Times* March 9, 1997, part 5, page 8.

30. "I Want to Be a Clone!" *The People* April 20, 1997, page 15.

31. Charles Arthur, "Don't Rule out Cloning Urges Ethics Body," *The Independent* March 6, 1997, page 5; Jeremy Laurance, "Embryo Watchdog Backs Trials in Human Cloning," *The Times* March 6, 1997, page 4.

32. Benedict Brogan, "Dolly's Creator: Humans *Can* Be Cloned," *Daily Mail* March 7, 1997, page 1.

33. Peter Conrad, "Constructing the 'Gay Gene' in the News: Optimism and Skepticism in the American, British and Gay Press." In *Markers and Links: Genetics and Behavior in the News*, Forthcoming.

34. *Lancet* ibid.

35. Tim Radford, "Researcher Hits Back at Clone Critics," *The Guardian*, February 28, 1997, page 13.

36. Personal communication from Sue Charles, chief executive of HCC De Facto, June 2, 1998.

37. Fifth report of the House of Commons Science and Technology Committee Session 1996–97. "The Cloning of Animals from Adult Cells." Volume II Minutes of Evidence and Appendices. HC 373-II HMSO, 1997.

38. This is the title of James Curran and Jean Seaton's classic text on the British media (Routledge, 1997). The full quotation ends: "The prerogative of the harlot throughout the ages." A comment by Rudyard Kipling on Lord Beaverbrook, *Journal*, 38, 180 (December 1971). Used by Stanley Baldwin in a speech in London on March 18, 1931.

39. Robin McKie, "Fears of a Clone," *The Observer* January 11, 1998, page 23.

40. Human Fertilisation and Embryology Authority and Human Genetics Advisory Commission, "Cloning issues in Reproduction, Science and Medicine," January 1998, Department of Health, London.

41. Nigel Hawkes, "Advisers Ask for Public's View on Cloning Benefits," London *Times*, page 4; Clive Cookson "Complete Cloning of Humans Is to Remain Illegal," *Financial Times*, page 18; Charles Arthur, "Laboratories Are Told They May Produce Embryo Human Clones," *The Times* page 3; all January 30, 1998.

CLONING

The Ethical Issues

13

Does Ethics Make a Difference?

The Debate over Human Cloning

Arthur L. Caplan

WHY TAKE ETHICS SERIOUSLY?

What does human cloning have to do with ethics? Or, more accurately, why should human cloning have anything to do with ethics? Once the initial frenzy over the cloning of Dolly the sheep had abated, a large number of people began to express skepticism or even outright hostility to the idea that ethics had anything of value to say about human cloning. Biologist Lee Silver spoke for many when he wrote, "In a society that values individual freedom above all else, it is hard to find any legitimate basis for restricting the use of reprogenetics" (Silver's term for genetic engineering including cloning and assisted reproductive technologies) (1997, p. 9). [Editor's note: For an elaboration of Prof. Silver's views, see chapter 5, this volume.]

The skeptics found a basis for their skepticism in three areas. Cloning had taken on a life of its own and had powerful supporters who were committed to seeing human cloning advance to serve their own agendas (Adler, 1997; Powers, 1998). Cloning should advance unhindered because every American has a fundamental and constitutionally guaranteed right to reproduce (Robertson, 1994; Silver, 1997; Wolf, 1997). Human cloning should advance because science must always be free to go where it wishes to go (Stolberg, 1998; Kolata, 1998).

None of these arguments is especially persuasive as a reason not to think about the ethics of human cloning. The fact that some may want to pursue their own agenda or their own self-interest is in fact a very good reason for thinking about the ethics of human cloning, especially since cloning involves the creation of new persons. The fact that persons do have a liberty right to reproduce says nothing about their right to an entitlement to technological aid in having children or whether it makes sense to limit that right if the mode used for the creation of children is not in the child's best interest (NBAC, 1997; Davis, 1997; Caplan, 1998). And it is simply not true that science and biomedical research enjoy open-ended, unbounded liberty when it comes to the pursuit of new knowledge. Anyone who has submitted a grant for peer review knows that the right to inquiry is almost always limited by the ability to command the support of the community to pay for it.

The strongest reason for skepticism about the relevance of ethics to human cloning was that a large number of people in positions of authority doubted that ethics would make any difference to the pace or path that cloning took. This form of skepticism is present in the commonly voiced concern of politicians, policy makers and scholars that ethics seems always to trail behind the latest scientific or medical breakthroughs (Fox and Swazey, 1992; Silver, 1997), and that there is no reason to presume ethics will prove more potent with respect to human cloning than it has in curbing, modifying, or stopping any technology in biomedicine in the years since the Second World War.

The phenomenon of the "ethics lag" has been accepted by many commentators on cloning as a fact (Adler, 1997; Silver, 1997). All one need do to see the depth of this belief is track any story about the ethics of any major new breakthrough in biology or medicine. It will not be many paragraphs before the writer notes either that ethics always seems to be lagging behind scientific advances or that biomedicine has outstripped the capacity of ethics and the law to keep pace. The "ethics lag" is a powerful presumption in American, European, and Japanese assessments of the future of biomedicine (Adler, 1997; Weiss, 1998).

One way to respond to the worry that ethics cannot keep up was to call for bans on human cloning (NBAC, 1997). The president of the United States moved quickly to ban the use of federal money to support research into human cloning. This was followed by many calls for Congress to enact legislation banning human cloning. More then 20 states were considering bills to ban cloning by the summer of 1998. [Editor's note: For a discussion of state legislative initiatives, see chapter 23, this volume.] Many nations quickly expressed grave concern about human cloning. But there was and remains a great degree of doubt that even bans would work (Wolf, 1997; Kolata, 1998).

Many people believe that it is a simple matter to evade a ban and conduct human cloning research secretly or in a Third World location. Some believe not only that it is simple but that it is inevitable. This is the only way to explain the elevation of Dr. Richard Seed, a retired Chicago physicist who announced at a conference in Chicago on December 5, 1997, that he intended to clone human beings, from obscurity to a figure capable of inspiring national anxiety. Seed was a pathetic figure who had absolutely no hope of cloning anyone or anything at any time. Still, his elevation for a few months early in 1998 to a national nightmare was the most obvious manifestation of the belief in the ethics lag. But there were many other manifestations of doubt that ethics would make any difference whatsoever to the future of the genetic revolution in the months after the birth of Dolly in Scotland became public knowledge.

Commentators and pundits went bonkers over the appearance of Dolly. Some fretted about the national security risk posed by clone armies in the hands of rogue regimes. Others wrung their hands over the use of cloning to create hordes of clones who might be mined to supply tissues and organs to those in need of transplants. A few commentators speculated on the societal implications of immortality achieved by means of cloning oneself sequentially. These sorts of speculations made little scientific sense, but they did reflect deep public doubt and mistrust of advances in the realm of genetics and genetic engineering (Caplan, 1998).

One legislator who spoke out very vociferously about human cloning on the basis of the Dolly experiment was Sen. Tom Harken of Iowa. In hearings on cloning he expressed the view that once science had started down the path toward new knowledge, there was nothing anyone could do to stop its progress. He ventured the opinion that no law, or moral rule or set of values, had ever deterred biomedicine from doing anything and that the best the world could hope for was that those working on cloning chose to do so in an ethical fashion (Lane, 1997; NBC News, 1997).

The view that biomedicine cannot be stopped, shaped, or changed by ethics might well be called *Harkenism*. The position holds that biomedical progress moves under its own momentum. It advances a supremely fatalistic and skeptical view about ethics: once science has made a key breakthrough and gets rolling, there is nothing anyone can do to stop it.

There is something terrifying about Harkenism. If accepted it means that there is really no point in debating or arguing about the ethics of any biomedical advance. That future will be what it will be, and there is nothing anyone can do about it. Worse still, if the unscrupulous or the crazy get their hands on biomedical advances—if a competent and rich Dr. Seed were to seek the sponsorship of a renegade regime to start his cloning company—there is nothing anyone can do to deter or stop this sort of thing. The only problem with both the invocation of the ethics lag and with Harkenism is that they are both wrong.

Has Ethics or Bioethics Ever Stopped Anything in Medicine?

Many years ago, in the late 1970s, when I was a graduate student just beginning a position at the Hastings Center, the nation's most influential private bioethics institute, Daniel Callahan, then the director, and I had a standing bet. We would ask of the various scholars, physicians, and researchers who came to the center to give talks or participate in seminars that they name a single technology that had been stymied, blocked or destroyed as a result of a bioethical objection or argument. Our bet was that no one would be able to do so. We agreed to provide a free lunch for all staff if someone ever came up with a single case of a technology that had been stopped because of ethical concerns or reservations. No one ever did.

Dan and I would use the inability to identify any scientific application or technology that had ever foundered on the rocks of ethics as a way to calm the worries of physicians and researchers that if they even talked about ethics they might somehow wind up being responsible for hindering inquiry. No act could have been seen as more treasonous, more incompatible with being a member of the biomedical community, then to permanently hinder scientific progress for ethical reasons. Reassured that they could not do permanent damage to their own research programs or those of colleagues, the visitors would then almost always dig in for a dialogue on bioethics, since they felt certain that talk of ethics would not put the practice of science at any real risk.

I have come to think that Daniel Callahan and I were wrong about the power of ethics. The problem was that when we asked for case examples we were looking for instances in the very recent past where someone's bright idea had gone up in smoke forever due to ethical worries. However, seeing the real impact of ethics on science is more akin to detecting the processes of evolutionary change, being aware of barometric pressure, or being alert to the presence of gravity.

Evolution is a phenomenon that is difficult to observe because it goes on very slowly all around us. It is hard for anyone to be aware of the weight of air or the pull of gravity, because they are present in our lives at all times. These forces are a part of our environment. We adjust to them. It is only in their absence, when humans travel into space or deep into the sea, that we realize the powerful force they constantly exert upon us.

Similarly, ethics is most noticeable with respect to the role it plays in shaping science when it is not present or present in a very different form. The inhumane experiments conducted in the German and Japanese concentration camps by competent scientists and physicians and public health officials during the Second World War show how very different scientific behavior is in the absence of the normal ethical restraints that dominate the practice of science and medicine. Research conducted on serfs and slaves in the United States and other nations in the nineteenth century, who were not seen as persons or even as human, or on animals in the eighteenth and early nineteenth centuries, give more tragic evidence of the role played by ethics in biomedical research today (Caplan, 1998).

It is simply not true that ethics has not had or cannot have an impact on what biomedicine does or what biomedicine becomes. While the influence is not always obvious or rarely is even detectable, once one looks closely it can be found.

For example we presume that doctors will reveal to potential subjects the nature of experiments they might want them to serve in and that they will obtain their permission before studying them. The requirement of informed consent in recruiting subjects to biomedical research is, however, a relatively recent innovation. As recently as the 1930s and 1940s, subjects were routinely lied to or deceived about the nature of human experimentation and consent was often not sought.

Prohibitions on research on retarded children living in institutions and upon fetuses except when it might be for their benefit have been in effect for decades, as have prohibitions against embryo research and fetal tissue transplantation. These moral bans have had the effect of bringing these areas of inquiry almost to a complete halt. Research on the total artificial heart and the use of animals as sources of organs for transplants was halted for more than a decade as a result of moral objections. The inclusion of women in clinical trials is a direct response to moral criticism. The decision to halt research involving recombinant DNA work in the 1970s until sufficient oversight could be applied to experiments was fueled by moral doubts on the part of basic scientists about the safety of early research with recombinant DNA (Singer, 1977). It is hard to maintain a strong allegiance to either the ethics lag or Harkenism once one takes a close look at the history of biomedical research.

True, ethics cannot always restrain or curb biomedicine's drive to know. Nor can it always provide a reliable safeguard against the actions of a fiend or a nut. But the fact that ethics is not omnipotent should not blind us to the fact that it is not impotent either.

The power of ethics in steering and even prohibiting certain kinds of conduct is not always easy to see. Just as Jane Goodall spent fifteen years observing chimpanzees without seeing them engage in killing before an all-out war broke out in the groups she had known and written about as peaceable, ethics may not be much in evidence until a true conflict of interest or scandal sends everyone scrambling for their code of ethics.

SHOULD ETHICS GUIDE HUMAN CLONING?

If there is no prima facie reason to doubt that human cloning does raise key ethical issues, and if it is not ridiculous to suggest that ethics might actually succeed in steering the direction of future research and application of knowledge about cloning to humans, what are the reasons for ethical concern about cloning?

The reasons are simple—safety and the best interest of the clone. There is insufficient verified knowledge available about the safety of cloning involving DNA obtained from adult cells. There is more knowledge about cloning involving the splitting of embryos to create clones (a subject that has drawn almost no moral commentary, even though it is probably the form of human cloning most within our reach).

To create a human clone based on the existing success record of animal cloning would be blatantly immoral. The clone could be born deformed, dying, or prematurely aging. There would be no basis for taking such risk unless there were some overwhelmingly powerful reason to clone someone. Safety alone justified moral concern in the form of clarifying the ethics of human experimentation. At what point will enough data from animals be on hand to justify a human trial? At what point would the risks involved still permit someone to try cloning? Who should attempt to clone, and for what reasons? People of goodwill can and do disagree about the answers to these questions but the very fact that disagreement exists shows the centrality of ethics to the enterprise of human cloning.

The other reason ethics is very relevant to human cloning is that it is not clear that cloning is a good way to make a person. If clones feel burdened by having a very close resemblance to one parent, if they feel that their future is not their own because they were made to conform to someone else's expectations and dreams (Davis, 1997), if they feel overwhelmed by the burden of knowing too much about their biological destiny because it is written in the body and appearance of the parent from which they came, if they elicit inappropriate or hostile reactions from parents and others, then it may prove to be too burdensome to ask someone to go through his or her life as a clone. It is not clear that cloning is too burdensome. But it is far from clear that it is not. Until that issue has been debated then there is no reason to think that ethics should be excused if it lags in any way behind the science of cloning.

REFERENCES

Adler, Eric. 1997. As Dolly, the first clone of an adult mammal, made her debut last week, a skeptical and doubting public asked . . . What's next? Technology inspires wonder and worry. *Kansas City Star*, March 2, A1, 5–6.

Caplan, Arthur L. 1997. *Due Consideration*. New York: John Wiley.

Caplan, Arthur L. 1998. *Am I My Brother's Keeper?* Bloomington: Indiana University Press, 1998.

Davis, D. 1997. Genetic dilemmas and the child's right to an open future. *Hastings Center Report* **26**: 6–9.

Fox R. C., and J. P. Swazey. 1992. *Spare Parts*. New York: Oxford, 1992.

Geller G., Botkin J. R., Green M. J., et. al. 1997. Genetic testing for susceptibility to adult onset cancer. *JAMA* **277**: 1467–74.

Kolata, Gina. 1998. With an eye on the public, scientists choose their words. *New York Times*, January 6, B12, F4.

Lane, Earl, 1997. Senator. Human clones ok but scientists tell panel: for animals only, *Newsday*, March 13, A8.

Lerman, C, Croyle R. 1994. Psychological issues in genetic testing for breast cancer susceptibility. *Archives of Internal Medicine* **154**: 609–16.

National Bioethics Advisory Commission. 1997. *Cloning Human Beings*. Rockville, MD, June.

NBC News Transcripts. 1997. *TODAY*, Senator Tom Harkin discusses his views on human cloning, March 13.

Powers, William, 1998. A slant on cloning. *National Journal* 30, 2 (January 10): 58.

Robertson, J. A. 1994. *Children of Choice*. Princeton: Princeton University Press.

Silver, L. 1997. *Remaking Eden: Cloning and Beyond in a Brave New World*. New York: Avon.

Singer, Maxine. 1977. Historical persepectives on research with recombinant DNA, in: *Research with Recombinant DNA*, Washington, DC: National Academy of Sciences Press.

Stolberg, S. G. 1998. A small spark ignites debate on laws on cloning humans. *New York Times*, January 19, A1, 11.

Weiss, Rick, 1998. Fertility Innovation or Exploitation? Regulatory Void Allows for Trial—and Error—Without Patient Disclosure Rules. *Washington Post*, February 9, A1, 16.

Wolf, S. 1997. Ban cloning? Why NBAC is wrong. *Hastings Center Report* **27**, 5: 12–14.

14

Cloning Humans and Cloning Animals

Peter Singer

If we were to judge by the amount of attention it has received, from the media, from political leaders, and from opinion makers, the ingenious technical breakthrough that enabled Dr. Ian Wilmut and his team at the Roslin Institute in Scotland to clone an adult sheep would have to be the most momentous scientific event since the first atom bomb was dropped on Hiroshima. The president of the United States demanded a report on whether the new procedure should be banned. Even the birth of the world's first "test-tube baby" did not provoke so swift a response. Nor did the subsequent cloning of mice, cows, goats, and pigs.

In the aftermath of the Dolly announcement it was no surprise that *Time* and *Newsweek* ran cover stories on the issue, nor even that *Newsweek*'s cover featured, not the most photographed sheep in history, but three identical human babies standing inside laboratory glass beakers. That story was spread over eight pages, plus an additional page of comment that concluded, only too predictably, with the question "Do we really want to play God?"

More remarkable is the fact that one of the world's most sober newspapers had articles on cloning virtually every day for more than a week. Cloning was on the front page of the *New York Times* on February 23 (with a continuation over 4 columns on a later page); February 24 (the continuation took an entire page); March 1, March 2, and March 3, when a special report began across the top of the front page and took up an additional three full pages.[1] Opinion articles were also published on February 26, 27, 28, and March 2. As Stephen Jay Gould put it, Dolly became the most famous lamb since John the Baptist designated Jesus the Lamb of God.

The media justified its unusually extensive coverage by quoting authoritative voices telling the reader that cloning is very scary stuff. The French minister for farming, Philippe Vasseur said "tomorrow someone could well invent sheep with eight feet or chickens with six legs." The German minister for science and research, Juergen Ruettgers, said that cloning of human beings "can never be allowed. . . . Each and every human being is a unique creation that cannot be the subject of manipulation." Robert Coles, a Harvard child psychiatrist and author, likened cloning to Eastern ideas of reincarnation. The most frightening comment of all came from Nobel peace prize winner Dr. Joseph Rotblat, who compared Ian Wilmut's breakthrough with the creation of the atom bomb.

Really, Dr. Rotblat? Will cloning cause the instant annihilation of tens of thousands of people, and the slow death from a debilitating illness of thousands more, as the atoms bombs dropped on Hiroshima and Nagasaki did? Will cloning ever have the power to destroy all life on Earth, as nuclear weapons do today? It hardly seems likely.

We might shrug this off as predictable media sensationalism, supported by a few off-the-cuff comments from politicians who are trying to win votes, and other commentators who are not expert in bioethics. What I find much more troubling is the fact that many bioethicists and others who have plenty of expertise in the field reacted in a very similar way. Daniel Callahan, one of the founding fathers of American bioethics, called cloning "a profound threat to what might be called the right to our own identity" and said that a parent who cloned him or herself "robs the child of selfhood."

Dr. Hiroshi Nakajima, director general of the World Health Organization, said that "WHO considers the use of cloning for the replication of human individuals to be ethically unacceptable as it would violate some of the basic principles which govern medically assisted procreation. These include respect for the dignity of the human being and protection of the security of human genetic material."[2] This statement was subsequently repeated in a resolution of the Fiftieth World Health Assembly. Frederico Mayor, the head of UNESCO, was even more sweeping in saying, "Human beings must not be cloned under any circumstances."[3]

The European Parliament passed a resolution on cloning which said in its preamble:

> The cloning of human beings . . . cannot under any circumstances be justified or tolerated by any society, because it is a serious violation of fundamental human rights and is contrary to the principle of equality of human beings as it permits a eugenic and racist selection of the human race, it offends against human dignity and it requires experimentation on humans.

In the first clause of the resolution, the European Parliament added another ground for prohibiting human cloning. It asserted that "each individual has a right to his or her own genetic identity."[4]

In my view these are hasty, follow-the-crowd judgments. Let us focus first on the prospect of cloning from an adult human being, since this is what set the media alight. I doubt that the ability to make clones from mature individuals will change the world in any dramatic way. What is it that we fear? Have we been so carried away by science fiction that we believe that megalomaniac dictators are going to try to make thousands of clones of themselves? Let us assume that these dictators have the scientific resources at their disposal to do this, and can find the thousands of women necessary to bear their clones.

It will still take 18 years for the clones to become adults—and then what? During these 18 years the clones will be growing up in environments totally different from those of the dictators from whom they were cloned. They will, for example, know that they are the clones of a dictator. It is impossible to tell what effect these differences will have on their personality, abilities, and views about the world. Megalomaniac dictators usually find easier—and much nastier—ways to leave their mark on the world. The European Parliament resolved that human reproductive cloning must be prohibited because each individual has a right to his or her own genetic identity. It would be hard to find a better example of the absurdity of the current fashion of plucking new rights out of thin air. From where does such a right come? On what is it grounded? And where does it leave identical twins? Does the mere existence of their twin violate their right to their own genetic identity? Could one twin use it as a defense to a charge of murdering his or her twin: "Your Honor, I acted in order to defend my right to my own genetic identity"?

What about the claim that cloning is contrary to the equality of human beings because it permits a eugenic and racist selection of the human race? It is true that cloning does permit this, but so do a host of other techniques. Artificial insemination (AI), for example, is already used to select cattle and other domestic animals for particular characteristics. There is nothing in the technique of AI that would rule out a similar use in humans. Should we therefore prohibit the use of AI among humans, for fear that it will be used in a racist or eugenicist way? Much better, I think, to prohibit specific morally objectionable applications of such techniques than to prohibit any use at all of them, no matter what the circumstances.

Since, at present, untrammeled free enterprise is a more realistic concern than fascism, perhaps we should worry not about state-promoted racist or eugenic uses of cloning, but rather about movie stars, sporting heroes, and Nobel Prize–winning scientists seeking to cash in on their fame by selling their DNA to people who would like to be the parents of their clones?

Even if cloning became a simple enough technique to make this affordable for some, I doubt that it would ever become widespread. The *New York*

Times tried to find people who might be interested in cloning themselves. They asked Donald Trump, whose ample desire to spread his name across the world has led to his practice of naming office towers and airlines he owns after himself. He wouldn't have any trouble in raising the cost of cloning himself, but he wasn't interested. The only person I know to have commented favorably on the idea of cloning himself is Richard Dawkins, who said that "it would be mind-bogglingly fascinating to watch a younger edition of myself growing up in the twenty-first century instead of the 1940s." Fascinating? Yes, I suppose it would be, especially to someone interested in the interplay of environment and heredity, but I doubt that many people would go to the trouble of cloning themselves just to see how it all turns out.

Roger Short, a distinguished researcher in the biological sciences, has relayed to me a story of another person who wanted to be cloned, a boy who suffered from cerebral palsy as a result of a tragic birth mishap. When he heard of Dolly he told his father: "Please have me cloned, so that you can see what I'd be like if I didn't have this dreadful condition." An unusual idea, perhaps, but surely not an evil one, nor one that, if carried out, would lead to any particularly bad consequences.[5]

Most couples prefer to have children who are genetically their own, if they can. Infertile couples will, if they can produce eggs and sperm, go through repeated cycles of IVF in order to conceive a child, even when adoption of the use of donor sperm would be a simpler means of having a child. And if a few people did give birth to clones of Mick Jagger, Madonna, Michael Jordan, or Jane Goodall, would that be such a terrible thing? We might pity the children, who could be under great pressure to live up to the talents of those from whom they were cloned, but to compare their problems with those of the victims of nuclear weapons, as Dr. Rotblat did, is grotesque.

If some politicians and bioethicists have made comments on these issues that were not well considered, research scientists have not done much better. Even Ian Wilmut has said:

> I am uncomfortable with copying people, because that would involve not treating them as individuals. And so I posed the question that I would like to ask anybody who is contemplating such a use: "Do you really believe that you would be able to treat that new person as an individual?"[6]

Does anyone think that people who are identical twins do not have an identity, or are not treated as individuals? If not, I would respectfully ask Dr. Wilmut why we should think such things of cloned human beings? I am sure that I do not need to remind him that an adult and his or her clone would be less similar than identical twins, because they would necessarily be brought up in different environments. We can expect them to have very different opinions, as identical twins, brought up together, often do.

Perhaps Dr. Wilmut would say that twins are not planned, but there is a special problem with the deliberate creation of a new person who is a clone of an existing person. Such remarks are reminiscent of the storm of criticism that greeted the news, in March 1990, that a Los Angeles couple, Abe and Mary Ayala, were having a baby in the hope—the odds were only one in four—that the child would be a bone marrow donor for their 17-year-old daughter, Anissa, who was dying of leukemia and for whom a two-year search had produced no suitable donor. This, medical ethicists thundered, is using a human being as a mere means! It was wrong, even "outrageous."

Despite the criticism, the Ayalas went ahead and, luckily for Anissa, the child was a match. Anissa's life was saved, and Marissa, the new addition to the family, though perhaps initially desired instrumentally, soon became a much loved child of the family. If what the Ayalas did was wrong, it seems to have been a remarkable kind of wrong, for it has greatly benefited at least three people—Anissa Ayala and her parents—and it has harmed no one. Indeed, if we can benefit a child by bringing her into existence and doing our best to ensure that she has a happy life in a loving family, that is exactly what the Ayalas have done for Marissa, so arguably it is not three, but four people who have been benefited.[7]

This last question—do we benefit beings by bringing them into existence, if their lives are not clearly awful?—is generally not raised in debates about cloning humans, yet it is clearly relevant to the ethics of cloning a human being. Are we going to condemn cloning if the life of a cloned human being might be somewhat more troubled than the life of a human being produced by the usual process? Or is cloning wrong only if we can show that the life of the cloned human being would be so bad as not to be worth living?

Think for a moment about the fact that many extremely premature newborn infants survive only because of the great skill and dedicated labor of highly trained health care professionals. Yet of any given baby with a birthweight of, shall we say, under 750 grams, we know that, if the baby survives at all, there is a risk of somewhere between 25% and 50% that it will have a moderate or severe disability. We do not see this as a sufficient reason for not trying to keep the baby alive (although where the disabilities are so severe that the life of the child will clearly be awful, we may do so). I would suggest that the situation ought to be similar with cloning a human being. In the absence of good evidence that the life of the cloned human would be awful, we cannot justify stopping cloning on the grounds that we are acting in the best interests of the cloned human.[8]

There may be genuine medical grounds for cloning a human being, and a situation like that of the Ayalas could be one of them. Cloning would eliminate the genetic lottery, and ensure that a future child could be a perfectly matched donor for an existing child. Of course, if there was a genetic component to the disease, we might well not want to clone from the existing child, but there may be circumstances in which this problem does not arise.

In any case, I don't think we are in a position to make any sweeping declarations about the wrongness of cloning a human being until we have carefully considered such possibilities. I can understand, however, why anyone in Dr. Wilmut's position would want to distance his work from human cloning and join in the chorus of condemnation of that practice. I do not expect any scientist to be oblivious to the threat of loss of funding for his or her research. Nevertheless I would hope that in time we may achieve a more nuanced discussion of the risks and benefits of human cloning.

I turn now to the ethical questions raised by the cloning of animals. Despite the fact that we are here faced not with a possible future scenario, but with something that has already happened, these issues have received relatively little discussion outside the animal movement. I find it curious that there is so little thinking about why some moral principles are supposed to hold for humans, but not for nonhuman animals. Why, for example, does the German minister for science and technology say that humans must not be cloned because every human being is a unique creation, when the nature of the genetic lottery ensures that every sheep, cow, pig, and dog is also a unique creation? Is it important that every human should be unique, but not important that every animal should be unique? Why?

I noticed this kind of thinking at a conference held in Melbourne in April 1998. A number of speakers referred to ethical concerns regarding techniques that may involve cloning humans, even when we are talking only about cloning human embryos. Dr. Wilmut, for example, referred to such potential problems when he suggested that cloning by nuclear transplantation from a patient with Parkinson's disease might be used to create an embryo from which differentiated cells could be taken to provide a therapy for the disease. Now I am aware that Dr. Wilmut was not himself saying that he thought this would be wrong, but rather that some societies might consider it wrong and not permit it.

As a prediction, that statement may prove to be correct. But I would hope that eventually people would be able to see that here we are talking about a very early embryo, far too undeveloped to be capable of feeling pain or suffering in any way. I see no serious ethical problems with the use of such an embryo. On the other hand, Dr. Wilmut also mentioned the possibility of using sheep as a model for cystic fibrosis. He did not suggest that there might be ethical concerns about this. I recognize, of course, that cystic fibrosis is a tragic condition, and I do not know enough about the proposal to say whether I might be persuaded that it is a defensible use of an animal. But it does seem to me obvious that there are serious ethical concerns about deliberately creating animals who will suffer from diseases such as CF.

Unfortunately, a dichotomy in our thinking regarding animals and humans is what we have to expect, given existing practices regarding animals. Is there really any reason for the French minister for farming to be so worried about a six-legged chicken when he shows no signs of any concern about the many millions of hens already on farms under his jurisdiction who can make no use of their wings because they are kept in cages too small to allow them to stretch even one wing at a time? Not, as far as I can see, as far as animal welfare is concerned.

In the comment about the six-legged chicken we see the operation of the "yuk" factor in ethics. In other words, think of something, and if your initial reaction is "yuk," then it must be wrong and we should stop it. But let us not forget how moral conservatives have appealed to the "yuk" factor to persuade us that we ought to prohibit abortion—remember those images of the tiny fetus being torn limb from limb—and almost every form of sexual activity that does not involve placing the penis in the vagina. "Yuk" ethics quickly slides over into the idea that something is only right if it is "natural," but that term is, as John Stuart Mill once wrote, "entangled in so many foreign associations, mostly of a very powerful and tenacious character," that it is now "one of the most copious sources of false taste, false philosophy, false morality, and even bad law.[9]

Though more than a century has passed since Mill wrote those words, they remain as true today as they were then. The term *natural* has no clear descriptive sense, and if it did have, then, since there are no values built into nature, it would not be possible to deduce any moral conclusions from it. If, to continue with the French minister's fanciful suggestion, we ought to prohibit six-legged chickens, then it is not because chickens in nature have only two legs, but because six-legged chickens would be in pain, or because producing six-legged chickens does not serve any worthwhile purpose.

This point was nicely put by Grahame Bulfield, the director of the Roslin Institute: the technology itself is a red herring. If an animal is lame because of genetic modification or selective breeding or poor nutrition or because I kick it, it is wrong that it's lame. So you have to pay attention to the phenotype—that is, to the animal itself—rather than the technique that produces the problem."[10]

Bulfield goes on to suggest that while there are some things one should never do to animals, there are others where you need to consider whether "the possible suffering is outweighed by the good." This is, he points out, a consequentialist analysis, and since I am a consequentialist ethicist, I can hardly disagree. But I want to insist that the weighing of the suffering and the good must be done without discounting the sufferings of the animals merely because they are animals, that is, not members of our species. Pain is pain, no matter what the species of the being who feels it.[11]

No doubt there are some forms of suffering and distress that require higher mental capacities not possessed by mice or sheep, but there are other forms of pain that they can feel, and which we must presume they feel in a manner similar to the way in which we would feel it. As the Australian Code of Practice for the Care and Use of Animals for Scientific Purposes says, "Investigators must assume that animals experience pain in a manner similar to humans. Decisions regarding the animals' welfare must be based on this assumption unless there is evidence to the contrary."[12] That seems to me a reasonable precautionary principle against which we should evaluate current and future research using animals.

In balancing our concern for animals against the possible benefits of the research, we should note that some cloning research has primarily commercial goals. Its aim is to produce more productive, and hence more commercially valuable domestic animals. Let us not pretend that we are talking about feeding the starving billions. Generally speaking, the research has been directed at improving the productivity of animals used in developed countries, and there is no way that the starving billions can afford these products.

There is, in any case, ample evidence that people in the developed world would be healthier on a diet that was lower in animal products. We should not be putting our research efforts into encouraging the population of the developing world to follow our own bad example of a diet that is high in fat, low in fiber, and environmentally disastrous. Hence, while I understand how commercially tempting it may be to seek to produce a faster-growing pig or steer, I do not regard this goal as justifying the production of sick or miserable animals, even for a transition period.

The cost/benefit equation is different when the goal is to produce therapeutic proteins in the milk of genetically modified and cloned sheep or cattle. If such products will be much less expensive than existing products, and will hence be affordable to people who otherwise would have died without them, then the infliction of a degree of distress on animals could be defended, if it were kept to the minimum necessary for the research.

Something similar might be said about the production of animals who could provide compatible organs for xenotransplantation, but here the question becomes much more complicated. We need to consider other possible sources of organs (for example, using an "opting out" system of organ donation rather than the "opting in" system still used in most countries). We must also take into account the resource implications of a possible dramatic increase in the number of organ transplants: is this the best use of our scarce health care resources? Nevertheless, I have to agree with those who point out that when pigs are being slaughtered in order to provide people with bacon, then, other things being equal, it is difficult to object to pigs being slaughtered to provide people with hearts or kidneys.

Cloning is one among many breakthroughs in biotechnology that have occurred in the past twenty years, and its significance should not be exaggerated. As with each of these breakthroughs, we need to make decisions about the many different issues it raises. I have suggested that we need to make these decisions with less haste, with more careful thought, and with a greater awareness of the ethical obligations we have to nonhuman animals, as well as to human beings.

NOTES

1. For a discussion of the American media response to the cloning of Dolly, please see chapter 12, this volume.

2. WHO press release (March 11, 1997), quoted from John Harris, "Goodbye Dolly? The Ethics of Human Cloning," *Journal of Medical Ethics* **23**, 6 (1997): 354.

3. *Ibid.*

4. *Ibid.*

5. Personal Communication.

6. See chapter 1, this volume.

7. Rachels, James. "When Philosphers Shoot from the Hip." *Bioethics* **5** (1991): 67–91.

8. Singer, Peter. *Practical Ethics*, 2nd ed. Cambridge: Cambridge University Press, 1993, chapter 4.

9. Mill, J. S. "Nature," in *Three Essays on Religion*. London: Longmans, 1885.

10. See chapter 1, this volume.

11. Singer, Peter. *Animal Liberation* New York: New York Review of Books, 1990, chapter 1.

12. Canberra: Australian Government Publishing Service, 1990, pp. 7, 1, 13.

15

Animal Cloning

The Pet Paradigm

Arlene Judith Klotzko

Two years after the birth of Dolly the sheep, and eight months after Dr. Richard Seed—the eccentric physicist and would-be cloner of humans for cash—took his brief turn on the world stage, cloning returned to the news. Not with a bang or even with a bleat. This time it was a bark. Scientists at Texas A&M University announced that they were poised to clone—for a $2.3 million reward—a pet dog named Missy, an 11 year old border collie/husky mix, adopted from the pound by an adoring and very rich American couple who have zealously guarded their anonymity.

While at first glance this so-called Missyplicity Project may seem like a natural extension of farm-animal cloning, it is actually a very different kind of enterprise. The imperatives driving people toward pet cloning are much more similar to those for cloning humans. They are deeply personal. In contrast, the cloning of sheep, cattle, and pigs is a quintessentially practical endeavor, devoid of sentimentality. Dolly—arguably the most famous farm animal in history—was the product of a series of experiments seeking to produce large transgenic animals (those carrying human genes) that secrete therapeutically useful proteins in their milk.

Because Dolly was cloned from an adult animal, she set the stage for cloning pigs and cattle from prototypes that have specific characteristics.

The ability to clone pigs—now demonstrated by several scientific teams—could be the most medically significant application of animal cloning if such pigs can be genetically modified in very precise ways to eliminate endogenous retroviruses and to address the process of acute vascular rejection of pig organs. When and if pigs with the desired characteristics can be produced they will still be basically interchangeable. As with farm animals generally, they will not be created to be loved—only to be used, as will their nongenetically modified farm friends cloned for agricultural as distinct from medical purposes.

But people and their pets are different. While cloning us and them may replicate our genetic identity, people and their pets are far more than the sum of our genes. (And in Missy's case there is even a genetic variable; the scientists are not using Missy's own eggs, but donor eggs, which contain their own mitochondrial DNA.) Environment and experience play a crucial role in the formation of personality.

What we prize in our friends and pets is that which makes them special—even unique. And these qualities are not merely a function of genetic identity: We may be acquainted with identical twins, yet find ourselves drawn to one and not the other. Mark Westhusin, the scientist who is overseeing the Missyplicity Project, believes environment is a major factor in the development of a dog's personality, and that the relative weights of genetics and environment will be clear only after they clone Missy.[1] Her clone would be raised in a different environment from Missy, who came from the pound—not from a loving home. Certainly, if Missy had been abused as a puppy and the clone were not, their personalities would be quite different.

To Westhusin, the Missyplicity Project is more than just a way to produce another Missy. He views it as real science, a chance to study basic reproductive physiology in dogs, and perhaps catch up with the work already done in this area with other species. But is the motivating force behind this $2.3 million project really the pursuit of science? Although five of the six project goals described on the Missyplicity website (www.missyplicity.com) do deal with hard science, cloning Missy is goal number one.

Even more telling are the "Missy Tales," written about Missy by her "human mother." "How Missy found a home" tells readers how she was selected at the pound: "I didn't want another dog whose bite . . . would bring blood. I wanted to know if she had a high ear-shattering bark (which I don't like), and if she could howl (which I do like). . . . I offered a howl to her and she raised her nose and howled at the roof. I barked at her and she barked right back, a low, rich-toned, businesslike bark. I whined, she whined." Missy seems to have met all the specifications.

There has been a long history of manipulating dogs to give them traits we like. Since 1884, the American Kennel Club has been registering pure-bred dogs that represent different ideals of beauty, taste, and style. Such predictability is the reason that one might choose a golden retriever over a mutt. But replacing a beloved dog that has died with another of the same breed will not bring back the lost dog. Neither will cloning.

The clear intention behind the $2.3 million expenditure by Missy's owner is the same as one of the reasons that has been advanced to justify the cloning of humans—to perpetuate a "unique" life that is drawing to its close by replicating it. For example, we are told, dying children could be cloned to produce replacements and thereby assuage their parents' grief.

The cloning of human beings for just such reasons may be a lot more imminent than we think. A cult that worships UFOs and is led by a former sports journalist named Rael, has recently announced its intention—and alleged ability—to clone a child from cells taken from a 10-month-old girl who died, tragically, as a result of a medical mistake. The unnamed parents are reported to have paid the cult $500,000 for its efforts. Rael claims to have 50 willing surrogates. Of course, this endeavor may be as impossible a dream—or nightmare—as the visions of Richard Seed, but the fact is that in the United States, there is no law to prevent such privately funded attempts.

The current price of a pet clone—assuming that the Missyplicity Projects's technique is shown to work—is $250,000. And there is now a new company formed to carry out pet cloning for those who can afford it—the delightfully named Genetics Savings and Clone (www.geneticsavings andclone.com). Until the cloning technique is perfected, pet owners can store their pet's DNA for a charge of $1,000. Cat owners should not feel left out in the cold. There's a project for them too, called—of course—"Copy-cat." Is there anything wrong with this sort of thinking? Well, it does seem to reduce people and pets to interchangeable—or at least replaceable—objects. And, of course, conferring immortality through cloning by serially perpetuating a unique and beloved human or pet is an impossible dream.

Cloning a pet is less morally problematic than cloning a human. One of the arguments against the latter rests on the possible psychological damage that being a clone might create. If I learned that I had been cloned to be "like" someone else, would I feel pressured to become that person? To excel in the ways that person excelled? To be a violinist, a baseball player, a physicist? Could I watch my older genetic double sicken and die prematurely without being paralyzed by the fear that this would be my destiny as well?

Cloned pets, of course, would be spared such knowledge.

A remarkable paradox lies behind the Missyplicity Project. Its rich benefactor has asked that scientists employ a technique that many see as a threat to uniqueness in a futile quest to perpetuate a unique life by reconstructing a new dog. Most of us would abhor the idea of a world in which there are a hundred people just like us. But, contrary to such fears, cloning could never produce such a world. Those who seek reincarnation must look to metaphysics—not genetics.

NOTE

1. Personal communication.
2. "The Missyplicity Project: Missy Tales: How Missy Found a Home," www.missyplicity.com

16

A Pragmatic Approach to Human Cloning

Glenn McGee

Human cloning presents a bewitching test of any bioethical method. One can scarcely imagine a worse mess to clean up. Public discussion of human cloning was promulgated by Dolly, a cloned Scottish ewe named after a country music singer,[6] and inflamed by Richard Seed, a Chicago scientist who played the Jack Kevorkian role of announcing on National Public Radio that he planned to clone himself several times "for fun." Public debate about cloning centered on stopping Seed, cloning pets and livestock, and the likelihood that a despot somewhere in the world would set about the task of breaking what seems to be an international consensus against the reproductive uses of cloning technologies. Virtually every philosopher with an interest in ethics was suddenly called on by television to play Solomon, or at least Nostradamus, to questions like "Is it ethical to clone a recently deceased child?" or "Would a clone have a soul?" Within a year of the birth of Dolly the odd, marginal, and unlikely problem of human cloning had been elevated to one of the most hotly debated issues in twentieth-century science and health.

Philosophical debate about cloning has been mounted but along fairly predictable lines, with scant examination of the implications of cloning for human nature, social institutions, or the practice of basic biological science. The question of the day remains narrow and consequentialist: Does anyone have the right to make a clone, and upon whose rights would such a process infringe?[1,2,12,24] This narrow question seems quite urgent as a result of two recent announcements: the news that a clinic in Korea may have made a human embryo from cloned adult DNA, and reports that a team at the University of Massachusetts has inserted human DNA into a cow egg, with plans to launch a humanlike fetus from that material for transplantation research.[14]

Given the hysteria and narrow moral debates, it would seem an odd time to attempt a comprehensive treatment of the philosophical issues in cloning. In fact that is exactly my intent. I believe that cloning is an ideal test of the usefulness of the uniquely American philosophical strategy called *pragmatism*. For those working in the classical American philosophical tradition of pragmatism, facts and values are intimately connected. As I argue in much more detail elsewhere, pragmatism works by finding the most satisfying solution to emergent social problems within the context of what is possible and what is demanded.[7,9,10,15] If a pragmatic approach to bioethics is to work anywhere it is the field of human genetic.[8,27,25] Genetic research is infused perhaps more than any other area of natural science with pragmatic aims.[8]

At bottom and in its implications, genetic science of the twentieth century affects the way we understand our capacity, meaning, and potential. Genetics is intimately tied to procreation, sexuality and reproduction, which are also the foci of our most intimate institutions, such as the family and church. When we make children and when we think of our inheritance, we are building our personal and communal understandings of loyalty, privacy, happiness, and growth. And, at the same time, human genetic information is rapidly becoming both a language of medical diagnosis and a commodity for licensure and ownership. Someone *owns* techniques for cloning mammals, including humans. It has become important to make social choices about the institutions that should be entrusted to reconstruct the family in an era of advancing reproductive technology, genetics, and cloning. Pragmatism is uniquely poised to address such questions but also to cope with the fog of current debates about cloning.[18]

Elsewhere I argue that the most elegant expression of a pragmatic social method is found in Dewey's *Logic: The Theory of Inquiry*.[7,11] In the *Logic*, Dewey offers a matrix for human inquiry into social problems, which I will use in this essay to frame my reflections on the implications of human cloning science and technology.[3] Pragmatic bioethics focuses on the biological, cultural, and commonsense dimensions. By selectively emphasizing and analyzing these three dimensions of the context of cloning, rather than rushing to more obviously normative aspects, we will see that human cloning is neither a special moral issue nor a radical step forward. Instead, human cloning is seen to be an element in a set of moral and scientific problems that compel us to reconstruct the enterprise of social thought about the embryo, the family, and future generations.[13,16]

BIOLOGICAL DIMENSIONS OF HUMAN CLONING

While there is an accepted biological definition of cellular cloning, and there are now well-understood (indeed patented) practices for the transfer of nuclei from embryos or somatic cells into enucleated eggs, it is still not possible to define a cloned mammal organism. That this is so has not gone unnoticed in the biological and philosophical literature of the latter part of this century. Yet now that mammalian cloned organisms are among us, and human clones seem imminent, we must ask again how we are to obtain semantic and scientific clarity about the meaning of a mammalian clone. Must it have all of its DNA from a single other creature? Must the donor of a clone's DNA be an adult? Can a clone's egg come from a source other than the DNA source? If the source DNA contains a slight mutation, is the resultant organism still a clone? To be a clone, must a clone act or sound or seem like its source organism, or perform that organism's role in the community or herd? These questions have not yet been answered, despite the use of "clone" as a descriptor for, at last count, more than 400 living mice, sheep, cows, and other mammals.

Received definitions of a human clone come from science fiction, not the lab. Stories of cloning have been used to illustrate the problems of nature versus nurture, the problem of defining the content of human character, and the problem of preserving our memories in future generations. Captain Kirk's transporter failed, splitting him into two Kirks, one aggressive and domineering, the other intellectual but indecisive. Fictional clones underwent "replicative fading" in *Brave New World* as they were copied one from another. Mostly, clones of our imagination have carried the memories, feelings, and ambitions of one generation into a next generation. Mostly, clones have been dupes and dopes, only occasionally rising above Frankenstein's guttural longings. When it was announced that Dolly had been constructed with DNA taken from the udder of her progenitor, American fear of cloning was motivated and circumscribed by the clones of a hundred years of imagination. President Clinton penned a letter within hours of the announcement calling his previously unfunded "presidential bioethics panel" into action to prevent abominations of the family, with exactly these fears in mind.

How one defines a clone seems to depend on to which side of the issue one stands. Those who see no problem with human cloning, such as Princeton geneticist Lee Silver[26] and Alabama philosopher Greg Pence,[20] matter-of-factly compare any cloned human embryo to a monozygotic twin, which contains the same genetic information as its womb-mate sibling. Twins, it is noted, happen frequently in human life, and it is common today after in vitro fertilization to keep one "sibling" embryo frozen in nitrogen long after the birth of a first. To avoid the pejorative overtone about clones and cloning, Pence suggests a new term: "somatic cell nuclear transfer." By contrast, those who disapprove of human cloning technology point to the centrality of sexual recombination in mammal reproduction, and argue that it would be extremely difficult to predict either the viability or risks associated with gestating or being born a human clone.[4,1]

Can there be a sober, commensurable definition of a clone? Not in this decade. While the brute techniques required to produce a clone are getting better, embryologists cannot state with absolute certainty the genetic or phenotypic identity of a clone. We think of the identity of mammals, including human beings, more and more in terms of the genetic code they bring into the world.

A variety of new, urgent, and troubling legal cases force adults to puzzle over the meaning of that code as it bears on parenthood and identity. Biologists and the broader culture would thus like to be able to at least define cloning in terms of something stable: genetic similarity. Cloning, after all, seems to raise the possibility of a wholly new kind of child, one made not from sex or sexual recombination, but rather from the transfer of genetic information from a single progenitor into its offspring. But in reality, while we do not know what sort of a human being a clone would be, neither do we have any real objective purchase on the variety of new kinds of children we make through new reproductive technologies and through new social mechanisms. We may be able to determine the origins of a child's DNA, but that only begins the process of reinventing ideas of relatedness and how relatedness conveys status and responsibility. We have amazing new ways to make children, and think of that process in increasingly design-oriented terms.[8,5,23]

That this is so is a function of the biological, political, and economic history of pregnancy and childbearing, which others have discussed in much more detail than I will attempt here.[5,23,28] Elsewhere I have drawn the conclusion that new genetic technologies and neonatal intensive care, as well as advancing diagnostic science, have changed the nature of the pregnancy experience from one of having to one of making babies.[8] By this I mean that our best ethnographic and qualitative studies suggest that parents of our time are able to identify with and care for a future child, and that their relationship to future children, including fetuses as well as those not yet conceived, is one that frequently feels like it includes an obligation to prevent future harm.

Despite our cultural insistence on the absolute right of a woman to terminate a pregnancy prior to the time a fetus is viable outside the womb, parents and social institutions are increasingly able to think of the fetus as a child for a variety of purposes. Thus, for example, parents who fail to care for their pregnancy, or physicians who fail to diagnose a fetal malady, are subject to sanctions or damages for the tort of harming a being that does not (at the time of pregnancy) have a right to exist per se, but seems to nonetheless have a right not to be brought into the world in a way that is harmful to it.[19]

The identification of a parental responsibility to future offspring has been long in coming and is tied to a variety of changes in what we mean by childhood and what we expect of children and childbearing. In the course of creating the most recent birth and genetic technologies we have found a way under the hood of pregnancy, radically increasing the ability of adults to take care in choosing the time and manner of pregnancy. We use ultrasound; we conduct amniocentesis; we mix and match genetic parents; we screen for the healthiest embryos all for this purpose.

For example, if my wife's eggs are in some way defective, and if I can take a second mortage or have a free credit card, we will be *treated* for infertility. Why? Because we now say that wholly apart from our need to make love to one another, we feel the separate need to have a biologically similar child, the need to do something to make such a child. The new tool of egg donation implies the possibility we might ameliorate a new kind of need. We want a child, we want it to feel like our child, we want to give birth to it. That need is old. But the need to have a child of such specific parameters is a new one, inspired by our culture's increasing tendency to think of fertility and parenthood as a state of affairs that includes both gestation and genetic relation.[16] Our imagination is of a child that is "mostly" ours.[29] But a baby from egg donation, we are told, is not 100% our genetic child. We are not going to be able to completely emulate the "fertile" state. So, electing to use a donated egg, we are under the hood, tinkering with what for most parents is just a shiny surprise. Our child is going to require more planning. No more will our sexual encounters be about making babies. Our baby will come from a dish. We control, or at least hope to control, what goes in the petri dish. Put more accurately, parents will feel responsible for what goes in the dish. We won't want to choose a donor who has a dangerous congenital anomaly. If we can choose a donor who is more likely to produce offspring with traits resembling our own (height, eye color), we might spare our child the feeling of being obviously different from us. And if we are under the hood anyway, we might also make sure that one of our children is male, and pay a small amount more for a young, Ivy League donor.

That it is odd to be under the hood is obvious. That it is a different kind of parental decision making, less subtle and more commodified, seems likely. But the point to be noted here is that our advancing reproductive technologies exascerbate the evolving problem of assigning and enacting parental responsibility. Where the abortion debate focused the attention of the Western world on the comparatively simple question of when an *in vivo* fetus takes on moral status, new reproductive technologies raise the problem of what it means to be a parent, and what value that experience has for those involved. In the case above, we will try to compensate for the 50% loss of parental DNA by making wise choices about the donor, choices that will both make us feel responsible and further assert our claim to dominion over the resultant child.

In the case of a cloned embryo, it is not at all obvious who are the parents. The person who donates DNA from a somatic cell is the progenitor, in that the child carries that person's DNA. But the mammalian parents of the cloned child are the grandparents, if what you mean by parent is that the person contributed 50% of the genes to the recombination process that formed the genome of the person in question. If the egg used to raise the clone comes from another person, as it would in the case of a clone of a male, there is in addition an "egg parent," a person who contributes mitochondrial DNA and RNA in the egg wall, the collective role of which on an organism is unknown but perhaps significant. If the progenitor of the clone is itself an embryo or aborted fetus, the parent would not only be a virgin, but also a nonconsenting nonperson that itself has no legally established standing apart from the wishes of its own progenitor. Cloning makes acute what is already true in many new technologies and for embryology more generally in our time: we do not know what is in the petri dish, and must make overtly stipulative claims about our relationship to the thing in the dish.

That this problem is acute in cloning results from the complex and engineered nature of that procedure. It is not obvious that a cloned embryo is an embryo. One part of what makes a mammalian embryo, after all, is conception. Sperm and egg fuse, and an embryo is formed. This is not so for a clone. An egg whose nucleus has been removed is fused with DNA from, for example, a human skin cell. The result is that the egg, in some cases, begins to behave much as an embryo. In the best of cases, that of the cloned mice from Hawaii, successful pregnancies of such embryolike things result in only about 4% of all attempts. This is, or we believe it to be, much less frequent than pregnancy rates for mice (or humans) attempting to have offspring through sex, though about the same as the rate of pregnancy from human sexual intercourse more generally. Put another way, a cloned mammalian embryo appears to be less viable than a noncloned embryo. What does it take to call the creature an embryo? Must there be fusion of egg and sperm? Must there be marked potential for gestation? Further, what is the bar for such a creature to count as a restoration of fertility, or as a therapy for infertility?

This last question is the most vexing part of the biological dimension of cloning. The felt need to have a child is undeniable in our society, and more than $1 billion is spent annually on the pursuit of biological parenthood through infertility medications and procedures. At one level, we need to know what sort of role individuals should be able to play in designing children; how far under the hood they should be allowed by our institutions to go. There surely are some negative rights against governmental interference in procreative activity,[21] and these perhaps include some right to experiment with technologies like cloning.[22] But more problematic is what it means to provide care for those who have a need to parent. Elsewhere I have noted that it is a common mistake to assume that it is species typical for human beings to have children that carry our own genes or are biologically similar to us.[16] Thus while it is fairly easy to establish that infertility includes an inability to contribute gametes or gestation to a child's birth, sequela to some organic dysfunction, the rub is that one cannot always cure the organic dysfunction itself. The therapy for infertility is often a technology aimed at providing as many children as are desired by some parent or parents. But is infertility cured by providing this therapy? Would adoption cure the condition of infertility as well? Does cloning present a cure? It seems clear that the answer requires us to turn to the way that the needs of biology as regards reproduction manifest themselves in our individual and cultural habits.

CULTURE AND CLONING

I was raised in the 1970s with a story about what it meant to be a child. The idea was that parents loved each other, got married, made love, and babies resulted. Parents loved each other so much that they raised those children as their own, and made sure that they could handle the responsibilities of parenting, marriage, and career by organizing life in such a way that only one of the parents would work, while the other raised the children. It is the story of the birds and the bees. Birds and bees, of course, do not live that way. But the story has powerful resonance for many Americans, representing what has taken on the name "traditional family values" in political discourse, despite the fact that such families are increasingly rare. It is a story that links sex, reproduction, and family in strict terms. While our technologies for making children have changed quite a bit, most aim at and are measured against the story of birds and bees.[17] In divorce and adoption, for example, the model of the birds and the bees is used by jurists to measure degrees of variation from the norm, and to aim at restoration of it.[17]

The data are fairly clear that tomorrow's children will not be raised in the world of birds and bees. Perhaps the most common model for parenthood in our time is that of the ants and termites, who live in large groups with distributed parental roles. We live in a culture in which children are frequently raised by some combination of nongenetic parents, or by those who are not parents at all. More than 40% of those born after 1998, we now believe, will have more than one mother or father by age 18. The majority of American children are effectively raised in day care, while all three or four of their parents pursue careers. Many in our society have long believed that a critical role one can play in the life of a child is that of godparent, or coach, or foster parent, and many families in many ethnicities have well articulated roles for these mentors.

New technologies will necessitate new stories. Octuplets and septuplets will be the first in our species to hear a story of the dogs and the cats, about being part of a litter. We need a story for a child whose entire first-grade class, and soccer team, consists of siblings. Children of postmenopausal pregnancy will need a new story more fitting than that of "the accidental" late-born child of yesterday. Children of sperm and egg donors will need a story. While today most parents do not tell their children of the presence of donor DNA, eventually it will not be optional. Perhaps these children will be told a story about the racehorse, bred from chosen samples of stud sperm. Lions represent a story for children who are gestated by one woman, with an egg from another and DNA from a third. As transgenic egg donation from monkeys or cows finds its way into human reproduction, stories for that technology, too, will be needed.

But what story can one tell a clone? Already we have noted that human cloning is unprecedented in the natural history of mammals. Twins are the closest existing phenomenon, and they are separated by at most a few hours. The stories of parental roles in cloning in the media are frightening in almost all cases. One has parents replicating a child who has died early due to an accident. Another has an infertile woman seeking a genetic link to her recently deceased husband through a clone from a tissue sample she happens to have lying around. Still a third has the parent raising a clone of his wife to realize his dream of seeing his wife as a child. The point of discussing children's stories is twofold. First, it is clear that whatever progress we make in infertility technologies, an important part of realizing the potential of such technology to satisfy the felt needs of adults is an account of what the technology will mean for the child. More, such family relationships are heavily textured by their social and institutional histories. Being tolerant of new kinds of family will have to begin with existing technologies and move out slowly and experimentally toward the margins.

Second, our children's stories—and the lack thereof—evidence the cultural manifestations of methods of satisfying parents' demands for children. The predominance of the story of birds and bees is symptomatic of our cultural and institutional commitments to genetic determinism, which in this case means our social faith that what matters about blood relation and about relatedness itself is programmed in and received through the genes of parents. People get married, make babies, and raise them in ways that seem normal to us because of our history, the habits passed down through the last three or four generations of Western families. It is only recently that we could consider the possibility of lesbian or gay reproduction, or ponder the relative value of different kinds of offspring or relatedness. So our efforts to squeeze every case into a standard of deviation from the normal model of birds and bees is merely a kind of collective dissonance with forming new habits about such an intimate matter. Moreover, we struggle in our new technologies to restore the apparent equilibrium of the "classical" family; we work to find technologies that give us as much of the birds and bees as is possible. This is one reason why, for example, most couples will use sperm injection rather than donor sperm. It is simply assumed that it is better, more normal to have a child that shares more identity with me. Thinking about and emphasizing the role of children's stories help to bring these two issues into focus.

But our habits in making our own families are only part of the culture of reproduction. Parenthood is for some purposes at the luxury of the community, and it is more than idle Platonic fantasy that children are raised by the state. We have already noted that economics, politics, and theology play roles in how infertility is understood and treated. The family is also only one among many institutions that raises children. In fact, when parents fail in a variety of tasks (from immunization to feeding to education), they can lose their parental rights, to be restored only at the discretion of representatives of our democracy. The upstream manifestation of this public concern for the welfare of children is manifest when, for example, it is argued that children in general ought not be born clones, or that research to clone humans is of a comparatively low priority in the existing array of choices for research spending. Even editors of scientific journals and newspapers have a choice about what they will send out for review and in what way they will publish findings about cloning. The goal of examining culture is to square the variety of contexts within which a tool comes to be with the ends it is actually capable of achieving. Dewey calls this the placing of means and ends in strict conjunction, and points to the continuum between means and ends for the purpose of seeing our social methods solve social problems.

COMMON SENSE AND CLONING

Common sense is the most misunderstood element of pragmatism. The goal of pragmatism is not to skip the difficult questions and move on to progress. As is already apparent, in the present case pragmatism unpacks the meaning of satisfying the complex and situated demands of a variety of people within a social context. More, though, pragmatism shows that ethical evaluation of social problems requires that we take seriously the challenges of science to social thought, and in this respect cloning is clearly a paradigm case. Cloning does not uniquely challenge what it is to make a child, but it has called attention to the vast array of new technologies that make new kinds of families whose parameters and relationships are neither pregiven nor socially sanctioned. It is insufficient to ask, as do most critics of cloning, whether a child of cloning would be deprived of a right to individuality.[17] No child has an open future, and even our cursory examination of the changing history of parenthood makes clear that it is not individuality but rather correct forms of responsible relation that are our goal.

I have not addressed, in this essay, the tough or exceptional cases. Richard Seed wants to make human clones. Greg Pence suggests the viability of cloning dead scientists. A Korean clinic may soon make the "first" human clone. The tough cases are interesting, and many will ask whether Seed should be stopped or Korea sanctioned. But the pragmatic question is more important: what institutions and arenas are right for situating the debate about human cloning and its ken? Elsewhere I have argued that the adoption procedure is a metaphor for what is possible: regional, localized evaluation of candidates for new procedures, accompanied by education and tolerance of new kinds of families and reproduction.[17] But other and more experimental methods too may be called for. The claim of this essay is that a

Deweyan, pragmatic approach to cloning demonstrates the need to reconstruct the entire enterprise of making children in the twenty-first century as a backdrop for debate about human cloning. Once this is accomplished, we can move beyond exceptional approaches to general problems and develop new institutional and personal habits for making and supporting families in the twenty-first century.

REFERENCES

Acknowledgments I acknowledge Arlene Klotzko, David Magnus, Arthur Caplan, Ezekiel Emanuel, Rosemarie Tong, Pilar Ossario, Andrea Gurmankin, Garth Green, and grants from the Commonwealth Foundation and the government of Britain in the form of an Atlantic Fellowship in Public Policy, and from the Greenwall Foundation to support study of the relationship between empirical and normative approaches to research in medicine.

1. Caplan A. 1997. *Am I My Brother's Keeper?* Indianapolis: Indiana University Press.

2. Davis D. March 1997. "A Child's Right to an Open Future," *Hastings Center Report.*

3. Dewey J. 1918. *Logic: The Theory of Inquiry.* New York: Free Press.

4. Kass L. 1997. "The Wisdom of Repugnance," *New Republic.*

5. Kitcher P. 1997. *The Lives to Come.* New York: Free Press.

6. Klotzko A. 1997. A Report from America: The Debate about Dolly," *Bioethics.* Reprinted and expanded in this volume, chapter 12.

7. McGee G. 1993. "Method and Social Reconstruction," *Southern Journal of Philosophy.*

8. ———. 1997. *The Perfect Baby: A Pragmatic Approach to Genetics.* New York: Rowman and Littlefield.

9. ———. 1997. "Parenting in an Era of Genetics," *Theoretical Medicine.*

10. ———. 1998. "Introduction to Pragmatic Bioethics," in *Pragmatic Bioethics,* ed. G. McGee. Nashville: Vanderbilt University Press.

11. ———. 1998. "Pragmatic Method and Bioethics," in *Pragmatic Bioethics,* ed. G. McGee.

12. ———. 1998. "Human Cloning: An Introduction," in *The Human Cloning Debate,* ed. G. McGee. Berkeley: Berkeley Hills Books.

13. ———. 1998. "Genetic Exceptionalism," *Harvard Journal of Law & Technology.*

14. ———. 1998. "A Cow's Egg," *Breaking Bioethics,* MSNBC Online.

15. ———. 1998. "Pragmatic Bioethics in Execution," response to questions, Annual Meeting of the American Society for Bioethics and the Humanities, Houston, Texas.

16. McGee G., McGee D. 1998. "Nuclear Meltdown: Ethics of the Need to Transfer Genes," *Politics and the Life Sciences,* March.

17. McGee G., Wilmut I. 1998. "Cloning and the Adoption Model," in *The Human Cloning Debate,* ed. G. McGee.

18. Moreno J. "Bioethics Is a Naturalism" in *Pragmatic Bioethics*, ed. G. McGee.

19. Parfit, D. 1986. *Reasons and Persons*. Oxford [Oxfordshire]: Oxford University Press.

20. Pence, G. 1997. *Who's Afraid of Human Cloning*? New York: Rowman & Littlefield.

21. Robertson, J. 1994. *Children of Choice*. Princeton: Princeton University Press.

22. ———. 1998. "Legal Issues in Human Cloning," *Texas Journal of Law*.

23. Rothman, B. K. 1998. *Genetic Maps and Human Imaginations: The Limits of Science in Understanding Who We Are*. New York: W. W. Norton & Co.

24. Roy, I. 1998. "Philosophical Implications of Human Cloning," in *The Human Cloning Debate*, ed. G. McGee.

25. Saatkamp, H. 1998. "Genetics and Pragmatism," in *Pragmatic Bioethics*, ed. G. McGee.

26. Silver, L. 1998. *Remaking Eden: Cloning and Beyond in a Brave New World of Genetic Engineering*. New York: Avon Books.

27. Shenk, D. 1997. "Biocapitalism: What Price the Genetic Revolution?" *Harpers*, December.

28. Steinbock, B. 1992. *Life before Birth: The Moral and Legal Status of Embryos and Fetuses*. New York: Oxford University Press.

29. Toennies, F. 1957. *Gemeinshaft und Gesellshaft*. Trans C. Loomis. East Lansing: Michigan State University Press.

17

Human Reproductive Cloning

A Look at the Arguments against It and a Rejection of Most of Them

Raanan Gillon

H uman reproductive cloning—replication of genetically identical or near identical human beings—can hardly be said to have had a good press. Banned in one way or another by many countries including the United Kingdom,[1] execrated by the General Assembly of the World Health Organization as "ethically unacceptable and contrary to human integrity and morality,"[2] forbidden by the European Commission through its Biotechnology Patents Directive,[3] by the Council of Europe through its Bioethics Convention,[4] and by UNESCO through its Declaration on the Human Genome and Human Rights,[5] clearly human cloning arouses massive disapproval. What are the reasons, and especially the moral reasons, offered as justifications for this wholesale disapproval? In brief summary these seem to be: "yuk—the whole thing is revolting, repellent, unnatural and disgusting"; "it's playing God, hubris"; "it treats people as means and not as ends, undermines human dignity, human rights, personal autonomy, personality, individuality, and individual uniqueness; it turns people into carbon copies, photocopies, stencils, and fakes"; "it would be dangerous and harmful to those to whom it was done, as well as to their families; it would particularly harm the women who would be bearing the

184

babies, and especially so if they were doing so on behalf of others as would probably be the case; it would harm societies in which it happened, changing and demeaning their values, encouraging vanity, narcissism, and avarice; and it would be harmful to future generations." "Altogether it would be the first massive step on a ghastly slippery slope toward"—here fill in the horror—'Hitler's Nazi Germany, Stalin's USSR, China's eugenic dictatorship; or, from the realms of literature, *Boys from Brazil*, mad dictators, and of course mad scientists in science fiction, Big Brother in *Nineteen Eighty-Four*, and the human hatcheries of *Brave New World*." "It would be unjust, contrary to human equality, and, as the European Parliament put it, it would lead to eugenics and racist selection of human beings, it would discriminate against women, it would undermine human rights, and it would be against distributive justice by diverting resources away from people who could derive proper and useful medical benefits from those resources." So, clearly, where it has not already been legally prohibited it should be banned as soon as possible.

Note that I have grouped these objections into five categories. The first constitute a highly emotionally charged group that includes yuk, horror, offence, disgust, unnaturalness, the playing of God, and hubris. Then come four clearly moral categories—those concerned with autonomy (in which for reasons given later I have included dignity); those concerned with harm; those concerned with benefit; and those concerned with justice of one sort or another, whether in the sense of simply treating people equally, of just allocation of inadequate resources, of just respect for people's rights, or in the sense of legal justice and the obeying of morally acceptable laws.

Two types of cloning have generated particular moral concern: the first involves taking a cell from a human embryo and growing it into a genetically identical embryo and beyond; the second, made famous by the creation of Dolly the sheep,[6] involves taking out the nucleus of one cell and putting into the resulting sac, or cell wall, the nucleus of another cell to be cloned. Strictly, the Dolly-type clone is not quite a clone because the cell wall also contributes a few genes, the mitochondrial genes, which are incorporated into the resulting organism, but the vast majority of the genes in a Dolly-type clone come from the nucleus, so that, for example, if a nucleus from one of my cells were implanted into a cell sac from someone else, and the resulting cell were grown into a human being, he would have a gene complement almost but not entirely identical to mine. On the other hand, clones that result from splitting off of cells from embryos and growing them have exactly the same gene content as the embryo from which they came. With either of these cloning techniques, the process can be carried to early stages of development for a variety of potentially useful purposes, without any intention or prospect of producing a developed human being (lumped together here as nonreproductive human cloning and referred to only in passing). The human cloning that produces the greatest concern, and is the main subject of this paper, is of course reproductive human cloning, which would aim to produce a human person with the same genes as some other human being.

HUBRIS, YUK, AND SO ON

First, then, the group of responses based on yuk, it's unnatural, it's against one's conscience, it's intuitively repellent, it's playing God, it's hubris—a group of responses that one hears very frequently. I have to admit that this sort of essentially emotional response tends to evoke a negative emotional response in me when it is used in moral argument, as it often is (by moral argument I mean, following David Raphael,[7] argument about what is good or bad, what is right or wrong, what ought or ought not to be done and about our values and norms). The trouble is that these gut responses *may* be morally admirable, but they may also be morally wrong, even morally atrocious, and on their own such gut responses do not enable us to distinguish the admirable from the atrocious. Think of the moral gut responses of your favorite bigots—for example, the ones who feel so passionately that homosexuality is evil, that black people are inferior, that women should be subservient to men, that Jews and Gypsies and the mentally retarded or mentally ill ought to be exterminated. People have existed—some still exist—who have these strong "gut beliefs" which they believe to be strong *moral* feelings, which indeed they believe to be their consciences at work; and my point here is that gut responses provide no way for us to distinguish those moral feelings that we know or strongly believe to be wrong, from the moral feelings that we ourselves have, which we know or strongly believe to be right. To discriminate between emotional or gut responses, or indeed between the promptings of deeply felt moral intuitions or of conscience, we must reflect, think, analyze in order to decide whether particular moral feelings are good or bad, whether they should lead to action or whether they should be suppressed (and yes, I think moral reflection shows that it is important to suppress, or even better reeducate so as to change, one's moral feelings when on analysis one finds they are wrong). Without such moral reflection the feeling itself, while it may be an important flag that warns us to look at the issues it concerns, is no more than that. With such reflection we may find that the flag is signaling an important moral perspective that we should follow; or we may find that the flag is signaling us to respond in a morally undesirable way.

An analogy which I like to use concerns medical practice. Doctors, especially surgeons, cut people up quite a lot; they (we) also stick their fingers in people's bottoms. Most of us, I imagine, would feel quite deeply that both of those activities are rather disgusting and not to be done; yet we know, through thought and reflection in our medical studies, that we had better overcome these deep feelings because in some circumstances it is right to cut people and in some circumstances it is right to put our fingers in people's bottoms. Both are extraordinary and counterintuitive things to do, but on analysis we find that they are sometimes the *right* thing to do. The same need for reflection, thought, and analysis applies to our deeply felt moral feelings in general. We need those deep moral feelings, those deep moral gut responses. Moral feelings are—here we may agree with Hume[8]—the mainsprings or drivers of our moral action. They lead us to action against social injustice and corruption, against the tyrant, the torturer, the sadist, the rapist, the sexual aggressor of children but—and now I part company with Hume—we need to reflect on and educate our moral feelings so as to select and develop the good ones, and deter and modify or preferably abolish the bad ones.

In the context of our deep feelings let's just remind ourselves about Huxley and his *Brave New World*.[9] It has become a trigger title, needing only utterance to provoke strong negative feelings, especially about the use of science and technology to control and predetermine people's feelings, attitudes, and behavior. From a rereading of *Brave New World* I was satisfied that Huxley's main target is not science and technology but rather their misapplication by the despotic state that systematically sets out to undermine the possibility of freedom—freedom in the sense of humanity's ability to make thought-out choices and live by them, autonomous freedom. But while Orwell, in his *Nineteen Eighty-Four* imparted to us a horror of tyrannical social control, of "Big Brother is watching you," Huxley with *Brave New World* is more commonly perceived to have patterned our thoughts and feelings against "runaway science" and especially against genetics and the artificial reproduction of embryos, fetuses, and babies away from their mothers, so as to control every aspect of their development. Recall, as an example itself of conditioning (of Huxley's readers), the ghastly example early on in which babies are naturally attracted to books and to flowers, and then, to ensure that the particular class of worker that the babies are destined to become will detest flowers and books, are subjected to nasty noises, terrifying explosions, sirens, alarm bells, and finally, to make sure, electric shocks. Two hundred repetitions of the Pavlovian conditioning would cause them to grow up with what the psychologists used to call an *instinctive* hatred of books and flowers. "Reflexes unalterably conditioned. They'll be safe from books and botany all their lives," declared the director of hatcheries.

Later on infants are socially programmed while they sleep through "hypnopaedia," in which they are conditioned to have predetermined attitudes. Thus children predestined to be in the Beta class of citizen hear over and over again:

"Alpha children wear grey. They work much harder than we do because they're so frightfully clever. I'm really awfully glad I'm a Beta, because I don't work so hard. And then we are much better than the Gammas and Deltas. Gammas are stupid. They all wear green, and Delta children wear khaki. Oh no, I *don't* want to play with Delta children. And Epsilons are still worse. They're too stupid."

So, the director concludes,

"At last the child's mind is these suggestions, and the sum of the suggestions *is* the child's mind. . . . The adult's mind too—all his life long. . . . But all these suggestions are *our* suggestions." The Director almost shouted in his triumph. "Suggestions from the State," he banged the nearest table. . . . "Oh, Ford! . . . I've gone and woken the children."

We will return to *Brave New World*, but I do not think we need Huxley's warnings about childhood conditioning of our attitudes and beliefs and prejudices to know that, even in our ordinary lives, many of our strong attitudes and prejudices and beliefs have emerged as a result of childhood patterning. We have been programmed to some extent into our attitudes. The big difference, of course, is that as we grow up and are educated we are able to reflect on these attitudes and beliefs and to *decide* whether to own them or reject them. Nonetheless many of our deep moral attitudes and beliefs are firmly embedded from early childhood (as the book of *Proverbs* reminds us "Train up a child in the way he should go: and when he is old he will not depart from it") and, even if we decide that some of them are wrong, we have to work very hard if we want to change them. Huxley in *Brave New World* warns against despotic *misuse* of science and technology that, by painfully embedding attitudes and feelings in early infancy, and by later social prohibition or discouragement of reflection about those attitudes, makes the development of moral agency impossible or at least extremely difficult.

However, we have not been the recipients of such state conditioning and control in our own societies, and I find it difficult to understand the strength and depth and origins of the contemporary widespread hostility to the very idea of cloning human beings. Certainly the existence of contemporary nature's own human clones, identical twins, seems harmless enough *not* to account for such deep hostility to the idea of deliberate cloning, though in passing it may be relevant to note that, down the ages, twins have been mysteriously subject to ambivalent prejudices. Thus, apart from literary and dramatic jokes about them from the comedies of Plautus via Shakespeare to Stephen Sondheim, twins and their mothers have been persecuted in some societies, revered but also feared in others, for example as unnatural miscegenated offspring of gods. On the other hand, there may be quite strongly positive attitudes to twins. Wendy Doniger in a review in the *London Review of Books*[10] quotes from Lawrence Wright's work on twins on "the common fantasy that any one of us might have a clone, a Doppelganger; someone who is not only a human mirror, but also an ideal companion; someone who understands me perfectly, almost perfectly, because he is me, almost me."

We will return to the issue of identity, because the myth that genetic identity equals personal identity lies at the root of much misunderstanding about cloning. First, let us pursue in more detail the argument that cloning is *unnatural* and therefore wrong. What role does "unnatural" play in moral argument? Our first requirement is to disambiguate the term—what do we *mean* by unnatural in this context? Anything that occurs in nature could be said to be natural, but that sense of natural is not going to do much moral work for us, for we and what we do are natural, not unnatural, in this sense. In any case, right and wrong, good and bad, insofar as they occur in nature, also are equally natural in this sense, so that to say that something is natural will hardly help us distinguish between the two. Another sense of natural means unaffected by human intervention. But unless we wish to argue that all human interventions are bad and or wrong and all states of nature are good or right, then this sense of natural too is not much help for moral judgment. Think of all the truly horrible and morally undesirable things that occur in nature uninfluenced by humans; think too of all the human interventions in nature that are clearly morally desirable, but "unnatural" in this sense—including all medical interventions, and all the other activities by which we help each other, including the provision of food, housing, clothing, and heating.

But there are two more senses of unnatural that are of moral relevance. The first is that it is part of human nature to be a moral agent (with perhaps a few exceptions) and thus human people who behave immorally or even amorally are acting unnaturally in this sense of acting against their human nature. I personally find this theme of enormous moral importance and a way of linking theological natural law theory with secular morality. But it does not afford us any simple basis or method for moral assessment—instead it demands assessment of what the moral part of our human nature requires of us. So "natural" in this sense, important though it is as a moral concept, does not give us a way of deciding whether cloning and the other genetics activities are good or bad—it simply requires us to make such distinctions. Like the objections based on the 'yuk response', the deep moral intuition, the moral repugnance and the claim of conscience, the objection that cloning is unnatural, when used in this morally plausible sense, requires moral reflection and judgment, but does not itself provide that moral reflection and judgment. If—but only if—such reflection and judgment lead us to conclude that cloning is immoral, *then* we can say that cloning is unnatural in this morally relevant sense of going against our moral nature.

There is another sense of unnatural which I think is also of potential moral relevance. If we do something that weakens, undermines, destroys, or harms our human moral nature, then this is immoral and unnatural in the sense of antinatural or against nature; and that of course is of enormous moral significance not just in relation to cloning but for the whole of the new genetics enterprise. So, to show that any activity, such as cloning, is unnatural in a morally relevant sense we need to give *reasons* that demonstrate why it is contrary to our human moral nature, or why it will undermine that human moral nature. Until we can give such reasons let us be particularly careful to avoid pejorative claims about cloning being unnatural, not simply for the reasons I have just given, but also because it must be very hurtful for the world's identical twins to hear that they, by association, are considered to be "unnatural" and therefore that their existence is morally undesirable.

To continue with this range of somewhat mysterious objections to cloning and sometimes to the new genetics as a whole, we need to look now at hubris and playing God. *Hubris* is a pejorative term meaning a contemptuous arrogance, especially against God or the gods. *Playing at God* combines both an implicit accusation of hubris with an implicit accusation of immaturity and lack of skill—as when children play doctors, somewhat inefficiently. Suffice it to agree that contempt, arrogance, puerile immaturity, and lack of skill are all morally undesirable in one fulfilling a responsible task. But are these accusations justifiably made against the whole enterprise of human cloning, or indeed against the whole enterprise of the new genetics? One would require specific cases and examples rather than sweeping generalizations, which otherwise boil down to mere abuse. There clearly is an important moral issue here, especially in relation to the question of whether at present we can safely and sufficiently skillfully carry out reproductive cloning, even if we wish to do so, and I shall return to this. But, without specific evidence, it seems straightforwardly tendentious to brand the whole enterprise of human cloning, let alone the whole of the new genetics, as "hubris" and "playing God."

AUTONOMY AND PERSONAL IDENTITY

The next set of objections against cloning concerns personal identity and dignity, the undermining of autonomy, of individuality, of personality, of uniqueness, the production of carbon copies, photocopies, stencils, and fakes of human beings.

Even if reproductive cloning were to produce a person identical with the person from whom he or she was cloned, it is not clear to me why this should be immediately condemned as morally unacceptable, though the idea so greatly strains the imagination that one might argue that it would be irresponsible to try any such trick even if it were possible. But of course reproductive cloning would *not* produce two identical people—only two people with identical (or in the case of Dolly-type cloning near identical) sets of genes. Genetic identity neither means nor entails personal identity.

Once again the proof of this exists all around us, for genetically identical twins are obviously different people, even though their genes are identical or near identical (in a type-type sense of identity, such that billiard balls are type-type identical, even though each billiard ball is token identical only with itself). But this genetic type-type identity of people who are clones does not make them identical as people, in either sense.

Some commentators make a different criticism. It is not only personal identity that must not be replicated; nor must genetic identity, for that itself is morally important, indeed even a right, according to the European Parliament. They assert that every one of us has a right to his or her own genetic identity. Here there seems to be serious confusion or conflation between token identity, type-type identity, and uniqueness. On analysis the claim surely cannot be that we all have a right to our genetic identity in the sense of token identity (every thing being token identical with itself and with nothing else), for that is simply an analytic truth. We all, including identical twins, necessarily do have that sort of genetic token identity, and if it is not incoherent to describe this definitional truth as a right, it is certainly pointless. But if the claim is made in terms of type-type identity, whereby we are claimed to have a right to type-type genetic identity, then identical twins and any other human clones *do* have such genetic identity; that is precisely the sort of identity that they have (or near identity in the case of Dolly-type cloning). So presumably it is not genetic identity that the European Parliament can sensibly be claiming as a right. Perhaps instead it is genetic exclusivity or uniqueness. If I have such a right, then no one else is entitled to have the (type-type) identical genes that I have. It might be described as a claim right that one's genetic identity, in the sense of token identity, must be unique—in other words, a claim right *not* to have type-type genetic identity. But if that is the European Parliament's claim it is not merely bizarre; if taken seriously it is morally malignant, for it implies morally malignant consequences for identical twins, nature's existing examples of people who are clones. If we have this right to genetic uniqueness, then somebody must have the corresponding duty—the duty to destroy one of each pair of existing identical twins, both born and *in utero*. Fortunately such counterexamples, plus the general tendency of morally reflective people to be morally and legally unconcerned about the lack of genetic uniqueness of identical twins, indicate that genetic uniqueness is unlikely, *pace* the European Parliament, to be of moral importance, let alone a moral right, and still less a right that ought to be enshrined in law.

But maybe there is a difference between cloning that occurs naturally and cloning that occurs by intention? Perhaps it is deliberate cloning that is the problem, rather than the cloning that occurs naturally, in the sense of unmediated by human beings? And perhaps the problem is that such deliberate reproductive cloning somehow demeans human dignity? Certainly both the World Health Organization and the European Parliament have stated that such cloning would offend against human dignity. Well, once again we need to know what we mean by human dignity. We all know that human dignity is good and ought to be promoted and respected, but most of us, I suspect, would find it very difficult to say what we actually mean by human dignity and to explain why its violation is wrong.

The *Shorter Oxford English Dictionary* offers definitions such as "the quality of being worthy or honourable . . . worth, excellence . . . high estate, position or estimation . . . honour, rank"—definitions that indicate that to say that someone has dignity is to say that he or she is valued. But with this sort of understanding of "dignity," to say that cloning violates or offends against human dignity is simply to assert that it diminishes worth without in any way explaining *why* this should be so. Interestingly the *Encyclopedia of Philosophy*, the *Encyclopedia of Bioethics*, and the *Dictionary of Medical Ethics* all lack entries under "dignity." Those who wish to use infringement of human dignity as an argument against human reproductive cloning thus need to explain what they mean by the term. For me the most plausible account of human dignity is Immanuel Kant's. For him human dignity resides in our ability to be autonomous, to will or choose to act according to the moral law.[11] I suspect that many uses of the term "human dignity" are consistent with this Kantian notion that our human dignity is our ability to make autonomous choices for ourselves according to what we believe to be right. If so, when we commit ourselves to respecting human dignity, to treating others in ways that respect their human dignity, we mean roughly that we should treat them in ways that they themselves on reflection and deliberation would believe to be good or right ways; and that when we make decisions on behalf of people who cannot make their own decisions, we should try so far as we can to replicate the decisions they themselves would have autonomously chosen (or if they have not yet become autonomous, can be expected and desired to make were they autonomous).

If we accept some version of the Kantian meaning of human dignity and its basis in autonomous choice, then it is not at all clear to me why reproductive cloning should in any way undermine such dignity. Of course there might be ways of destroying or damaging that dignity by damaging the underlying genetic basis for such autonomous choice—and any such activity should be morally condemned precisely because of the damage to human dignity, a version of the reputable anti–human nature argument above. But no reason has been offered for accusing reproductive human cloning of damaging human dignity in this way.

Another objection to cloning that may also reside in the notion of human dignity is that we must never treat other people merely as means to an end, but always as ends in themselves—one of the versions of Kant's categorical imperative.[11] This claim is frequently misrepresented as a moral obligation never to use each other as means to an end, or as instruments or as tools or as objects. That misrepresentation is plainly wrong, for of course we morally can and morally do frequently use each other as means to an end, as tools, and it is highly desirable that we continue to do so. If I ask you to bend forward so that I can climb on your back in order to get over my garden wall to let myself in through the kitchen window because I have lost my key, I am using you as a means to my end, as an object, a sort of stepladder, an instrument or a tool. But I am not treating you *merely* as an object or a tool or an instrument. By asking and obtaining your permission I am treating you as an end in yourself as well as treating you as a means to my end. The issue is complicated with embryos because it is a matter of unresolved and passionate moral debate whether embryos and fetuses are

HUMAN REPRODUCTIVE CLONING ■ 193

within the scope of the Kantian requirement to treat each other as ends in themselves. Many of us believe that they are not, and thus would permit, for example, the cloning of human embryos for research purposes with disposal (i.e., destruction) of the experimented-on embryo at an early stage in its development. In the United Kingdom the law allows this sort of thing. On the other hand, many others would say that this is morally outrageous because the human embryo *does* fall within the scope of the Kantian categorical imperative, being itself a human person from the moment of its creation. I am not going to address that argument, but it is important to see how it complicates the issue of cloning, both sorts of cloning. For if in creating an embryo, by whatever method, we have created a person, then of course we must treat it as a person, and thus not use it merely as a means to an end. If, on the other hand, it is not yet a person then we may use it merely as a means to an end, as a research tool for example, and destroy it after such use. That is an unresolved philosophical or theological problem.

Suppose, however, we put aside that piece of the argument and revert to human reproductive cloning. Then the requirement always to treat people as ends in themselves, even when we also treat them as means, is entirely compatible with reproductive cloning. The issue surely turns, not on the method of reproduction, cloning or otherwise, that one may choose, but rather on how one actually treats and regards the child that results. Take the example of parents who seek to clone a child because they want to have another child with the same blood or marrow type, so that they can transplant some marrow from the new child into an existing child mortally ill with leukemia. Such a process would necessarily involve, it is often claimed, treating the new child merely as a means. Not at all, I would counterargue. The argument needs to be broken down into two parts. The first part concerns the question, why do and ought people decide to have children? In particular, is there any moral obligation to have a child only for the sake of the child-to-be? If so, then surely the vast majority of parents have behaved immorally, for while there must be many different reasons for having a child, I doubt that there exist very many parents who have decided to have a child because they decided there was now a need to have a new person in the world to whom duties were owed that he or she should exist. Much more commonly (and yes, this is mere supposition) people decide to have a child because they *want* a child for their own reasons or, perhaps more commonly, instincts. They feel like it, or they are ready to have children, or they want to fill a gap in their lives, or perhaps they want an heir, or someone to take over the business, or someone to look after them in their old age; all sorts of personal selfish reasons may operate, or none at all. My argument is that, until shown otherwise, we should accept that there is nothing wrong with making either a self-interested or an instrumental decision to have a child.

Having implemented such a decision, the second stage of the argument applies, for now, of course, mere self-interest can no longer be justified. Once there is another person created as a result of one's decision, then that person must be accorded the same moral respect as is due to all people and must not be treated merely as a means to an end, an object, a tool, an instrument. So while one may perfectly properly decide to have a child in order to provide a source of life-saving cord blood or marrow for one's existing child, one must of course then respect the new child as an end and never treat him or her merely as a means to an end. I can see no reason for the parents' instrumental motivation for having a child in any way necessitating their treatment of the new child merely as a means and not an end. If anything, I suspect that human psychological nature would tend to lead parents to treat such children even more lovingly and respectfully than usual.

I have given reasons for doubting that cloning would infringe the human dignity and autonomy of the cloned person. Let us now consider the dignity and autonomy of those who wish to engage in reproductive cloning. Such considerations favor noninterference on the grounds that in general people's autonomous choices for themselves should be respected, unless there are very strong moral reasons against doing so, and that this is particularly true in respect of those rather personal and private areas of choice, notably those concerning reproduction, sexuality, choice of partners, and decisions about babies. Intervention by the state, or anyone else, in these areas of private morality undermines the human dignity/autonomy of those people. Moreover, respect for people's dignity/autonomy in these areas is not only right in itself, but is also likely to lead to far greater overall good and far less harm than if we start erecting state apparatuses for intervention in these private areas.

I think this is Huxley's main message in *Brave New World*. Do not let government start to control our private decisions, our autonomy or our development. Do not let the apparatus for state control in these areas be developed. By leaving such choices decentralized not only will people's dignity/autonomy be respected—a good in itself—but human welfare generally will benefit. Similarly, beware state control of science and technology, for in the name of social order it will lead to the end of liberty. As Huxley later admits, this is an overstated case, and I am certainly not arguing for total libertarianism and absence of state controls either of citizens' behavior or of science and technology. But I am arguing against excessive state control, and in favor of a substantial zone of respect by the state for private autonomous choices where such respect does not entail harm to others. And so, I think, was Huxley.

HARMS AND BENEFITS

Which brings us to the next group of arguments, based on the harms and benefits of cloning. Let us briefly examine these in relation to the people cloned, their families, their societies and future generations. It is in the context of the social and personal harms of human reproductive cloning that *Brave New World* (and also the Ira Levin book of 1976, *Boys from Brazil*, in which clones of Hitler are bred in an attempt to rekindle the Nazi enterprise) has done so much to turn us against cloning, even succeeding in rendering the term pejorative. What was common to both of those books, but was especially evident in Huxley's novel, was that the cloning involved either selection of already impaired humanity for cloning (e.g., Levin and the cloning of Hitlers) or the deliberate impairment of human embryos before they were cloned, as in Huxley's notion of "Bokanovsky's Process."

"Essentially," the director of hatcheries and conditioning explains, bokanovskification, or cloning, "consists of a series of arrests of development"—arrests by chilling the embryo, by X-raying it, by adding alcohol, and by oxygen starvation. When one of the students bravely asks what the benefit was of this process,

> the Director wheeled sharply round, "can't you see, can't you *see*?" He raised a hand; his expression solemn. "Bokanovsky's Process is one of the major instruments of social stability. . . . Ninety-six identical twins working ninety-six identical machines! . . . You really know where you are. For the first time in history." He quoted the planetary motto: "Community, Identity, Stability." Grand words. "If we could bokanovskify indefinitely, the whole problem would be solved" . . . standard Gammas, unvarying Deltas, uniform Epsilons . . . "But alas," the Director shook his head, "we *can't* bokanofskify indefinitely."

Note that Huxley has here combined and conflated three quite separate ideas. One is reproductive cloning; the second is a crude and simplistic genetic determinism (ascribed to the rulers of course and rejected by Huxley himself) whereby genetic identity equals personal identity; and the third is intervention in the cloning process to impair the normal development of the human embryo. But we have seen that cloning does not entail personal identity, and so far as we know cloning need not harm or impair the embryo.

But of course we do not yet know. Cloning by nuclear substitution has only just begun in mammals, with Dolly the sheep being one successful outcome out of 277 attempts to produce such a clone. Imagine that being done in human beings and the harms to the women producing the eggs and undergoing the unsuccessful implantations (as stated above, I am leaving unargued the issue of whether it is permissible to harm the human embryo itself for the purpose of such research).

Claims about the potential harms caused by human reproductive cloning are extensive. Animal experiments are reported to have produced many abnormal embryos and fetuses, many spontaneous abortions, and many abnormal births. Theoretical reasons are claimed to indicate that the offspring will be particularly prone to various diseases including those associated with premature aging. Psychological harms are predicted for individual children thus born, including resentment at having their genetic structure predetermined by their parents, resentment at having been conceived merely as means to benefit others (for example as blood or bone marrow sources), a sense of overwhelming burden if they have been cloned from someone with great achievements that they are supposed to emulate; confusion about their personal identity and relationships (if, for example, they are clones of one of their parents). Physical and emotional harms are also predicted for women bearing cloned embryos, including the high rate of failure and abnormality of the pregnancies; and if the women are also surrogates even more emotional harms can be anticipated (a point made by my colleague Donna Dickenson).

In addition to the general social disasters noted above, as envisaged in books such as *Brave New World* and *Boys from Brazil*, contemporary concerns include undermining of social values by opening the doors to racist eugenics, encouragement of "vanity, narcissism and avarice" as the U.S. Commission on Bioethics has reported,[12] and the creation of "a means of mass destruction" with "science out of control" as the Nobel Laureate scientist Joseph Rotblat is reported to have claimed.[13] Add to all this the concern of the European Parliament about "violation of the security of human genetic material," anticipated reduction in the genetic variability of the human race and the consequent threat to human evolution that the WHO is concerned about, and also worries about the geometric increase through germline inheritance of any mistakes that are created by cloning, and we see a wide range of anticipated harms that may result from cloning.

Once again we need to look at this range of harms rather more precisely. Let it be acknowledged immediately that *at present* the technique of human cloning is not well developed enough to be safely used in humans for reproduction, but this is not to acknowledge either that each of the preceding harm arguments is valid, or that the harm arguments that are currently valid are sufficiently strong to prevent further research into ways of reducing such harms—for example, by animal experimentation.

What, then, of the formidable lists of harms, mainly psychological harms, anticipated to affect children? In brief, I think we need to set against these purported and anticipated psychological harms of being a clone child the very important counterconsideration of what is the alternative for that particular child? This argument commonly irritates, sometimes enrages, but rarely convinces. Yet it seems valid, and I have not encountered plausible *counterarguments*. The alternative *for those children* is to not exist at all, so if we are genuinely looking at the interests of those children who are anticipated to have the various psychological problems of being clones, and the difficulties that undoubtedly we can anticipate those will raise, and if we are genuinely looking at those problems from the point of view of the child, then the proper question to ask is: What is preferable for that child? To exist but to have those problems, or not to exist at all? It is an argument that I learned from the so-called pro-life movement, though I suspect this is not a use that pro-lifers themselves would wish to make of it. I found that the argument radically changed the way I thought about anticipated harms. Of course, it in no way stops one from deciding, for example, not to have a baby, or to have an abortion, or not to pursue reproductive cloning.

But it does force one, or should force one, to realize that one's reasons are unlikely to be the best interests of the child whom one is thinking of not having, but are instead one's own reasons and preferences, largely about the sort of world one wishes to participate in creating. And if that is the case, why should one's own reasons and preferences prevail over the reasons and preferences of those who do wish to carry out reproductive cloning? After all, they do not claim a right to prevent us from reproducing according to our preferences; why should we claim a right to prevent them from reproducing according to their preferences?

As for the arguments about the potential social harms of cloning, other than those based on safety of the techniques, it seems to me that they are either frankly implausible (the argument that cloning is a threat to further human evolution surely falls into this category, given the likely numbers of cloned versus more conventionally produced people), too weak to justify imposition on those who reject them (for instance the arguments that reproductive cloning encourages vanity, narcissism, and avarice), or powerful but misdirected. Thus it is not cloning, nor the techniques of the new genetics more broadly considered, that might lead to the social harms of racism, eugenics, mass destruction, or the violation of the security of genetic material, but rather social structures that permit dictatorships and other forms of immorally enforced control of people's behavior by their rulers. Those are the harms that we need to be concerned about; and the most important way of avoiding them—of avoiding oppression of all those who are oppressed by the strong, including the widespread oppression of women by men—is not to ban cloning or to become obsessed with the new genetics, but rather to reform those social structures that result in such harms and to maintain in good order those social structures that do largely avoid these harms.

What about the germ-line argument of dangers to future generations? Well certainly the genome resulting from reproductive cloning is germ-line transmissible, and any mistakes that occur can be passed on to future generations. But so too, of course, can any benefits. If, for example, a cloning technique results in the elimination of some genetic abnormality that would otherwise have been transmitted through the germ line, then the cascade effect is geometrically beneficial, just as, if a mistake results and is passed on through the germ line, that, too, is geometrically inheritable. Clearly, care is needed to minimize the chances of the latter and maximize the chances of the former. But in general, with ever increasing voluntary personal control over reproduction, it seems likely that even if genetic mistakes do occur, if they are severe people will be reluctant to pass them on to their offspring, thus reducing the risks of a cascade of negative genetic effects down the generations. On the other hand the precedent of deciding to prevent certain sorts of reproduction on the basis of the risk to future generations of deleterious genetic effects is itself one of the social harms—enforced eugenics—that opponents of the new genetics are usually very keen to avoid. The current orthodoxy that somatic genetic interventions that are beneficial can be accepted but that germ-line interventions, even if clearly beneficial, should be forbidden seems to be one of those undefended taboos that need to be rejected. If we develop a genetic intervention that helps one offspring we need very good reasons for denying it to that offspring's offspring.

In general, and in relation to possible harms of new techniques, we need to beware excessive concern with the "precautionary principle." Insofar as it tells us to avoid doing harm, it is an important moral concern to balance against our continuing search for new ways of doing good—of benefiting others. In other words, the principle of beneficence should always take into account the principle of non-maleficence, and the objective should be an acceptable probability of doing good with minimal and acceptable harm and risk of harm. But sometimes the precautionary principle is used as a sort of moral blunderbuss, like the use of *primum non nocere* when this is translated as "above all do no harm." That way lies a beneficence moratorium, with all applied medical research, indeed all new medical interventions, being banned, for whenever we seek to benefit we risk harming. The morally desirable use of the precautionary principle is to weigh anticipated benefits and their probabilities against anticipated harms and their probabilities, always aiming at a likely outcome of net benefit with minimal and acceptable harm and risk of harm.

So what about the benefits? I have been able to find less in the published work about the potential benefits of reproductive human cloning than about its potential harms. The same is not true about nonreproductive human cloning, for which a wide variety of impressive potential benefits has been claimed. These include production of useful pharmaceuticals from cloned transgenic animals; basic research into DNA and aspects of genetics, human reproduction and infertility, aging and *oncogenesis*; as well as the possible production of cloned human tissues and organs (for use, for example, in transplantation).[14] But even human reproductive cloning can be anticipated to provide certain benefits. For example, in rare cases Dolly-type cloning techniques could prevent inheritance of rare and disabling mitochondrial genetic disorders. The genetic abnormality being in the mitochondria, these cloning techniques make it possible to replace the cell membrane containing the defective mitochondria with an unaffected cell membrane and then to insert into that the unaffected genetic material in the cell nucleus. For the affected people such reproductive cloning could be of major benefit. A second potential benefit could arise where parents wish to have a further child, as already suggested, in order to provide, for example, compatible bone marrow or cord blood for an existing child who needs it to survive (I owe the cord blood example to Dr. Matjaz Zwitter). A third example of potential benefit might be where a car crash has led to the death of a husband and the fatal injury of the only child and where the surviving woman wishes to have a clone from the child as the only means of raising a child who is her husband's biological offspring. A further potential benefit of reproductive cloning might be to a couple who are carriers of a fatal recessive gene and prefer to clone a cell from one of them to avoid the genetic danger, rather than reproduce by means of other people's genetic material.

Given the limited potential benefits of reproductive human cloning, the benefit/harm analysis does not seem at present to create much moral pressure to undertake this activity (though in nonreproductive cloning there certainly seem to be a large number of potential benefits, with far fewer potential harms). Nonetheless, given that there are some benefits that may be anticipated from reproductive human cloning, given the counterarguments offered above to many of the claims that this would create major harms, and given the arguments from respect for reproductive and scientific autonomy, then at the very least we should thoroughly question contemporary absolutist proposals to ban human reproductive cloning for ever and a day—even if prudence and precaution indicate a temporary ban until the safety of such techniques can be researched and developed.

JUSTICE

But do the last set of moral arguments against human reproductive cloning—those based on justice—lead us to require a permanent ban on the technique? Justice arguments can usefully be considered from the point of view of rights-based justice; of straightforward egalitarian justice (according to the European Parliament, cloning is contrary to the principle of human equality because it leads to eugenics and racism), of legal justice (the requirements of morally acceptable laws); and finally, and perhaps in this context most important, of distributive justice—the fair or just distribution of scarce resources, including consideration of the opportunity costs of using such resources for one purpose rather than another.

The only rights-based arguments that I have found against reproductive human cloning are based on the right to have a genetic identity—a claim that I have examined above and found morally unacceptable in regard to identical twins. On the other hand, in favor of reproductive cloning are rights-based arguments claiming rights to reproductive autonomy and privacy and rights to carry out morally acceptable scientific research.

Egalitarian theories of justice are fine (everyone should be treated equally) provided they pass the Aristotelian test for theories of justice—notably, that it is equals who should be treated equally, while those who are not equal in a morally relevant sense ought *not* to be treated equally but treated unequally in proportion to the morally relevant inequality. Thus, cloning does not treat everyone as equal if it is not done for everyone, but that is not unjust; for not everyone *needs* cloning and not everyone *wants* cloning. However, the European Parliament has claimed that cloning is contrary to human equality because it leads to eugenics and racism. Suffice it to say that, while both racism and imposed eugenics are morally unacceptable (though not the sort of eugenics that stems from uncoerced reproductive choice, against which there are, I believe, no convincing moral arguments), there seem to be no reasons for believing that cloning is or entails either of these morally unacceptable phenomena. At best this is an empirical "slippery slope" type of argument; and there seems no reason to believe that the slipperiness of this slope is so uncontrollable that we should never start down it.

Legal justice arguments require us at least *prima facie* to obey morally acceptable laws. I will simply assert that by this is meant laws that have been created in a morally acceptable manner, rather than laws whose content one morally approves. From this point of view we should obey the many laws that have now been passed in morally acceptable ways which ban reproductive human cloning. That in no ways settles the question of whether their moral content *is* morally desirable, and the bulk of this paper has been arguing that permanent bans on such cloning are *not* morally desirable and should be reversed.

Distributive justice arguments seem to offer the most plausible case against development of human reproductive cloning, or at least against funding such development from community funds, simply because the anticipated benefit-harm ratio does not seem to justify the undoubted costs and especially the opportunity costs. But this argument does not rule out private funding of such research, nor does it result in a permanent ban on provision of state funding, should the anticipated benefits become substantially greater.

CONCLUSION

And so I conclude that all the arguments for a permanent ban on human reproductive cloning fail and that most of the arguments for even a temporary ban fail. However, four arguments in favor of a temporary ban do, I have indicated, currently succeed. The first is that at present the technique for human reproduction by cloning is simply not safe enough to be carried out in human beings. The second related argument is that, given these safety considerations, the benefits including respect for the autonomy of prospective parents and the scientists who would assist them are at present insufficient to outweigh the harms. The third is the argument from distributive justice, but this is only sufficient to prescribe a low priority for state funding for human reproductive cloning. And finally, respect for autonomy within a democratic society requires adequate social debate before decisions are democratically made about socially highly contentious issues so a moratorium is also needed to provide time for this full social debate and, with luck, for more informed, more deliberated, and less frantic decisions.

The issues that underlie, and in my view are far more morally important than the cloning debate and indeed much of the contemporary opposition to the new genetics are those that Huxley pointed to in *Brave New World*—notably that both science and government must be used as servants of the people and not as our masters. This is something that Huxley explicitly addresses in the foreword to his 1946 edition of *Brave New World*, where he points out that if he had written the book again he would not have had just two alternatives—essentially either the madness of state control or the madness of the savage's emotional and unreasoned lifestyle. He would also have included a middle way in which reason was used in pursuit of a reasonable life, in which science was applied for the benefit, for the *eudaemonia* or flourishing, of humankind.

The second issue that underlies the cloning debate, and indeed the overall debate about the new genetics, and is of deep moral importance, is the need to protect the genetic *underpinning of human autonomy and free will*. Some such basis, however complex, there must be; and in pursuing any sort of human genetic research and development, safeguarding and protecting that genetic kernel of what, to embroider on Aristotle, is humanity's specific attribute, notably our autonomous rationality—*that* must be the underlying moral challenge and moral imperative for the new genetics, along with its and our shared obligation to protect ourselves against the predations of the control freaks, whether they are the control freaks of state or religion or science or big business, or simply of crooked gangsters who seek to use us for their own ends. Those I think should be the central moral concerns in developing the new genetics, and that is the lesson that I have most vividly taken to heart from this excursus into the ethics of human reproductive cloning and from my rereading of Aldous Huxley's *Brave New World*. But let us also not forget the origin of Huxley's title—in Shakespeare's *Tempest*. There the phrase does not have the negative connotations that Huxley has given it. Rather, Miranda is excited by the prospect of a new world of new people (and especially one wonderful new man), and a new life away from the tiny island on which she had been brought up and on which the only people she had ever seen till then were her father and Caliban. It is to the prospect of a new, more varied, and fuller world and life that she wonderingly refers when she exclaims "O brave new world, That has such people in't." Like her and like Huxley in his 1946 preface I think we should look more positively at our brave new world—the brave new world of genetics. We should learn Huxley's lessons, protect ourselves against the depredations of those who would unjustifiably control us, and realize that the potential problems lie less in cloning and genetics and more in politics and political philosophy—and of course in *their and our* underlying ethics.

REFERENCES

While the ideas and arguments in this are my own, many are similar to arguments put forward by others. In particular I would like to acknowledge the pioneering work in this area by Professor John Harris (See reference 13 for example), and the thorough analysis of the issues by the U.S. National Bioethics Advisory Commission (reference 12).

202 ■ THE ETHICAL ISSUES

1. Human Genetics Advisory Commission (HGAC). *Cloning Issues in Reproduction, Science and Medicine.* London: HGAC, 1998.

2. World Health Organization. Resolution on cloning in human reproduction (50th World Health Assembly). Geneva: WHO, 1997.

3. European Parliament and Council. *Draft Directive on the Legal Protection of Biotechnical Inventions.* Brussels: European Parliament, 1997.

4. Council of Europe. *Protocol to the Convention on Human Rights and Biomedicine on the Prohibition of Cloning Human Beings.* Strasbourg: Council of Europe, 1997.

5. United Nations Economic and Social Council. *Universal Declaration on the Human Genome and Human Rights.* Paris: UNESCO, 1997.

6. Wilmut I, Schnieke A E, McWhir J, Kind A J, Campbell K H. Viable offspring derived from fetal and adult mammalian cells. *Nature* 1997; 385: 810–8.

7. Raphael D D. *Moral Philosophy*, 2nd ed. Oxford: Oxford University Press, 1994:8.

8. MacIntyre A, ed. *Hume's Ethical Writings.* London: Collier-Macmillan, 1970.

9. Huxley A. *A Brave New World.* Harmondsworth: Penguin, 1955.

10. Doniger W. What did they name the dog? (Review of Wright L. Twins: Genes, *Environment and the Mystery of Identity.* London: Weidenfeld, 1997). *London Rev Books* March 19, 1998: 32.

11. Kant I, Groundwork of the metaphysic of morals. In: Paton H J, ed. *The Moral Law.* London: Hutchinson, 1964.

12. National Bioethics Advisory Commission (USA). *Cloning Human Beings.* Rockville, Maryland: NBAC, 1997 [http://bioethics.gov/pubs.html].

13. Harris J. "Goodby Dolly?" The ethics of human cloning. *J Med Ethics* 1997; 23:353–60.

14 Winston R. *The Future of Genetic Manipulation.* London: Phoenix, 1997.

18

A Life in the Shadow

One Reason We Should Not Clone Humans

Soren Holm

One of the arguments which is often put forward in the discussion of human cloning is that it is in itself wrong to create a copy of a human being. This argument is usually dismissed by pointing out that (a) we do not find anything wrong in the existence of monozygotic twins even though they are genetically identical, and (b) the clone would not be an exact copy of the original, even in those cases where it is an exact genetic copy, since it would experience a different environment which would modify its biological and psychological development (throughout this chapter I will assume that clones are perfect genetic copies, even though the present cloning techniques do not in most instances produce perfect clones because they do not clone the mitochondrial DNA).

In my view both these counterarguments are valid, but nevertheless I think that there is some core of truth in the assertion that it is wrong deliberately to try to create a copy of an already existing human being. It is this idea which I will briefly try to explicate here.

THE LIFE-IN-THE-SHADOW ARGUMENT

When we see a pair of monozygotic twins who are perfectly identically dressed some of us experience a slight sense of unease, especially in the cases where the twins are young children. This unease is exacerbated in cases where people establish competitions where the winners are the most identical pair of twins. The reason for this uneasiness is, I believe, that the identical clothes could signal a reluctance on the part of the parents to let each twin develop his or her individual and separate personality or a reluctance to let each twin lead his or her own life. In the extreme case each twin is constantly compared with the other and any difference is counteracted.

In the case of cloning based on somatic cells we have what is effectively a set of monozygotic twins with a potentially very large age difference. The original may have lived all his or her life and may even have died before the clone is brought into existence. One of the scenarios where cloning might be desired is in a family in which a child is dying or has died, and the parents wish to replace this child with a child who is genetically identical. In such a case there would not be any direct day-by-day comparison or identical clothing, but I think that a situation even worse for the clone is likely to develop. I shall call this situation "a life in the shadow," and I shall develop an argument against human cloning that may be labeled the "life in the shadow argument."

Let us try to imagine what will happen when a clone is born and its social parents begin rearing it. Usually when a child is born we ask hypothetical questions like "How will it develop?" or "What kind of person will it become?" and we often answer them with reference to various psychological traits we think we can identify in the biological mother or father or in their families, for instance "I hope that he won't get the kind of temper you had when you were a child!"

In the case of the clone, however, we are likely to give much more specific answers to such questions. Answers that will then go on to affect the way the child is reared. There is no doubt that the common public understanding of the relationship between genetics and psychology contains substantial strands of genetic essentialism, that is, the idea that genes determine psychology and personality.[1] This public idea is reinforced every time the media report the findings of new genes for depression, schizophrenia, and every time a novel or a film portrays a link between the "criminal traits" of an adopted child and the psychology of its biological parents. Therefore it is likely that the parents of the clone will already have formed in their minds a quite definite picture of how the clone will develop, a picture based on the actual development of the original. This picture will control the way they rear the child. They will try to prevent some developments and promote others. Just imagine how a clone of Adolf Hitler or Pol Pot would be reared or how a clone of Albert Einstein, Ludwig van Beethoven, or Michael Jordan would be brought up. The clone would in a very literal way live his or her life in the shadow of the life of the original. At every point in the clone's life there would be someone who had already lived that life, with whom the clone could be compared and against whom the clone's accomplishments could be measured.

That there would in fact be a strong tendency to make the inference from genotype to phenotype and to let the conclusion of such an inference affect rearing can perhaps be seen more clearly if we imagine a hypothetical situation. Suppose that in the future new genetic research reveals that there are only a limited number of possible human genotypes, and that genotypes are therefore recycled every 300 years (i.e., somebody who died 300 years ago had exactly the same genotype that I have). It is further discovered that there is some complicated, but not practically impossible, method whereby it is possible to discover the identity of the persons who 300, 600, 900, and so on years ago instantiated the genotype which a specific fetus now has. I am absolutely certain that people would split in two sharply disagreeing camps if this became a possibility. One group, perhaps the majority, would try to identify the previous instantiations of their child's genotype. Another group would emphatically not seek this information because they would not want to know and would not want their children to grow up in the shadow of a number of previously led lifes with the same genotype. The option to remain in ignorance is not open, however, to social parents of contemporary clones.

If the majority would seek the information in this scenario, firms offering the method of identification would have a very brisk business, and it could perhaps even become usual to expect of prospective parents that they made use of this new possibility. Why would this happen? The only reasonable explanation, apart from initial curiosity, is that people would believe that by identifying the previous instantiation of the genotype they would thereby gain valuable knowledge about their child. But knowledge is in general only valuable if it can be converted into new options for action, and the most likely form of action would be, that information about the previous instantiations would be used in deciding how to rear the present child. This again points to the importance of the public perception of genetic essentialism, since the environment must have changed considerably in the 300-year span between each instantiation of the genotype.

WHAT IS WRONG ABOUT A LIFE IN THE SHADOW

But what is wrong with living your life as a clone in the shadow of the life of the original? It seems to me that it diminishes the clone's possibility of living a life which is in a full sense of that word *his* or *her* life. The clone is forced to be involved in an attempt to perform a complicated partial reenactment of the life of somebody else. In our usual arguments for the importance of respect for autonomy or for the value of self-determination we often affirm that it is the final moral basis for these principles, that they enable persons to live their lives the way they themselves want to live these lives. If we deny part of this opportunity to clones and force them to live their lives in the shadow of someone else we are violating some of our most fundamental moral principles and intuitions. Therefore, as long as genetic essentialism is a common cultural belief there are good reasons not to allow human cloning.

FINAL QUALIFICATIONS

It is important to note that the life-in-the-shadow argument does not rely on the false premise that we can make an inference from genotype to (psychological or personality) phenotype, but only on the true premise that there is a strong public tendency to make such an inference. This means that the conclusions of the argument follow only as long as this empirical premise remains true. If ever the public relinquishes all belief in genetic essentialism the life-in-the-shadow argument would fail, but such a development seems highly unlikely. It could be suggested that the argument does not speak against cloning, but for a much more active information campaign to change peoples erroneous beliefs in genetic essentialism. This is absolutely correct, but there are two important things to note about this suggestion. The first is that it is probably unrealistic to expect much change in public perceptions about genetics, even if we mount a very strong information campaign. The beliefs concerning the relationship between "blood" and character traits are culturally very old and deeply entrenched in our way of thinking. We may therefore never reach a situation where the life-in-the-shadow scenario disappears. Second a belief in genetic essentialism is definitely wrong, but it is not in itself inherently ethically problematic, as is, for instance, a belief in the natural superiority of one specific race. It is thus not certain that we can draw an analogy from the sound argument that the negative consequences of racist beliefs should be combated by changing the beliefs not by accommodating them, to a similar argument about beliefs in genetic essentialism.

In conclusion I should perhaps also mention that I am fully aware of two possible counterarguments to the argument presented above. The first points out that even if a life in the shadow of the original is perhaps problematic and not very good, it is the only life the clone can have, and that it is therefore in the clone's interest to have this life as long as it is not worse than having no life at all. The life-in-the-shadow argument therefore does not show that cloning should be prohibited. I am unconvinced by this counterargument, just as I am by all arguments involving comparisons between existence and nonexistence, but it is outside the scope of this chapter to show decisively that the counterargument is wrong. It is, however, perhaps worth nothing that if the argument is accepted it also entails that every kind of intervention I perform which in some way harms a future human person is ethically innocuous, as long as I make sure that either (a) it is part of a series of interventions leading to the only life the person can have, or (b) make sure that the intervention harms the embryo, fetus, on newborn sufficiently for it to become a different person from the one it would have been without the intervention (thereby ensuring the fulfillment of criterion a). I find this conclusion profoundly counterintuitive and morally pernicious.

The second counterargument states that the conclusions of the life-in-the-shadow argument can be avoided if all clones are anonymously put up for adoption, so that no knowledge about the original is available to the social parents of the clone. I am happy to accept this counterargument, but I think that a system where I was not allowed to rear the clone of myself would practically annihilate any interest in human cloning. The attraction in cloning for many is exactly in the belief that a person can re-create himself. The cases in which human cloning might solve real medical or reproductive problems are on the fringe.

NOTE

1. Nelkin D, Lindee MS. *The DNA Mystique: The Gene as a Cultural Icon.* New York: W. H. Freeman and Company, 1995.

19

Clones, Harms, and Rights

Rosamond Rhodes

As the possibility of cloning humans looms on the horizon people are worrying about the morality of using the new technology. They are anxious about the ethical borders that might be crossed when virtual genetic duplicates can be produced by creating a zygote from an existing person's genetic material. They are apprehensive about eugenics, concerned about creating humans as sources of spare parts for others, uneasy about producing humans in the absence of the intention to allow them to live and develop, and uncomfortable about using human clones in business ventures.

The religiously inclined are concerned about meddling with the "sanctity of life". As Paul Ramsey has explained, "The value of human life is ultimately grounded in the Value God is placing on it. . . . [The] essence [of human life] is [its] existence before God and to God, and it is from Him."[1] For believers, cloning sounds dangerously close to playing God, trespassing in his domain, or treading on the sanctity of life.

Those who are sensitive to environmental issues are worried about meddling simpliciter.[2] Having witnessed so many problems created by the shortsighted use of new technologies, they are concerned at the prospect of upsetting the delicate balance of nature. They imagine that cloning could have serious implications for limiting reproductive diversity and so could harm the survivability of the species.

Philosophers have been busy forecasting ethical problems that will be hatched by cloning. For example, in a note on the moral problems of cloning, Frances Kamm flagged five such problems, three of which would be unique to cloning: (1) By cloning we could develop a standard (or multiple standards) for an ideal person which could, in turn, diminish the appreciation for other types of people. (2) The availability of genetic multiples could make us careless about those who already exist since they could be replaced like interchangeable (fungible) parts. And (3) government or business could control the cloning process and breed qualities for their purposes, such as compliant soldiers or workers with great endurance and a high tolerance for monotony.[3]

In response to such concerns coming from so many disparate perspectives there has been a call for a moratorium on the research and a demand for legislation to limit or outlaw the use of cloning techniques.[4] For many the specter of cloning is so awful that they want cloning banned or contained even before it begins. Yet, in the face of these premonitions of disaster and the bans that have already been put in place around the world, this paper will argue that we must resist the movement to proscribe or prohibit cloning. Our society's commitment to liberty requires that we allow individuals to make choices according to their own lights, and absent actual substantial evidence that such practices cause serious harm or at least a demonstration of a significant likelihood of untoward repercussions, we are not justified in denying individuals the option to clone themselves.

LIBERTY

From its inception, our society has embraced the value of liberty. Freedom has been our creed and the foundation for building our government. While there has been a range of interpretation offered for the concept of liberty, John Stuart Mill's account has been given the greatest weight in moral and political philosophy, and in this discussion, because of the strength of his arguments and the analytic power it yields, I will follow his account. As Mill has explained the commitment, for people who extol liberty, "the sole end for which mankind are warranted, individually or collectively, in interfering with the liberty of action of any of their number is self-protection."[5] This principle of limiting legislation, which has become known as the "harm principle," demands that no action be forbidden unless it can be shown to cause harm to others in the enjoyment of their rights.

> As soon as any part of a person's conduct affects prejudicially the interests of others, society has jurisdiction over it, and the question whether the general welfare will or will not be promoted by interfering with it becomes open to discussion. But there is no room for entertaining any such question when a person's conduct affects the interests of no person besides himself, or need not affect them unless they like.[6]

Although anything one person does may give another affront, upset, or sadness and thereby cause some harm, only those actions which "violate a

distinct and assignable obligation to any person or persons"[7] may be pro-
scribed by legislation. For example, the outfit I am wearing might offend
the aesthetic sensitivities of some of the people I meet. But since I have no
specific obligation to dress in accordance with their taste, and since I have
made no one any promise about what I would refrain from wearing, I have
violated no one's right. Therefore, society has no grounds for limiting my
freedom of self-expression through dress by legislating against what I
choose to wear.

Clones, Rights, and Harms

With respect to reproduction by cloning, the question relevant to Mill's
criterion is whether anyone's rights would be violated by cloning humans.
To answer we must consider all of those whose rights we could anticipate
might be violated. Those who might be harmed by the production of cloned
offspring would include the prospective cloned children, their peers (according to Kamm's scenario), and those in the community who would be
upset by people overstepping the line into God's domain. None of these,
however, would suffer any violation of rights.

Obviously multiple children with the same genetic inheritance produced by cloning would not be genetically unique. Having a genetic twin,
or several of them, however, is neither a clear harm nor a clear benefit. It
might be psychologically harmful because genetic twins could be so easily
confused or compared. But it could also be psychologically beneficial because of the special sharing, support, and intimacy that might develop between the genetically identical individuals. Furthermore, while clones
would be virtual genetic duplicates, identical twins or other children from
a multiple birth and, theoretically even some natural siblings, also have
common genetic material. No one has ever, on that account, charged that
their parents had violated their rights by having more than one child with
the same DNA. Apparently, even though most humans happen to be genetically unique, no one has a right to be unique. Therefore, no one who seeks
to produce multiple children by cloning should be prohibited from doing
so in order to avoid violating the rights of others.

Kamm's worry that we could develop a standard for an ideal person
which would diminish our appreciation for other types of people does not
meet Mill's standard for prohibiting the choice either. First, it seems that
no one has a right to prevent the existence of others who might be superior
to themselves and, just by living, make the inferior feel unappreciated. That
cloning technology might be the means to enable some superior individuals
to be born does not, therefore violate anyone else's rights. Second, Kamm's
concern with the development of a new "ideal" depends on a large and
significant number of people being produced by cloning. Since the cost,
inconvenience, discomfort, and loss of privacy entailed by the procedure
would be likely to make cloning a rarely employed technology, and since
there would be a variety of motivations and procreators, the numbers of

individuals produced by cloning would not be great enough or similar enough to have any significant impact on the social ideal of a person. Likewise, other objections to producing large numbers of cloned individuals could be discounted for now because that prospect is importantly different from what is presently being considered.

The religious concern over cloning interfering with the sanctity of life also fails to meet Mill's criterion for legislating against a practice. While liberty allows individuals the freedom to choose a religious perspective and the freedom to live according to the religious views they embrace, it limits individuals' infringement on the similar rights of others. In other words, no one may impose his own religious views on others. The religious liberty guaranteed by the harm principle does not extend rights to control the lives of others; thus, those whose religious sensitivities are upset by the prospect of other people meddling with the creation of human life cannot claim that harm as grounds for limiting others' procreative practice.

The Cautious Skeptic's Objection

Cautious skeptics, who might otherwise be guided by Mill's harm principle, nevertheless object to cloning because they can see no good reason for resorting to cloning technology.[8] Being hesitant about forging into uncharted territory in the face of well-known omens of disaster, they plead for applying the brakes to the momentum of technological capability. Skeptics imagine egoists using cloning technology to reproduce themselves and, limited only by the extent of their wealth and narcissism, cloning themselves many times. This consideration, which invokes neither harms nor rights, nevertheless deserves a response.

First, there are "good" reasons for employing cloning technology, such as allowing an infertile woman to have a genetically related child. Also, cloning could enable a couple to have offspring biologically related to both partners, although genetically related to only one. By using an enucleated egg from the woman (one from which the genetic material had been removed) and nuclear transfer technology to insert a cell nucleus from the man, they could have a cloned offspring with links to both parents (i.e., cytoplasm from the mother and DNA from the father). And, in the rare cases where both the man and woman lack gametes, cloning would be the only means of their having a child who is genetically related to one of them (the donor of the nucleus).

Putting these "good" reasons aside, it is important to point out that we do not question the reasoning which motivates non–technology-assisted reproduction. The ordinary desire to have biologically related offspring is not challenged even in the face of overpopulation and the large numbers of orphaned children around the world. Without aid and without society's interference people have children in order to pass along their genes, or to pressure a partner into marriage, or to get an apartment, or to keep a marriage together, or to get an inheritance, or to have a real live doll to play with, or to have somebody to love. It is not even clear which reasons are "good" reasons and which are not. But it is clear that privacy and respect for autonomy require that people be allowed to make reproductive decisions. So reasons for procreation should be irrelevant to policy makers. And, at least since Hobbes's writings in the seventeenth century, it has been understood that law can govern only action, not thought or belief. Considering the skeptics' objections to cloning, therefore, leads to limiting social impediments to cloning rather than supporting its prohibition.

An Alternative Approach

In America, legislation has become the most popularly advocated solution to every problem of managing medical technology. This may be attributable to the social awareness of medical needs that accompanied President Clinton's putting access to health care back on the political agenda. But whatever the reasons, whenever some medical misconduct is brought to light, the announcement is immediately accompanied by calls for legislation to correct the problem. For example, the revelation of the radiation experiments of the 1950s was followed by calls for laws to more carefully govern such research. The demand for more laws came despite the fact that the presently existing Institutional Review Board (research ethics committee) requirements, which went into effect after these unacceptable clinical trials had occurred, would have been adequate to have prevented those outrageous abuses of human subjects.

Then again, according to this pattern, when a U.C.L.A. study of treatment for schizophrenia was brought to light in March 1994 and publicly criticized, the discussions again called for more legislation to protect research subjects even though most of the critical commentary pointed to problems of inadequate implementation of existing rules.[9] And most recently, with the announcement that scientists have developed methods for cultivating human embryonic stem cells from germ cells of aborted human fetuses or human blastocysts, or by nuclear transfer involving the use of enucleated bovine eggs, there were new calls for banning the technology.[10–12] It seems that we have become like the fellow with a hammer who sees a nail as the solution to every problem. We have developed a practice of leaping at every situation with legislation and federal guidelines that are tantamount to legislation before the need has been demonstrated and before we have adequate empirical evidence to form the basis of desirable and effective laws. Legislation need not be proactive; there are other alternatives.

Two examples will illustrate a better approach to the development of social policy related to medicine. One comes from the Netherlands, the other from the history of American nursing. In the Netherlands today, while euthanasia is still officially prohibited under old laws designed against murder, the society is carrying on a program for deciding whether or not to accept the practice as they also try out standards for regulating it. They began in the late 1980s by formulating guidelines to govern physician-assisted dying. Then they undertook studies to gather empirical evidence on what was in fact being done to whom, when, and how, and to see how that practice was changing over time. One was an official government study, the Remmelink Commission Report of 1991–92, the other an independent research project by Gerrit van der Wall, 1991–92. They are now in the process of assessing their study findings so that the guiding policy can be amended based on what they have learned. Eventually they expect to draft a law that will regulate euthanasia but only after they have tried to assess the actual harms that they want to avoid.[13] As Mill advised, they are allowing that their ultimate decisions will be informed by "different experiments of living."[14]

In 1898, Isabel Hampton Robb, the first president of the American Nursing Association, advised the association to go slowly in pursuing one of its stated goals, "to establish a code of ethics." She argued that "it will be better to wait to learn the mind of the greater number on what shall constitute our national code of ethics. This code should be formulated to meet our own special needs in our own special way." First the American Nurses Association Committee on Ethical Standards worked on developing a professional statement of ideals. Then it took time to gather content from nurses, to reevaluate its ideals, and to formulate the wording. The committee finally recommended a code for nurses at the annual convention in 1926. The code was then referred back to the committee for further work and the suggested code was published in the *American Journal of Nursing* so that further comments could be elicited. The code was ultimately accepted in 1956. [It was revised in 1975, 1985, and again in 1989.][15]

This fifty-eight year history of the development of a nursing code of ethics illustrates again the alternative model of preceding legislation with a period of thoughtful deliberation, gathering empirical data, and assessing the real need for rules. Were such a model to be followed in guiding cloning practice, the process would begin by allowing physicians and their patients to use the new technology directed by the existing recommendations for the employment of technologies (like assessing genetic risk)[16] in "different experiments of living." Existing guidelines typically include requirements to establish precautions against conflict of interest and to provide for the informed consent of participants which would require giving the person information about the risks, benefits, efficacy, and alternatives. Surrogacy programs also typically include psychological screening for participants.

Obviously, such safeguards should be part of any program using cloning technology. Counseling and screening practices, however, should be seen as educational enhancements of liberty rather than intrusions on it. Again citing Mill, "human beings owe to each other help to distinguish the better from the worse, and encouragement to choose the former and avoid the latter."[17] Prospective patients being counseled about cloning should be encouraged to seriously evaluate all of the caveats raised by the objectors and to examine their own motives for pursing the technology. Psychological implications, not only for the resulting children but also for the parents who may be disappointed by their offsprings' lack of success or jealous of their superior achievements, should be raised. But the primary goal for the early stages of employing the new technology should be gathering the empirical evidence about the harms that might ensue and trying to learn the best way to avoid them.

CONCLUSION

The arguments above all suggest a single conclusion. Instituting legislation to bar or limit the employment of cloning technology would seriously violate our commitment to liberty. Similarly, panels consisting of representatives of particular political agendas which issue constraining policies that actually circumvent the democratic process and function as law, and also the tyranny of the majority, and also the blaring voice of vocal minorities, can all be enemies of liberty. Instead of attempting to arrest advances in cloning technology, those who are uneasy about proceeding should be tolerant of cloning. Once the technology is in use we can begin to experiment with rules that would help avoid any harmful consequences.

And in the meanwhile, those engaged in research and early applications of the new technology must also study the consequences and guidelines of research and practice in related areas of reproductive medicine. Conscientious care is required in scientific research, and conscientious care is also required for ethical implementation of new technology. And just as scientific advance must be based on careful evaluation of the relevant data, ethical rules that reflect our strongest moral intuitions about the value of liberty must also be based on the experiential data of actual harms and violations of rights.

NOTES

1. Paul Ramsey, "The Morality of Abortion," in *Life or Death: Ethics and Options* (Seattle: University of Washington Press, 1968), pp. 72–73.

2. William Safire, op-ed *New York Times* 1997; Feb. 27.

3. Frances Kamm, "Moral Problems in Cloning Embryos," *American Philosophical Association Newsletter on Philosophy and Medicine* (spring 1994, vol. 93, no. 1), p. 91.

4. See chapter 23.

5. John Stuart Mill, *On Liberty* (Indianapolis, Indiana: Hackett Publishing Company, 1978) p. 9.

6. Ibid, pp. 73–74.

7. Ibid, p. 74.

8. The skeptics' argument has been forcefully presented by Joe Fitschen in conversation, 1994.

9. *New York Times*, March 10, 1994, pp. A1 and B10.

10. Nicholas Wade, "Scientists Cultivate Cells at Root of Human Life: Hope for Transplants and Gene Therapy—Ethics at Issue," the *New York Times*, Nov. 6, 1998; pp. A1, A24.

11. Rick Weiss, "Human-cow cloning claim stirs questions," *Washington Post,* Nov. 13, 1998.

12. Richard Powers, "Too Many Breakthroughs," the *New York Times*, Nov. 19, 1998 (op-ed) p. A35.

13. This discussion draws on a paper by Margaret P. Battin, "Seven (More) Caveats Concerning the Discussion of Euthanasia in the Netherlands," *American Philosophical Association Newsletter on Philosophy and Medicine* (spring 1993, vol. 92, no 1), pp. 76–80.

14. Mill, *On Liberty*, p. 54.

15. This section follows a paper by Elizabeth Murphy, "A History of Nursing Ethics," delivered at the Oxford-Mount Sinai Conference, New College, Oxford, April 6, 1994.

16. Committee on Assessing Genetic Risks, Division of Health Sciences Policy, Institute of Medicine, "Assessing Genetic Risks: Implications for Health and Social Policy," Washington, D.C., 1993.

17. Mill, *On Liberty*, p. 74.

IV

CLONING & GERM-LINE INTERVENTIONS

The Policy Issues

20

Reflections on the Interface of Bioethics, Public Policy, and Science

Harold T. Shapiro

This chapter discusses the role of ethical considerations in the formulation of public policies aimed at shaping the scientific agenda. Specifically, new and controversial public policy issues will confront us in the twenty-first century as the result of developments on the frontiers of biomedical science. Some of the anxieties in the ethical arena generated by the rapid pace of these developments are likely to result in efforts to place constraints on the shape of the scientific agenda and the application of new knowledge.

The relationship between bioethics and public policy has become a rather broad subject that asks a rather simple question, namely: Which moral imperatives that arise out of the study and consideration of bioethical issues should be reflected in public policies that govern us all? Such policies are, after all, one of the end points of the ethical debate. The question is simple enough, but the answers are difficult because Americans live in a society where even the most thoughtful citizens do not share a moral consensus on many bioethical issues. Indeed, despite the rich, inspiring, and diverse array of current thinking in moral philosophy and bioethics, we continue to lack a moral consensus on some of the most profound ethical claims that some believe ought to be more fully reflected in actual public policies. This, of course, is not surprising, since it has long been recognized that no set of abstract rules can be expected to satisfy the particular contingencies represented by the cultural traditions and uncertainties that must be accommodated in real public policies. Nevertheless, since humans are a social species, all human societies continue to seek to establish rules of conduct that govern relationships between individuals and are thought to serve their collective interests. Furthermore, this search goes on within an evolving cultural context, and these collective rules of conduct must be constantly reviewed and perhaps revised and updated.

Thus, the first lesson is that in a society such as ours there will be perpetual uncertainty regarding which of the many competing ethical concerns ought to shape particular public policies. The second lesson follows from the first; namely, we cannot escape the anxiety that characterizes a situation where the justifiability of many ethical claims remains uncertain, or at least unconvincing, to important segments of the community. Finally, in my view, the set of "optimal" ethical views—those that will generate the most reliable and redeeming developmental outcomes—are unlikely to remain fixed in view of our dramatically changing circumstances. Our collective interests are likely to change, for example, given our new capacity to more radically control our future gene pool. Similarly, if we should gain the ability to transform cells at will from one form of gene expression to another, our concept of the moral status of various human biological materials may be altered. At a more macro level our collective interests and, therefore, our ethical obligations relating to the sharing of the earth's resources may well shift as we face radically new environmental problems. Indeed, such shifts in values might well be required if the survival of the species homo sapiens is important to us.

At times the many uncertainties surrounding all of the above concerns seem both important and almost boundless. We are, for example, uncertain about the nature of the limits to impose collectively on individual autonomy. We are undecided about the merit of the claims that various notions of distributive justice imply for our public policies, either across the generations or at any given time within particular societies or across the nations. We are uncertain which bioethical principles, theories, or frameworks should have priority in the formation of public policy. We are increasingly uncertain about what it means to be human either in a biological sense or a cultural sense. In addition, there continue to be areas of moral disagreement, particularly surrounding issues of birth and death, where it remains difficult in our society to have thoughtful and mutually empathetic conversations among those with opposing views. This is unfortunate since such conversations are the normal vehicle whereby a peaceful and morally pluralistic society learns to live together. Finally, even while we strive for greater moral agreement or understanding, we at least try to resist any temptation to exert moral tyranny over others and instead place our faith in the difficult task of momentarily stepping outside of our own commitments, histories, and circumstances to help define and redefine a robust core of moral propositions that can be incorporated into the narratives of most people.

Returning to the issue of bioethics and public policy, it seems certain that many new controversial public policy issues will arise out of ongoing developments in biomedical science and their associated bioethical considerations. Indeed, it is inevitable that the rapid pace of development of new knowledge and, therefore, of new opportunities—that is, applications—is certain to generate new issues and new anxieties in the ethical arena. We can anticipate, therefore, a continued search for those social processes or controls, possibly public policies of one type or another, that will improve our chances of selecting the most ethically acceptable applications of our expanding knowledge base. As a result, just as we expect that new science will gain its moral relevance from the nature of the uses we make of new knowledge, we should understand that our moral propositions—old and new—are themselves about to be tested and retested in their application to our evolving social, cultural, and historical circumstances and the changing technological context.

CONCERNS

Although signs of immense human accomplishments are all around us, no previous century has produced such a high level of apprehension about the future. Perhaps the reason for this is that as science generates a larger set of opportunities for us all, it simultaneously raises the level of moral responsibility that falls on our shoulders, and it is this moral or ethical challenge about which we are so uncertain. This nervousness, ethical malaise, anxiety, or even foreboding reflect, I believe, in addition to the fragility of traditional reference systems, a shared understanding that humankind's destiny will not be decided in full in the laboratory or at the genetic level, where we have a lot more confidence in our ability to find solutions. Whatever else we may need to address these concerns and the many new issues and opportunities rushing toward us, we need new sources of reflection to enrich the ongoing ethical debates.

Perhaps all this explains why a short time ago I was asked to address a conference with the rather "arresting" title of "Stopping Science." Although concerns regarding the transformation of our lives and our prospects by the application of new scientific discoveries have a very ancient pedigree, the stunning pace and nature of recent scientific advances, particularly in biology, have been disquieting to some. Indeed, let me quote from two different letters I received from President Clinton in my capacity as chair of the National Bioethics Advisory Commission:

> While this technological advance [somatic cell nuclear transfer cloning] could offer potential benefits . . . it also raises serious ethical concerns. (February 1997)

This week's report of the creation of an embryonic stem cell that is part human and part cow raises the most serious of ethical, medical, and legal concerns. . . . I am therefore requesting that the National Bioethics Advisory Commission consider the implications of such research . . . and report back to me as soon as possible. (November 1998)

Over the course of a year and a half, therefore, the president of the United States has expressed some serious ethical concerns arising out of developments on the scientific frontier. As always, of course, it was not the scientific developments themselves that caused the concern, but their potential applications, as well as, perhaps, concerns regarding the limits (if any) of technology and the pace of discovery in the absence of a moral compass. However, in both cases there was at least some suggestion that public policy should address itself to the moral or ethical content—the application—of scientific developments and perhaps try to arrest, or to shape in some way, further developments in certain scientific or technological areas. I would like to reflect, therefore, on the role of ethical considerations in the formation of those public policies specifically aimed at shaping the nature of the scientific agenda.

PUBLIC POLICY, THE SCIENTIFIC AGENDA, AND ETHICS

The overall impact of public policy on the shape and scope of the scientific agenda has become so widespread that, for the most part, we scarcely take any special notice of it. The influence of public policy operates day in and day out not only through the magnitude and distribution of government support for various aspects of the scientific enterprise (including the education of scientists), but also through a broad series of laws and regulations dealing with such diverse issues as the control of toxic substances, tax laws, the use of animals in research protocols, environmental regulations, patent policy, the design and conduct of clinical trials, and so on. In summary, public policy influences science through the problems it deems worth solving, the methods it finds acceptable, and the resources it makes available for the task. Moreover, behind some of these public policies lie ethical (and other) commitments and values that, for the most part, we scarcely bother to articulate. It is my own view that a more systematic articulation of the particular ethical commitments, if any, that help inform these existing and long-established policies would not only be salutary, but would prepare us better for the new and perhaps more complex ethical challenges that will face certain public policies in the future. It is also true that developments in science and technology and the views of the science and technology community influence public policy in general and science and technology policy in particular.

Although many of these ethical issues go unnoticed in the rush of our day-to-day concerns, we do take much greater notice of the stance of science policy when it directly concerns morally contested areas such as human cloning—the potential use of somatic cell nuclear transfer techniques to create infants—and embryo research or genetic engineering, which raise important moral concerns for some. With this in mind, I will narrow the scope of my paper to consider the special issue of how public policy concerns sometimes become focused on efforts to place *negative* constraints on the shape of the scientific agenda and the applications of the new knowledge. By negative constraints, I mean instructions not to do something, as opposed to positive constraints, which are designed to encourage one to proceed in a particular direction. More specifically, how did we ever become concerned, for example, with "stopping science"? For reasons that are not fully clear to me, it is also the case that public policy in the arena of bioethics seems primarily focused on restraints, but perhaps all of ethics is in this category!

STOPPING SCIENCE

Stopping phrases or expressions like *stopping science* have a rather special resonance to them. They seem to have a certain bracing quality that not only beckons us to purposeful action, but often contains an obvious imperative of some sort, such as "stop the violence," or an understandable wish, such as "stop AIDS" or "stop cancer," or even "stop inflation." Indeed, we are now stopping so many things that there is a new literary genre dealing with "the end": the end of empire, end of history, end of God, end of affluence, end of the nation state, and perhaps most relevant here, the end of science, and so on. The argument suggesting the "end of science" is stunning in its claims, but it is equally unconvincing. The claims are that all unique discoveries have already been made and that only increasingly specialized "cleanup" operations remain. The argument does, however, allow for a rather endless stream of further developments in the application of the "final" corpus of scientific knowledge—that is, technology. Therefore, even if science is at an end, its moral content would remain an ongoing challenge through the applications of the technologies that are selected.

I will focus, however, on a special subclass of "stopping" phrases; namely, those that urge us to stop, or slow down, something that most people think, on balance, is a very good thing. Here we are presented with a much more subtle matter. Like the more general class of "stopping" phrases, these rather distinct rallying cries are intended, once again, to promise us a better world if only we will act, but this time the promised prize is available only if we have the wisdom to see the ultimate futility of contemporary beliefs and the dangerous, but somewhat hidden, dynamic of our present circumstances. Thus, for example, however positive developments in science or technology may seem, someone is suggesting that a deeper look into their full impact would generate some concerns regarding such issues as the meaning of these new developments for sustaining our capacity to live together and for comprehending the place of human societies in the grander scheme of things.

These more subtle "rallying cries," therefore, not only call us to purposeful action, but call us to account, in a biblical sort of way, for our intellectual shortcomings and the lethargy that together prevent us from being agents of positive change. Such phrases are intriguing (even if often misleading) because they represent, on the one hand, an effort to mobilize us and, on the other, an effort to scold us for our inability to see things as they really are. It is almost as if one is calling on the blissfully ignorant to solve a complex problem! In any case, it is into this category that phrases such as "stop science" or "stop progress" (whatever that means)[1] or "stop technology" fall, and since they dare us to think anew about some central matters in our lives our initial reactions are often, to borrow a phrase from psychology, "flight (refuse to face the issue) or fight (refuse to accept the premises of the new idea)."

The reaction of flight or fight is understandable in the sense that to look below the surface of things is to risk the discovery of a new truth and, therefore, the need to replace previous beliefs with a new set of ideas. Fortunately, contemporary observers have many intellectual resources for attending to such situations, since much of modern thought, from Darwin, Freud, Marx, and Einstein to the ideas that surround the more recent notion of the social construction of reality (to say nothing about the continuing avalanche of scientific discoveries), deals with the "uncovering" of truths that replace a set of former beliefs that finally have been revealed to be badly mistaken or convenient myths.

The stopping-science crowd might suggest, for example, that our prevailing confidence in human competence, technology, and science may, in the fullness of time, turn out to be just another myth that needs to be modified or set aside! After all, the continuing popularity of the Faust legend suggests pretty much the same thing. Scientists should not feel too defensive about the existence of such views for at least two reasons. First, most people, as well as most policy makers, do not find this perspective compelling. Second, in addition to any concerns some might have with respect to science, there are other defining elements of twentieth-century life that cause despair in certain quarters. Many would say, for example, that there is or should be a widespread revolt against modernism in all its various forms, since its overintellectualization of life, values, and art undermines much older and more sustaining beliefs. I leave this latter issue for another time.

Irrespective of one's views on the ultimate impact of science and technology on the evolving human condition, there seems to be a clear need for all thoughtful citizens to consider the ongoing and changing impact of science and especially of technology on those institutions, values, and other cultural commitments that sustain our individual and common lives, since it is in these areas that science and technology gain moral relevance. It is important, therefore, not just to celebrate discovery and its many benefits but to consider as well the different possible moral repercussions of new knowledge. Indeed, the more dependent we become on new science and technology, the more essential it is to generate renaissance after renaissance in moral philosophy. For example, I believe it is the explosive growth in the development of new knowledge that has generated the renaissance in interest in moral philosophy of the last two or three decades.

SCIENCE AND THE HUMAN NARRATIVE

It is critical to remind ourselves that scientific theories have nothing very interesting to say about either the value of a human being or the meaning of the nature of the lives we lead together. As John Maynard Smith (1984) has pointed out, the purpose of "myth" or other constructed or revealed narratives is to define our place in nature and to give us a sense of purpose and value. It is these special "myths" or narratives that give us moral guidance or suggest how we should act. Stated another way, scientific theories say nothing about what is right in a moral sense, but speak only of what is possible. The source of our values, therefore, must come from outside science. Thus, although myths and narratives, on the one hand, and science, on the other, are both constructs of the human mind, their functions are quite different, and meet different needs. In particular, deciding how we should act, including what public policies we should enact, is a negotiated social decision that necessarily involves resources outside of science such as our cumulative cultural traditions and other historical contingencies.

Thus, although crass self-interest or unexamined fears may cause some to want to stop or redirect a branch of scientific investigation or to interfere with the existing dynamic underlying the development of a new technology, such concerns could, from time to time, also be motivated by the desire to try to understand the moral content of what is about to happen. After all, everyone could agree that what is scientifically and technically possible must be parsed into those applications that are desirable and those that are undesirable either on ethical or other grounds. As we march into our future, we should, in short, never confuse what we *can* do with what we *should* do.

We need to acknowledge that many persons with genuine respect for the continuing contributions of science and technology also have serious commitments to various ontological positions, having to do, for example, with deeply held views regarding the limits on appropriate human behavior or activity. Equally important, however, there are the more secular concerns about the continued capacity of human institutions and nature itself to survive both advancing science and technology and the associated desires to control and possess all. As Havelock Ellis (1923, p. 352) commented long ago in *The Dance of Life*: "The sun, the moon, the stars would have disappeared long ago, had they happened to be within reach of predatory human hands."

Concerns regarding such issues are found throughout the historical record of Western civilization and are widespread in the Western literary and cultural tradition. Indeed, uncertainty has always existed within the Western tradition about whether the relationship of humans to the natural world is one of control and exploitation or of praise, celebration, and awe. This uncertainty raises interesting questions. Does our survival as humans require a shift in our values and aspirations? Do we need new ethics to moderate our desires for dominion, acquisition, and power? Is it true that the only way we save our souls is to find some moral compass, or moral limits, to our desire to conquer and control all and to possess all? Perhaps only such limits can save us from the moral ambiguity of our own cleverness and help us realize that even we cannot transcend nature without nature.

Within the human narrative of the West, the notion that advances in science and technology are Janus faced—both friend and foe—and can bring both vast good and catastrophic evil is a truly ancient one, deeply embedded, for example, in classical Greek culture, where science and technology are often characterized as bringing both promise and peril, hope and despair. Moreover, even in those early days, the focus of concern was on the implications of new knowledge for the meaning of being human and on what new and perhaps dark human desires, which might distort the human journey, would be released by the new power generated by that knowledge. Since the earliest days, therefore, the issue has been how we understand the nature of what it means to be human within the context of our new knowledge about the natural world and how these developments will influence the future of the human condition. Listen, for example, to the voices of Ovid and Sophocles, which speak directly to the issue of mankind's evolving role and whether limits to our power are an essential aspect of our humanity.

What you want, my son, is dangerous, you ask for power beyond your strength and years: your lot is mortal. But what you ask is beyond the lot of mortals. ("Story of Phaethon," in Ovid's *Metamorphoses*, Book II)

Many things are formidable, and none more formidable than man. . . . And he wears away the highest of the gods, Earth, immortal and unwearying, as his ploughs go back and forth from year to year. . . . Skillful beyond hope is the contrivance of his art, he advances sometimes to evil and other times to good. . . . May he who does such things never sit by my hearth or share my thoughts. (Sophocles's choral "Ode to Man" in *Antigone*)

In this same vein, the "Tower of Babel" story portrays a massive but fruitless building project and epitomizes what some believe is the folly of human technological action taken in complete autonomy from God. Likewise, one could ask today if various genetic engineering proposals, for example, are a similar act of hubris as we try to alter what some believe God has wrought. Once again such questions might suggest that we should consider the natural world to be more than merely instrumental to human purposes. Elsewhere, of course—Ecclesiastes, for instance—the Bible warns that with new knowledge can come much grief. It goes almost without saying, however, that alternative views about science and technology also have an ancient lineage. Aspects of the Judeo-Christian ethic, for example, particularly the notion of humankind's perpetual progress within the divine unfolding of history (e.g., the Exodus story), have been responsible for, some would say, transforming a reverence for nature to a view in which nature is a "mere" resource to support humankind's efforts to achieve their (and God's) program of upward progress. Consider the following verses from Genesis.

> God said: Let us make humankind, in our image, according to our likeness! Let them have dominion over . . . all the Earth . . . Bear fruit and be many and fill the Earth and subdue it! (Gen. 1: 26–28)

The same verses, of course, not only have a variety of interpretations, but also provide support for the widely held Western notion of the moral superiority of human life over other forms of life.

These verses, and the countless others that have become part of the Western literary tradition, reflect that, in addition to our great enthusiasm for technological progress, often there exists just below the surface of our consciousness a certain amount of pent-up anxiety regarding the impact of science and new technology both on a wide variety of honored practices, important values, and other long-standing cultural commitments and on the inherent limits on the ability of science and technology to address important aspects of the human condition.

At some level, therefore, we have always known that the final crisis of science and technology is the realization that they cannot be relied upon to deal with some of the more important issues of our lives as individuals and our lives as members of coherent communities. Science and technology, despite their protean strengths, cannot help us decide how human beings should act or how they are to construct a coherent and secure narrative of their place in the grander scheme of things. It is not surprising, therefore, that early myths and narratives from almost every culture foretell the inevitable crises of knowledge; namely, that after we know all about the natural world and possess all material things, we will still find, as I have already noted, that many of our most important needs remain unaddressed.

Anyone who has actually studied the evolving condition of human societies over time cannot help but be impressed by the contribution of science and technology not only to a fuller understanding of the natural world and a fuller expression of our humanity but to the reduction of individual human suffering. Moreover, science cannot be regarded merely as a generator and storehouse of facts, but surely must be ranked as one of the great human endeavors of all time. Nevertheless, as with the ancients, we must continue to acknowledge that while there does not seem to be anything particularly convincing about calling new scientific and technological developments dehumanizing, these developments can bring in their wake a certain understandable disquiet and perplexity to many. We need to acknowledge that these concerns, even if ultimately proved groundless, need to be taken seriously, not simply because they reflect real worries, but because new scientific and technological developments often raise important moral and ethical issues that need to be confronted.

Whatever else one may say about science, one must allow that it can be quite subversive, since its focus on revealing the previously unseen reality of things works against the stability of current beliefs and our trust in or even reverence for certain values that may be required to sustain certain valuable human institutions. Although it may seem fine to have the "real truth" out (it certainly seems better than sustained ignorance!), our social institutions also rely on trust—as opposed to an unbridled skepticism—and, as I have already noted, even a reverence for a particular set of beliefs or cultural arrangements. In addition, we have to wonder if there is anything outside of ourselves and our efforts that is worthy of reverence and awe.

IN OUR OWN TIME

In our own time, of course, we must both celebrate and contend with the fact that science and technology are advancing at an unprecedentedly rapid pace. To the extent, therefore, that science sometimes works to undermine (often quite justifiably) our faith in existing arrangements for our individual and common life, these rapid advances also generate a somewhat elevated level of concern regarding the full meaning of this new knowledge for our existing cultural commitments. Hence, it should not surprise us that, in our own time, we are simultaneously concerned with "stopping science" and fully celebrating the new scientific discoveries that arrive daily.

Consider, for example, the case of genetic engineering. On the one hand, an early type of genetic engineering was the foundation of one of humankind's greatest achievements; namely, the successful domestication of certain plants and animals. On the other hand, these ancient processes of selective breeding were imprecise and unpredictable, and successful results took a very long time to evolve. In contrast, however, contemporary genetic engineering—the capacity to manipulate genes or to isolate and transfer genes both within and across species—has become the center of both stunning scientific advances and a good deal of ethical controversy.

It is easy to understand both the excitement and the consternation generated by our new capacities in this arena. This new technology could play a decisive role in helping to meet the nutritional requirements of the world's rapidly growing population, in assisting efforts to sustain the health of our environment, and in making a wide spectrum of clinically useful medical discoveries. But there are those, perhaps a small minority, who have deep concerns about the ethical and social implications of all these potential developments. Some worry that we are proceeding at too fast a pace along a path whose implications for our sense of what it means to be human are unclear and may be overwhelming. A favorite example is that we might, because of our quite unprecedented ability to choose for and against certain types of people as our descendants, be tempted to take some dark and foreboding journey into eugenics territory. Thus, there is both great enthusiasm and some uneasiness about contemporary advances in genetics.

THE CHALLENGE AHEAD

It is easy for many of us to dismiss the uneasiness with scientific and technological developments as the mark of Luddites and others who are always upset by change and always getting in the way of progress. It would probably serve us all better, however, to separate the unjustified and unexamined fears of critics from other of their concerns, which may have some useful warnings for us all, especially in those areas of our lives that still lie beyond the reach of science itself. At the very least, we have to concede that new knowledge does not ensure that our moral wisdom or our public policies will rise to the occasion.

Humans have always practiced technology by adapting the natural world to serve their own ends. Indeed, such momentous developments as the domestication of plants and animals enabled human societies to turn some of their energies to the creation of the great cultural artifacts that both define civilization and enable humans to realize more fully their potential. Nevertheless, as new technology was developed and incorporated into human societies, it has always remolded our societies in some way. At times these changes are rather minor. Often they enhance and enrich our human potential. At other times, however, the changes challenge our assumptions about life, our self-understanding, and our ways of relating to one another and the rest of the natural world, and these types of challenges continue to need our thoughtful attention.

One of the great responsibilities facing us in the twenty-first century, therefore, is to consider the social and human repercussions of our rapidly accumulating new knowledge and the appropriate stance of public policies with respect to these matters. For scientists, ethical reflection must become an integral part of the scientific agenda. This obligation is especially acute given our enhanced capacity to transform the lives of all manner of plants and animals, including ourselves. Perhaps it is not enough to use these powers to benefit humanity directly by relieving human suffering; in addition, we might wish to understand the resulting impact of these developments on the social and cultural institutions that are critical to supporting our individual and collective lives.

For many, the impact of these developments has been so startling and perplexing that some observers believe that a considerable level of hostility to developments in science and technology has developed. To some extent this hostility is sustained by the general lack of serious conversations between scientists and other thoughtful citizens. It may be, as one humorist noted, that modern science is practiced by those who lack a flair for conversation. This has not been my experience, but serious conversations between scientists and other thoughtful citizens are becoming more and more essential. By serious conversations, I refer to dialogues between individuals where the persons involved expect, through the reflective and thoughtful engagement with the work and ideas of others, to expand their imaginations, enlarge their awareness, deepen their understanding, and hone their ability to perceive new possibilities of all kinds. Both the dividends and the risks of such conversations are their ability to transform one in important ways. These conversations are not, therefore, for individuals who think they have nothing more to learn or who are frightened by new ideas and new opportunities. All such conversations, if their potential for personal growth is to be realized, will demand both an openness to new ideas and a focused effort from all who participate. Once again, to participate in such conversations one must risk that one's own ideas might not be accepted by others or that their ideas might be upsetting. Such exchanges, however, can build new worlds of meaning and lasting connections of all kinds. They can lead all participants to a deeper understanding of themselves and the human narrative of which they are a part, as well as provide an enhanced capacity to help all of us meet the challenges that lie before us.

The practical issue is not whether one set of concerns and attitudes should prevail over an alternative set, but how various strategies for prioritizing or addressing human needs and desires can have a deeper and healthier interaction with each other in the public marketplace of ideas, in public policy discussions, and within evolving cultural arrangements. Within such a discourse, scientists and others should understand that one of the great unresolved challenges of the scientific revolution is our ongoing failure to adequately consider its implications in moral terms. We may or may not need new moral philosophies, but we are more likely to get to the moral high ground if scientists and others work together to help shape the uses of new knowledge. The fact that the moral context of scientific discoveries is defined by the objectives and applications of this new knowledge gives all of us additional responsibilities and, inevitably, new anxieties.

Unfortunately, scientists, much more frequently than others, show considerable disdain for what I would call "preventive ethics programs," which are based on the perfectly rational and morally defensible principle of moral prudence. This principle simply recommends a certain caution in order to avoid leaving difficult questions to the lottery of future circumstances. We should recall the story of Ulysses, who in an action of considerable moral strength strapped himself to the mast of his ship in order to avoid a decision he would later regret!

My own view is that the human condition will continue to be well served by sustaining the vitality of our scientific enterprises. The continuing health and robustness of the scientific and technological enterprise itself, however, depend, in part, on sustaining the intellectual authority of the scientific community, and this requires both a willingness of scientists to work with others in order to take appropriate action as new scientific understandings emerge and a continued commitment for society to sustain its belief in and incentives for the development of new and challenging ideas. This is a social and political as well as a scientific process, and social or political decisions cannot be left to scientists alone. Scientists and other thoughtful citizens must work together not only to distinguish between self-interest and community interest, sentimentality and careful thought, learning and imagination, but also to understand the power and limitations of knowledge. We all must face the fact that one of our greatest responsibilities is to consider the full implications of our new knowledge not only for relieving human suffering and distress but for the social and cultural institutions that are as critical as DNA to supporting our individual and collective lives and informing the process of evolution through natural selection.

Concern for our social and cultural institutions is a critical matter, not simply because I have some considerable affection for many of our cultural traditions, but because I believe it is impossible to maintain the sharp distinction we have gotten used to between our biological evolution as a species and our cultural evolution as families and communities. Although I am a firm believer in the theory of evolution under natural selection, I believe, as Griffith and Grey (1994) and Oyama (1985) have articulated, that reliable developmental outcomes occur because of a wide variety of reliable interactions between developing organisms and their *total* environment, where the latter is composed of organisms, genes, and cytoplasm (along with a certain randomness and contingency), on the one hand, and quite different factors such as language, traditions, and even such prosaic matters as libraries, courts, universities, gravity, and sunshine (again with some random and contingent factors), on the other. It seems clear to me, for example, that the plasticity of the brain's complex network makes it quite accessible to a wide variety of social and cultural stimuli. Thus, there seems to be scant basis for privileging the gene over, for example, other developmental resources, both physical and cultural, that make a critical contribution to the reliable developmental outcomes that are the observed product of the evolutionary process.

A gene, after all, cannot even replicate itself without the help, for example, of ribosomes and proteins. It turns out that only much more complicated living systems are self-replicating. It is not clear to me whether prayers, a cultural resource, or F-18s, a technological resource, will be the most decisive element in the outcome of some future battle. Moreover, as Bateson (1978) observed a generation ago, if we say a bird's nest—a cultural resource for birds—is a gene's way of making another gene, we could just as easily remark that a gene is a bird nest's way of making another bird's nest. It is more critical than ever that we expand our moral imagination beyond the thrall of the truly stunning developments in contemporary biology and enlarge our horizons regarding our understanding of what resources of all kinds—scientific, cultural, ethical, and environmental—are required to continue to be human. In short, our lives are even more complicated than we thought, and it is only by integrating the special chemistries of our biological and cultural selves that we prove the following verse entitled "The Choice" by Yeats (*1933) to be quite wrong:

> The intellect of man is forced to choose
> Perfection of the life, or of the work,
> And if it take the second must refuse
> A heavenly mansion, raging in the dark.
> When all that story's finished, what's the news?
> In luck or out the toil has left its mark:
> That old perplexity an empty purse,
> Or the day's vanity, the night's remorse.

NOTE

1. Although many believe the notion of progress to be an idea of the eighteenth-century Enlightenment, biblical/religious sources both embrace it (e.g., the Exodus story) and reject it (e.g., we are mired in sin, misery, and pain). Moreover, there are contemporary observers who deny that any progress has taken place and maintain that even the process of evolution is meandering nowhere. To others it seems unclear whether the notion of progress is an appropriate criterion with which to measure ourselves. At the cosmological level, of course, the world as we know it seems to be a result of a rather chaotic series of events with direction but no real purpose.

REFERENCES

Bateson, P. 1978. Review of Dawkins' "The Selfish Gene." *Animal Behavior* **78**: 316–18.

Ellis, Havelock. 1923. *The Dance of Life*. Cambridge, MA: Riverside Press.

Griffith, P. E., and Grey, R. D. 1994. Developmental Systems and Evolutionary Exploration. *Journal of Philosophy* **21**: 277–304.

Oyama, S. 1985. *The Ontogeny of Information*. New York: Cambridge University Press.

Smith, John Maynard. 1984. Science and Myth. *Natural History* **93** (11): 10–24.

Yeats, W. B. 1951. The Choice. In *Collected Poems*. New York: The Macmillan Company.

21

The Regulation of Technology

Mary Warnock

Everybody recognizes that most of the problems in medical ethics arise, these days, from innovations in medical technology. We would not have had to lay down laws or ethical guidelines about assisted reproduction had it not been for the new technology of in vitro fertilization, which produced the first IVF baby in 1978. We would not be currently anxious about the ethics of possible human cloning had it not been for the production in Edinburgh of Dolly, the lamb whose birth resulted from the removal of a mammary gland cell from an adult sheep. So the question is whether there is some research into developing technology that is too dangerous, that will lead to consequences too dramatic for humanity, for the research itself to be permitted. Should there be control over what technological innovation should be permitted?

Put like this, the question looks absurd. It is not the discovery of new technological possibilities that is alarming, but the use to which these possibilities may be put. Control should not be over research, but over the uses of research. After all, even Plato, centuries ago, recognized that any skill, or *techne*, could be put to either good or bad use; the skilled doctor could also be a skilled poisoner.

However, the distinction between research and the uses of research is by no means easy to draw. First, it may be argued that if a procedure is shown to be possible (such as, for example, the transplant of organs from one human to another, or, transgenically, from one animal to another), then

233

someone, somewhere, will want to use this technique for therapeutic, not merely for research purposes. Second, the very possibility of such a technique may have been established only by means of its use on subjects, animal or human. There is no way of definitively distinguishing new and untried treatment from research. All treatment is, in some sense, a contribution to research, or may be such. Equally, in the field of medicine all research is undertaken with at least a vague hope that it will one day be used to improve treatment. Medical research is seldom entirely "pure." So the development of new technology cannot be fenced off from its use.

Nevertheless, it could be argued that the development of certain techniques is simply in itself too dangerous to be permitted. In the 1970s, when the genetic manipulation of plants became a widely recognized possibility, a moratorium was, for a time, called on such research, on the grounds that the research itself was too dangerous, carrying as it did a risk to those engaged in it, and a risk of the accidental release into the environment of genetically modified organisms, with unknown consequences. The moratorium did not last; and it is probably true to say that the safety of research workers and of the environment as a whole is better protected than it was because of a greater realization of the risks that may exist unless due care is taken. So the dangers of research are these days seen to be dangers of outcome rather than of the processes themselves. A parallel story could be told of the fears surrounding research in nuclear physics.

Thus the question must be asked again, are there some technologies to develop which would be so threatening that they should be subject to regulation, or even be prohibited by law? The technique of cloning is obviously a candidate for such prohibition. The public reaction to the birth of Dolly was little short of hysterical. On Sunday, February 23, 1997, the *Observer* carried the story of the cloning, to be published with proper scientific dispassion in *Nature* the following Thursday. The press reacted instantly, both in the United Kingdom and abroad. For some reason, Philippe Vasseur, the French minister of agriculture, warned Europe of the possibility of six-legged chickens. But, unsurprisingly, most concentrated on the possible use of the technique on humans. The German newspaper *Die Welt* called attention to the political implications of human cloning, saying that Hitler would have used it if it had then been possible; Jaques Santer, the president of the European Commission, instructed commission officials to investigate whether there was need for EU regulation of cloning; and the German Socialist MEP, Dagmar Roth-Behrend, called for a worldwide moratorium on the technique, on whatever animals it was used. Fortunately, no one rushed out instant regulative legislation. The scientific press managed to come up with explanatory and generally reassuring accounts of the procedure, and the Edinburgh team themselves very sensibly announced that they were not in favor of the use of the technique on humans.

There is, however, a lesson to be learned from the case of cloning. It is tempting for the press (and they are certain to fall for the temptation) to turn the announcement of any new biomedical technique into a shock/horror story; and the public will probably accept what they read and put pressure on Parliament to take steps either to prohibit further research altogether or at least to subject it to nonscientific regulation. To legislate in such circumstances, in response to popular feeling, is almost always a mistake. But in any case there is a fundamental objection to the regulation of scientific research, and in the excitement of the moment it must not be forgotten. It is the need to preserve academic freedom. By this I do not mean an absolute right of scientists or other academics to receive public funding for whatever they want to do, or to teach; I mean rather that academics themselves must be recognized as those who can decide what is or is not worth pursuing. Research, or indeed the content of teaching for that matter, must not be controlled by those who are ignorant. Parliament, and the general public, must trust those who actually know what they are talking about, and must be taught by them. We are all too likely to think that anybody is entitled to hold a moral view, either about what research is or is not worth pursuing, or about what the possible outcomes of such research may be. But this is a false belief; it is not possible to hold a responsible moral opinion on a matter of which one is ignorant. We need to learn the facts, and the probabilities, first, and then form a judgment upon them. Legislation based on popular indignation or fear, then, is nearly always going to be bad legislation that will be later regretted.

However, this is not to say that technology and the search for new technologies must never be subject to legislation. If only to allay public alarm (fear, that is, that scientists are too powerful, and that they like to "play God") it is often necessary that the use of technology should be, if not prohibited in certain cases, then at least regulated. And if necessary the criminal law must be invoked in the case of nonobservance of regulation. For example, so horrendous did people find the idea of fertilizing sperm and egg in the laboratory, and keeping the resulting embryo alive in its "test-tube" indefinitely, that a new criminal offense was invented in the legislation of 1990—that of keeping an embryo alive for more than fourteen days after the completion of fertilization, an offense that carries the penalty of up to ten years' imprisonment. Some would argue that the creating of human clones by the technique that produced Dolly should likewise become a criminal offense, though I believe that this is unnecessary, at least for the foreseeable future.

If regulation of the uses of new technology is ever to be thought desirable, then the question must arise of who is to take that decision. It may simply be a matter of professional self-regulation, with published guidelines. But, if there are not sufficient grounds to trust the professionals themselves either to follow the guidelines or to submit themselves to inspection to ensure that they are doing so, then it must be a matter for Parliament, and there will, as I have suggested, need to be legislation. Whether or not, in cases of biotechnology, the professionals are to be trusted will depend on the issues involved. But increasingly, it has to be said, there are huge sums of money to be made by pharmaceutical companies, who may develop their own research teams; increasingly patents are taken out for new techniques, and the competition between companies and consequential secrecy makes any kind of inspection or monitoring nearly impossible. We may therefore see more legislation in the field.

Parliament must obviously be well-informed if it is to produce legislation; for the issues involved will be moral issues, involving public policy of a particular kind, namely, that nothing shall be permitted that is genuinely outrageous to the value that ought to be accorded to human beings. And, as I have said, one cannot make proper moral judgments on a basis of ignorance. Here there is, I believe, a genuine role for a committee of inquiry, or royal commission, composed partly of scientists, partly of practicing doctors, partly of lawyers, and perhaps philosophers, or other reasonably level-headed persons, who will make the outcome of their deliberations public, will seek evidence and opinions from as many people as possible before reaching their conclusions, and above all will have the task of educating the general public.

This last point is of the greatest importance. Anonymous departmental civil servants—even if, as one hopes, they are strictly impartial and not under pressure from their ministers—cannot take on the educative role that is necessary in such cases. They cannot write articles or take part in broadcasts or lecture tours to explain the conclusions that they recommend and the arguments on which they are based. Without this fairly lengthy process, no regulatory legislation can be satisfactory. In the field of biotechnology, indeed, it is all too likely that legislation, even if it is not hastily cobbled together to allay public fears, will be overrestrictive and will tend to inhibit valuable research.

There is a difficult balance that must, if possible, be achieved between allowing new technologies to be developed which may have quite unforeseen beneficial uses, and on the other hand offending widely held and deepseated moral feelings. The best that can be hoped for is that by understanding more of the issues in each particular case people may come to feel that the freedoms allowed, and the restrictions imposed, are acceptable, even if not exactly what they would have personally liked to see. It is only if this balance can be achieved that the regulation of technological research can be compatible on the one hand with academic freedom, and on the other hand with a democratic regard for the moral views of people at large.

22

Cloning and the Regulative Dilemma

David Magnus

Reproductive and genetic technologies have developed far more quickly than our ability to cope with their implications. That is especially clear when it comes to dealing with the regulation of these technologies. Eugenic anxieties burble up into public discourse; yet little has been done to alleviate those anxieties.

The level of public interest over the prospect of human cloning provides a welcome opportunity to engage in a serious discussion of how to regulate reproductive and genetic technology. Much of the cloning discussion has been disappointingly superficial largely addressing two questions: Is the act of human cloning itself inherently immoral and problematic (Leon Kass and a few others see it as inherently repugnant, while Gregory Pence, Lee Silver, and others have devoted most of their efforts to rebutting this view), and Should cloning be banned?[1] Both of these questions ignore the social and economic context in which cloning will occur. The deeper, more critical, and largely ignored issues have to do with the likely uses and abuses of the technology in the situations within which it will be developed, how the technology is to be regulated and controlled within these contexts, how regulatory mechanisms in different countries will interact, and what social meaning will ultimately attach to the technology. My remarks in this chapter will largely focus on regulation in the United States because it is in this political and social context that the regulative dilemma I will discuss is most acute.

It might seem that the best way to deal with cloning is simply to pass legislation that curbs its inappropriate use. However, as Bonnicksen and others have argued, there are deep problems with this approach to regulation.[2] First, it sets a dangerous precedent. Congress would be passing legislation to ban or restrict a technology before it had even been developed, in response to a largely visceral public reaction. This risk of a serious setback for scientific and technological development is one of the primary reasons that most scientists have opposed an outright ban. A temporary ban could preempt discussion of some of the very real problems that human cloning presents. If we look to the past, it is clear that reproductive technology at one time considered dangerous (particularly if broadly used) came to seem routine over time.[3] When the public grows accustomed to human cloning, the visceral response may vanish. Without a serious discussion of the potential abuses of cloning (and how to regulate it) a temporary ban may lead to a completely unregulated technology in the long run.

Moreover, if the legislation is not crafted very carefully, it may preclude existing, widely used and extremely important techniques and therapies. At the same time, the wrong language could create loopholes that would allow human cloning to take place. As Bonnicksen has argued, legislating in the area of new, developing technologies is usually a bad idea precisely because the technology is evolving. New developments could render even carefully crafted legislation obsolete. If it is possible to transfer the nucleus of a human cell to a cow's egg, will that be included within a ban? The possibilities can never be completely anticipated—therefore legislation is the wrong mechanism for dealing with cloning.

There are other problems with regard to state regulation of reproductive and genetic technologies generally. These problems apply broadly even to nonlegislative regulation. Many would argue, for example, that state interference in reproductive decision making constitutes an unwanted and unwarranted intrusion into private matters. Schwartz-Cowan argues that once we allow state involvement in reproductive decision making, we are on a slippery slope to the loss of abortion rights. Indeed, it could be politically dangerous for advocates of abortion rights to allow any state regulation of reproductive decision making. Philip Kitcher has argued that no matter what regulation takes place, we will have to rely on the decision-making ability of individuals.[4] It is inappropriate and unwise to allow state involvement in decisions that are best left to prospective parents. Garland Allen has maintained that the history of eugenics in this century shows the dangers of state involvement in reproductive matters. Decisions that initially seem designed to further reasonable and humane purposes can be turned into dangerous measures during hard economic times. In the United States the eugenics movement culminated in the forced sterilization or more than 60,000 individuals.[5]

Partly as a result of these considerations, though probably due more to the tremendous political resistance in the United States to state intervention in almost anything, most reproductive and genetic technologies have been left to the market. However, this option seems to be as fraught with difficulty as state regulation. Troy Duster, Diane Paul, and others have warned of the dangers of "back-door" eugenics.[6] Individual decisions have collective consequences. These critics worry that obsessive concern for individual rights and patient autonomy may leave us without adequate state regulation. Communal and social welfare must be protected, and the market may be a poor way of doing that. Kavka shows how individually reasonable choices can collectively be disastrous. Feminists point to India and China, where aborting unwanted female fetuses identified through amniocentesis has resulted in unbalanced sex ratios.[7]

As Kitcher has argued, back-door eugenics is to some extent inevitable and unproblematic.[8] One of the benefits of genetic testing is the potential for parents to make more informed decisions. Those who are at risk for some of the most serious genetic disorders previously had a choice between having no children or risking the creation of a horribly diseased child. Genetic testing allows parents to abort fetuses and reduce the incidence of many of the worst diseases imaginable.

While technology can provide parents with the power to avoid bearing children with these terrible diseases, difficult questions also arise. At what point do we draw the line between those in India who abort unwanted female fetuses, and parents who (perhaps aided by a genetic counselor) choose to abort a fetus destined to die an early, painful death? And people often disagree about what constitutes a life worth living. Some patients with Huntington's disease, for example, feel that the several healthy decades of life that they have is what really matters. Other genetic traits many cause lesser health problems and risks. Will testing eventually stigmatize all those who are "unhealthy" or "abnormal" in any way? Will parents choose to test for socially important traits, such as being thin or tall? Will they test for homosexuality along with a propensity to develop heart disease?

Interestingly, defenders of the market respond to these problems by positing the existence of a well-informed, well-meaning public (Kitcher further assumes a large, public, well-financed single-payer medical system and strong socialist-styled social programs). Unfortunately, what is known about public knowledge of science in general and genetics in particular makes this assumption highly dubious. If such knowledge base is the requirement for the market to be unregulated, this solution has little bearing on the situation in the United States.

We already can see several examples of abuses that have resulted from the lack of regulation of reproductive and genetic technologies. IVF clinics in the United States regularly engage in practices that would land the "offending" clinician in prison in Britain. Genetic testing is performed in cases where it is not yet clear what the information means, potentially leading to bad decisions by patients who misunderstand their risks. Genetic testing is also often offered without adequate counseling. The specter of similar abuse is one of the reasons Dr. Richard Seed caused such a furor. The public came to the realization that despite near universal opposition there was nothing preventing him from attempting to clone (assuming, of course, that he could find reproductive clinicians to sign on). The FDA implausibly stepped into the breach to claim regulative authority. Clearly, some regulation of cloning (and other reproductive and genetic technologies) must take place.

We are left with a regulative dilemma. Neither state regulation nor an unregulated market seems to be a desirable way of dealing with reproductive and genetic technologies, including cloning. What is called for is a midlevel solution that allows us to steer a course between the Scylla of state regulation and the Charybdis of the free market.

McGee and Wilmut advocate one approach.[9] They argue that one of the problems with much of the discourse on reproductive technologies is the way it uses normal procreation as a model—making any sort of interference or regulation seem an unwarranted intrusion. Instead, they urge that adoption be considered as a model—the one area of "procreation" where it is legitimate for regulation which allows the welfare of the child be taken into account in deciding who gets to be parents. This would lead to regulation at a more local level, just as there is in adoption.

Such an approach has much to commend it, particularly as a way of conceptualizing reproduction; any discussion of regulation requires that the picture of reproductive technology as a mere extension of the bedroom must be challenged. However, there are problems with the adoption model. First, there are many flaws in the U.S. adoption system. Indeed, the decentralized nature of the system often causes problems such as the "Baby Jessica" case when different states have very different views about who the parents of a child should be. Second, it would seem that implicit in this approach is an assumption about the status of the nonexistent person. Adoption requires that we ask questions of prospective parents so that we can protect existing children. An account needs to be given to explain how this applies to regulating reproductive technologies. After all, the "children" being protected in this case have not been born (or even conceived). How can these unconceived children have rights and interests? Wouldn't this line of argument also threaten abortion rights? If we can protect the interests of these unborn children in general, it would seem that we would be hard-pressed to justify termination of a pregnancy.

The approach I will take does not necessarily preclude the adoption model or other forms of regulation. It is merely a way of taking advantage of existing institutions that can potentially help to regulate some of the more problematic consequences of the uses of genetic and reproductive technologies. Midlevel institutions, such as professional societies, patient advocacy groups, ethics advisory boards, and so on, represent the locus of the practices that are to be regulated. What I propose is a kind of self-regulation, where those with diverse interests and an actual stake in the practice reach consensus on at least some of the fundamental issues which these technologies present.

There are at least four obstacles to this approach to regulation, but each can be overcome. First, there is a risk that consensus about key issues cannot be reached. While that is certainly not unexpected when confronting controversial topics, one of the problems with the current situation (where technology is left entirely to the market) is that even particularly egregious violations of nearly universal standards are tolerated. However, institutional consensus has been reached on many issues in genetic testing, IVF implantation, and cloning, and with effort, consensus on a wider range of issues could be reached.[10]

A second problem with this approach is that guidelines put forth by the midlevel institutions often lack clarity, making them difficult to implement as a regulatory mechanism.[11] Again, however, guidelines that already exist provide evidence that at least some behaviors fall outside the range of what is acceptable. It is clear that in order for this approach to work, the institutions involved need to work harder to develop clearer recommendations, but with sufficient motivation, this would seem to be a surmountable task.

A deeper worry is the potential for conflict of interest. Many professional organizations reflect the interests of practitioners, as well as research scientists, which may not always coincide with the interests of patients or the public at large. For example, those scientists developing and running companies that sell clinical genetic testing might not put forward the best criteria for utilizing genetic testing.

Although conflict of interest is a serious concern, there are ways of blunting its force. Consensus among a number of different kinds of institutions helps to mitigate this risk. Particularly helpful will be recommendations of at least some independent midlevel institutions such as the (now defunct) National Advisory Board for the Ethics of Reproduction.

The most significant problem facing a midlevel approach to regulation is that such institutions typically lack regulative authority—there are already many areas where consensus has been reached without having any impact on behavior, which has continued to be controlled by the market.

There are two ways that midlevel institutions can have a greater role in regulating behavior. First, they can take action against their members. While that might sound like a token punishment, it is potentially much more significant. Under the Health Care Quality Improvement Act of 1986 that established the National Practitioners Data Base, all actions taken by professional societies against physicians are reportable. Any time a hospital, HMO, or licensing group does a routine check, the sanction by the professional society will be revealed, providing an important disincentive to flout the group's standards. Unfortunately, professional societies have refused to utilize the power they have. Between 1992 and 1996, there were nearly 20,000 "malpractice payments" per year made by physicians. Several thousand reportable actions and licensures took place each year. In contrast, there were fewer than 59 actions taken by professional societies in any of these years (NPDB).

State medical licensing boards could also take action against clinicians that act in ways that are universally acknowledged to be unethical. Conforming to ethical standards must be seen as an integral part of good medical practice. It is just as vital that clinicians police themselves over their ethical conduct as over conforming to universally recognized standards of care. If no standard of care has been established, there can be no breach. Yet it seems that too often, if a standard of care has not been violated, and no law is broken, no action will be taken against a clinician. This is especially problematic in genetic and reproductive technologies where no clear standards of care have emerged and where (as we have seen) there is little regulation, hence little legal restriction.

It is important that medical licensing boards be more aggressive in their enforcement of ethical standards. There are too many prominent cases where universally agreed-upon standards are flouted with no action taken.[12] Imposition of sanctions by state medical licensing boards and professional societies would help to reinforce the notion that ethical standards are as integral a part of the practice of medicine as prescribing appropriate dosage of a drug or ordering the right tests. It is unrealistic and undesirable for self-regulation of the profession to be taken to extremes or turn into self-policing. But once it is clear that ethical standards will be enforced, informal pressure—which plays such an important role in the governance of other aspects of clinical behavior and enforces norms about the standard of care—will similarly be brought to bear. For example, IVF practitioners who implant a large number of embryos would face pressure from their colleagues to conform to universally recognized standards of behavior and should fear loss of their medical license.

How would this approach deal with the issues raised by cloning? There are many points of controversy in the cloning debate. But again, there are certain areas where consensus has already been reached and areas where it could presumably emerge with respect to both short-and long-term issues. In the short term, there is agreement that human cloning is unsafe and constitutes unacceptable risks at the present time and presumably for the next five to ten years. Minimal preconditions for an ethically defensible attempt at human cloning would include more experience and success with animal cloning and a better knowledge of possible risks. This was the one issue on which NBAC was able agree. NBAC recommended that there be a moratorium on attempting to create a child using somatic cell nuclear transfer. This moratorium would be mandatory for any institutions using federal funds and voluntary for all others. The ban would have a sunset clause so that once the short-term safety issues are addressed, cloning could be allowed.

The midlevel approach could address such safety concerns. If Dr. Seed or some other ambitious person recruited a physician for the purpose of cloning a human—while it is still unsafe—that clinician would face the risk of action by professional societies and by state licensing boards, up to the loss of medical license. The possibility of such sanctions should be sufficient to prevent any clinician from engaging in such obviously problematic behavior.

With respect to long-term issues there are at least two different sources of concern. First, there are potential problems that emerge if the technology were to become commercialized. What would happen if a company were set up to clone humans? Would there be an existing market? If not, would cloning be advertised in order to create one? Would companies prey on the fears and pain of parents who have lost a beloved child, by making promises—such as being able to bring that child back—that cannot of course, be realized ("we can bring your child back"). Midlevel institutions would need to reach consensus on ethically appropriate advertising techniques. And far beyond the marketing realm, midlevel institutions could build consensus as to the limits of acceptable behavior. And they could take action against clinicians who choose to be a part of unethical organizations. Again, the goal would not be to police each practitioner, but to create a climate of informal social enforcement.

Another long-term concern involves harm to children produced by cloning. For example, a man or woman seeking to be cloned with the goal of creating an exact duplicate of themselves may not be a fit parent. In addition, the burden of the expectations being imposed on a child created to be a "copy" might be unduly burdensome. If agreement can be reached with respect to limitations on access to cloning technology, then guidelines could be developed for clinicians.

The midlevel approach could address and potentially solve a number of problems raised by the application of cloning technology to humans. But the approach does have limitations. Issues about which there is no hope of reaching consensus cannot be regulated by midlevel institutions or by any network of social enforcement that these institutions foster. These more intractable problems may require other regulative solutions. Or they may simply remain intractable.

NOTES

1. Kass, L. *The Wisdom of Repugnance*; Pence, G. *Who's Afraid of Human Cloning?* Oxford: Rowman and Littlefield, 1998; Silver, L. *Remaking Eden: Cloning and Beyond in a Brave New World*. New York: Avon Books, 1997.

2. Bonnicksen, A. Chapter 25 of this volume; and Bonnicksen, A. Procreation by Cloning: Crafting Anticipatory Guidelines. *Journal of Law, Medicine, and Ethics* 25, 1977: 273–282.

3. Schwartz Cowan, R. Genetic Technology and Reproductive Choice: An Ethics for Autonomy. In *The Code of Codes*, Ed. Kevles and L. Hood, Cambridge, Mass.: Harvard University Press, 1992.

4. Kitcher, P. *The Lives to Come: The Genetic Revolution and Human Possibilities*. New York: Simon and Schuster, 1996.

5. Allen, G. Eugenics and Genetics. *Technology Review*, September 1996.

6. Paul, D. *Controlling Human Heredity*. New Jersey: Humanities Press, 1995; Duster, T. *Backdoor to Eugenics*. London: Routledge, 1990.

7. Kavka, G. Upside Risks. In *Are Genes Us?* Ed. C. Cranor. New Brunswick, N.J.: Rutgers University Press, 1994.

8. Kitcher, P. *The Lives to Come: The Genetic Revolution and Human Possibilities*. New York: Simon and Schuster, 1996.

9. McGee, G., and Wilmut, I. Cloning and the Adoption Model. In *The Human Cloning Debate*, Ed. G. McGee. Berkeley: Berkeley Hills Books, 1998.

10. Moreno, J. Deciding Together: *Bieothics and Moral Consensus*. Oxford: Oxford University Press, 1995.

11. Sankar, P., Wolpe, R., and Cho, M. Unpublished manuscript.

12. Magnus, forthcoming.

23

Mom, Dad, Clone

Implications for Reproductive Privacy

Lori Andrews

On July 5, 1996, a sheep named Dolly was born in Scotland, the result of the transfer of the nucleus of an adult mammary tissue cell to the enucleated egg cell of an unrelated sheep, and gestation in a third, surrogate mother sheep.[1] Although for the past ten years scientists have routinely cloned sheep and cows from embryo cells,[2] this was the first cloning experiment which apparently succeeded using the nucleus of an adult cell.[3]

Seventeen months after Dolly's birth was announced, researchers at the University of Hawaii announced they had produced 50 cloned mice, representing three generations.[4] The first of these cloned mice, Cumulina—named because the adult cell used to create her was a cumulus cell—was born on October 3, 1997, by using a technique similar to that used to create Dolly. Of significance, however, is that the team at the University of Hawaii used a chemical bath instead of an electric shock to fuse the cell and the enucleated egg cell.

Though human cloning has yet to occur, President Clinton, as well as several states and foreign nations,[5] took action to prevent its potential use. In 1997 President Clinton issued an executive order banning the use of federal funds for human cloning.[6] The president also asked the National Bioethics Advisory Commission ("NBAC") to submit a comprehensive report on the scientific, ethical and legal issues raised by human cloning. That report, issued in July of 1997, recommended that Congress enact federal legislation to ban the creation of a child through cloning for three to five years, regardless of the source of funds.[7]

The bill President Clinton subsequently introduced failed but four states—California, Louisiana, Michigan and Rhode Island—passed legislation that prohibits human cloning.[8] The California statute makes it a violation of the Medical Practices Act for any corporation, firm, clinic, hospital, laboratory or research facility to clone a human being with civil penalties of up to $1,000,000[9] and loss of license.[10] An individual who violates the cloning prohibition in Michigan may be guilty of a felony punishable by imprisonment for up to ten years and/or a fine of up to $10,000,000[11] in addition to a $10,000,000 civil fine.[12] While the cloning statues in California, Louisiana, and Rhode Island are temporary,[13] the statute in Michigan does not include a sunset provision.

All four of the state cloning statutes contain language that would allow cloning for research purposes.[14] Using nearly identical language, the Louisiana and Michigan statutes both assert that the prohibition on human cloning does not apply to "scientific research" or "cell-based" therapies not specifically prohibited elsewhere.[15] The California legislation, after which the Rhode Island statute was modeled, specifically stated that it did not intend the moratorium on cloning to apply to the cloning of human cells, tissue, or organs that would "not result in the replication of an entire human being."[16]

The state laws banning human cloning may be outpaced by advances in technology. California's statute, for example, prohibits cloning based on a definition of cloning that specifically uses the term "human" when referring to enucleated eggs.[17] This may prove problematic in light of University of Wisconsin researchers' purported success in using enucleated cow eggs as incubators for the nucleic DNA of other mammalian species.[18] If a cow's enucleated egg were used instead of a human's, the California law, as written, might be evaded.

As of August 2000, six states had cloning bans pending—Illinois, Massachusetts, New York, New Jersey, Ohio, and Oklahoma. In New York, for example, a bill to amend its public health law has been introduced that would ban human cloning and make a violation of such a prohibition a felony.[19]

At the federal level, Representative Stearns introduced H.R. 2326 to prohibit the use of federal funds for cloning human beings, and Representative Paul introduced H.R. 571 to prohibit federal payment to entities that engaged in human cloning in the past year. Senator Specter, in S. 2015, in a bill that would allow the use of federal funds for embryo stem cell research, but prohibiting reproductive cloning. However, laws that ban human cloning might be challenged as violating an individual's or a couple's right to create a biologically related child. This article explores whether such a right exists and whether, even if it does, a ban on creating children through cloning should be upheld.

THE RIGHT TO MAKE REPRODUCTIVE DECISIONS

The right to make decisions about whether or not to bear children is constitutionally protected under the constitutional right to privacy[20] and the constitutional right to liberty.[21] The U.S. Supreme Court in 1992 reaffirmed the "recognized protection accorded the liberty relating to intimate relationships, the family, and decisions about whether or not to beget or bear a child."[22] Early decisions protected married couples' right to privacy to make procreative decisions, but later decisions focused on individuals' rights as well. The U.S. Supreme Court, in *Eisenstadt* v. *Baird*, stated, "If the right of privacy means anything, it is the right of the *individual*, married or single, to be free from unwarranted governmental intrusion into matters so fundamentally affecting a person as the decision whether to bear or beget a child."[23]

A federal district court has indicated that the right to make procreative decisions encompasses the right of an infertile couple to undergo medically assisted reproduction, including in vitro fertilization and the use of a donated embryo. *Lifchez* v. *Hartigan*[24] held that a ban on research on conceptuses was unconstitutional because it impermissibly infringed upon a woman's fundamental right to privacy. Although the Illinois statute banning embryo and fetal research at issue in the case permitted in vitro fertilization, it did not allow embryo donation, embryo freezing, or experimental prenatal diagnostic procedures. The court stated:

> It takes no great leap of logic to see that within the cluster of constitutionally protected choices that includes the right to have access to contraceptives, there must be included within that cluster the right to submit to a medical procedure that may bring about, rather than prevent, pregnancy. Chorionic villi sampling is similarly protected. The cluster of constitutional choices that includes the right to abort a fetus within the first trimester must also include the right to submit to a procedure designed to give information about the fetus which can then lead to a decision to abort.[25]

Procreative freedom has been found to protect individuals' and couples' decisions to use contraception, abortion, and existing reproductive technology. Some commentators argue that the U.S. Constitution similarly protects the right to create a child through cloning.

There are a variety of scenarios in which such a right might be asserted. If both members of a couple are infertile, they may wish to clone one or the other of themselves.[26] If one member of the couple has a genetic disorder that the couple does not wish to pass on to a child, they could clone the unaffected member of the couple. In addition, if both husband and wife are carriers of a debilitating recessive genetic disease and are unwilling to run the 25% risk of bearing a child with the disorder, they may seek to clone one or the other of them.[27] This may be the only way in which the couple will be willing to have a child that will have any genetic relationship to them.

Even people who could reproduce coitally may desire to clone for a variety of reasons. People may want to clone themselves, deceased or living loved ones, or individuals with favored traits. A wealthy childless individual may wish to clone himself or herself to have an heir or to continue to control a family business. Parents who are unable to have another child may want to clone their dying child.[28] This is similar to an existing situation in which a couple whose daughter died arranged to have a cryopreserved in vitro embryo created with her egg and donor sperm implanted in a surrogate mother in an attempt to re-create the daughter;[29] that particular attempt was unsuccessful.

Additionally, an individual or couple might choose to clone a person with favored traits. Respected world figures and celebrities such as Mother Teresa, Michael Jordan, and Michelle Pfeiffer have been suggested as candidates for cloning. Less well-known individuals could also be cloned for specific traits. For example, people with a high pain threshold or resistance to radiation could be cloned.[30] People who can perform a particular job well, such as soldiers, might be cloned.[31] One biologist suggested cloning legless men for the low gravitational field and cramped quarters of a space ship.[32]

Cloning also offers gay individuals a chance to procreate without using nuclear DNA from a member of the opposite sex. Clone Rights United Front, a group of gay activists based in New York, have been demonstrating against a proposed New York law that would ban nuclear transplantation research and human cloning. They oppose such a ban because they see human cloning as a significant means of legitimizing "same-sex reproduction."[33] Randolfe Wicker founded the Clone Rights United Front in order to pressure legislators not to ban human cloning research because he sees nuclear transplantation cloning as an inalienable reproductive right.[34] Wicker stated, "We're fighting for research, and we're defending people's reproductive rights. . . . I realize my clone would be my identical twin, and my identical twin has a right to be born."[35]

Ann Northrop, a columnist for the New York gay and lesbian newspaper *LGNY*, says that nuclear transplantation is enticing to lesbians because it offers them a means of reproduction that allows the potential for both women to contribute genetically in the form of mitochondrial DNA and nuclear DNA and has the potential of giving women complete control over reproduction.[36] "This is sort of the final nail in men's coffins," she says. "Men are going to have a very hard time justifying their existence on this planet, I think. Maybe women may not let men reproduce."[37]

The strongest claim for procreative freedom is that made by infertile individuals, for whom this is the only way to have a child with a genetic link to them. However, the number of people who will actually need cloning is quite limited. Many people can be helped by in vitro fertilization and its adjuncts; others are comfortable with using a donated gamete. In all the other instances of creating a child through cloning, the individual is biologically able to have a child of his or her own, but is choosing not to because he or she prefers to have a child with certain traits. This made-to-order child making is less compelling than the infertility scenario. Moreover, there is little legal basis to suggest that a person's procreative freedom includes a right to procreate using *someone else's* DNA, such as that of a relative or a celebrity. Courts are particularly unlikely to find that parents have a right to clone their young child. Procreative freedom is not a predatory right that would provide access to another individual's DNA.

The right of procreation is likely to be limited to situations in which an individual is creating a biologically related child. It could be argued that cloning oneself invokes that right to an even greater degree than normal reproduction. As lawyer Francis Pizzulli points out, "In comparison with the parent who contributes half of the sexually reproduced child's genetic formula, the clonist is conferred with more than the requisite degree of biological parenthood, since he is the sole genetic parent."[38]

John Robertson argues that cloning is not qualitatively different from the practice of medically assisted reproduction and genetic selection that is currently occurring.[39] Consequently, he argues that "cloning . . . would appear to fall within the fundamental freedom of married couples, including infertile married couples, to have biologically related offspring."[40] Similarly, June Coleman argues that the right to make reproductive decisions includes the right to decide in what manner to reproduce, including reproduction through, or made possible by, embryo cryopreservation and embryo twinning.[41] This argument could also be applied to nuclear transplantation by saying that a ban on cloning as a method of reproduction is tantamount to the state denying one's right to reproductive freedom.

In contrast, George Annas argues that cloning does not fall within the constitutional protection of reproductive decisions. "Cloning is replication, not reproduction, and represents a difference in kind, not in degree, in the way humans continue the species."[42] He explains that "[t]his change in kind in the fundamental way in which humans can 'reproduce' represents such a challenge to human dignity and the potential devaluation of human life (even comparing the 'original' to the 'copy' in terms of which is to be more valued) that even the search for an analogy has come up empty handed."[43]

The process and resulting relationship created by cloning are profoundly different from those created through normal reproduction or even through reproductive technologies such as in vitro fertilization, artificial insemination, or surrogate motherhood. In even the most high-tech reproductive technologies available, a mix of genes occurs to create an individual with a genotype that has never before existed on earth. In the case of twins, two such individuals are created. Their futures are open, and the distinction between themselves and their parents is acknowledged. In the case of cloning, however, the genotype has already existed. Even though it is clear that the individual will develop into a person with different traits because of different social, environmental, and generational influences, there is evidence that the fact that he or she has a genotype that already existed will affect how the resulting clone is treated by his family, social institutions and even himself.

In that sense, cloning is sufficiently distinct from traditional reproduction or alternative reproduction to not be constitutionally protected. It is not a process of genetic mix, but of genetic duplication. It is not reproduction, but a sort of recycling, where a single individual's genome is made into someone else.

ASSUMING CONSTITUTIONAL PROTECTION

Let us assume, though, that courts were willing to make a large leap and find that the constitutional privacy and liberty protections of reproduction encompass cloning. If a constitutional right to clone were recognized, any legislation which would infringe unduly upon this fundamental right would be subject to a "strict standard" of judicial review.[44] Legislation prohibiting the ability to clone or prohibiting research would have to further a compelling interest in the least restrictive manner possible in order to survive this standard of review.[45]

The potential physical and psychological risks of cloning an entire individual are sufficiently compelling to justify banning the procedure. There are many physical risks to the resulting child. Of the first reported successful somatic cell nuclear transplantations cloning, of 277 attempts, only one sheep—Dolly—lived. Although the University of Hawaii team achieved a slightly higher success rate—ranging from 1 in 40 to 1 in 80 survivors for every cloned embryo implanted in different experiments,[46] Ryuzo Yanagimachi, whose laboratory produced the cloned mice, cautions that although his team achieved a higher success rate, the efficacy rate is still modest and noted that the same technique did not achieve any male mice clones, only female mice clones.[47]

The high rate of laboratory deaths may suggest that cloning in fact damages the DNA of a cell. In addition, scientists urge that Dolly should be closely monitored for abnormal genetic anomalies which did not kill her as a fetus but may have long-term harmful effects.[48]

For example, all of the initial frog cloning experiments in the 1950 and 1960s succeeded only to the point of the amphibian's tadpole stage.[49] In addition, some of the tadpoles were grossly malformed.[50] More recently, when the Grenada Corporation in Texas began the cloning of cows[51] from differentiated embryonic cells, some of the cloned calves were abnormally large.[52] Some weighed up to 180 pounds at birth, more than twice the normal 75-pound birth weight of this breed.[53] Also, some of these calves were born with diseases such as diabetes and enlarged hearts, and 18% to 20% of these calves simply died after birth.[54]

The scientific team that created Dolly has also met with unsatisfactory results. After cloning Dolly, they used fetal cells to create cloned, transgenic animals.[55] In this experiment, the team successfully transferred the DNA from fetal sheep cells into 425 enucleated sheep eggs.[56] Of those attempted fusions, however, only fourteen resulted in pregnancy, and only six lambs were born alive.[57] Labor was artificially induced in all of the surrogate ewes, and in some instances the lambs were delivered by cesarean section.[58] Some of these cloned lambs weighed nearly twice the average amount.[59]

Initial trials in human nuclear transplantation could also meet with disastrous results.[60] Ian Wilmut and National Institutes of Health director Harold Varmus, testifying before Congress, specifically raised the concern that cloning technology is not scientifically ready to be applied to humans, even if it were permitted, because there are technical questions that can be answered only by continued animal research.[61] Dr. Wilmut is specifically concerned with the ethical issue which would be raised by any "defective births" which may be likely to occur if nuclear transplantation is attempted with humans.[62] Dr. Wilmut responded to the announcement that Dr. Richard Seed intended to clone human beings within the next two years by stating: "Let me remind you that one-fourth of the lambs born in our experiment died within days of birth. Seed is suggesting that a number of humans would be born but others would die because they didn't properly develop. That is totally irresponsible."[63]

In addition, if all the genes in the adult DNA are not properly reactivated, there might be a problem at a later developmental stage in the resulting clone.[64] Some differentiated cells rearrange a subset of their genes. For example, immune cells rearrange some of their genes to make surface molecules.[65] That rearrangement could cause profound physical problems for the resulting clone.

Also, because scientists do not fully understand the cellular aging process, scientists do not know what "age" or "genetic clock" Dolly inherited.[66] On a cellular level, when the *Nature* article was published about her, was she a normal seven-month-old lamb, or was she six years old (the age of the mammary donor cell)?[67] Colin Stewart believes that Dolly's cells most likely are set to the genetic clock of the nucleus donor, and therefore are comparable to those of her six-year-old progenitor.[68] One commentator stated that if the hypotheses of a cellular, self-regulating genetic clock are correct, clones would be cellularly programmed to have much shorter life spans than the "original," which would seriously undermine many of the benefits which have been set forth in support of cloning—mostly agricultural justifications—and would psychologically lead people to view cloned animals and humans as short-lived, disposable copies. This concern for premature aging has lead Dr. Sherman Elias, a geneticist and obstetrician at the

Baylor College of Medicine, to call for further animal testing of nuclear transplantation as a safeguard to avoid subjecting human clones to premature aging and the potential harms associated with aged cells.[69]

The aging process of clones is being examined in various studies. While Dolly appears to be normal and healthy, a study found her telomeres to be unusually short.[70] Unlike Dolly, however, six calves that were cloned using a slightly different method than the one used to create Dolly, appear to have telomeres that are actually longer than normal.[71] It is currently unknown whether either shorter or longer telomeres will ultimately have any impact on longevity or whether any such findings would be transferable to humans.[72]

The hidden mutations that may be passed on by using an adult cell raise concerns as well. Mutations are "a problem with every cell, and you don't even know where to check for them," notes Ralph Brinster of the University of Pennsylvania.[73] "If a brain cell is infected with a mutant skin gene, you would not know because it would not affect the way the cell develops because it is inactive. If you chose the wrong cell, then mutations would become apparent."[74]

WHEN PHYSICAL RISKS DECLINE

Three of the four state bans on cloning—those of California, Louisiana, and Rhode Island—put a five-year moratorium on creating a child through cloning, which will expire in 2003. During that time period, though, the physical risks of cloning will probably diminish. Animal researchers around the world are rushing to try the nuclear transfer technique in a range of species. If cloning appeared to be physically safe and reached a certain level of efficiency, should it then be permissible in humans?

The NBAC recommendations left open the possibility of continuing the ban on human cloning based on psychological and social risks.[75] The notion of replicating existing humans seems to fundamentally conflict with our legal system, which emphatically protects individuality and uniqueness.[76] Banning procreation through nuclear transplantation is justifiable in light of the sanctity of the individual and personal privacy notions that are found in different constitutional amendments, and protected by the Fourteenth Amendment.[77]

A clone will not have the ability to control disclosure of intimate personal information.[78] A ban on cloning would "preserve the uniqueness of man's personality and thus safeguard the islands of privacy which surround individuality."[79] These privacy rights are implicated through a clone's right to "retain and control the disclosure of personal information—foreknowledge of the clonant's genetic predispositions."[80] Catherine Valerio Barrad argues that courts should recognize a privacy interest in one's DNA because science is increasingly able to decipher and gather personal information from one's genetic code.[81] The fear that potential employers and health insurers may use one's private genetic information discriminatorily is not only a problem for the original DNA possessor, but any clone "made" from that individual.[82] Even in cases where the donor waives his privacy rights and releases genetic information about himself, the privacy rights of the clone are necessarily implicated due to the fact that the clone possesses the same nucleic genetic code.[83] This runs afoul of principles behind the Fifth Amendment's protection of a "person's ability to regulate the disclosure of information about himself."[84]

If a cloned person's genetic progenitor were a famous musician or athlete, parents might exert an improper amount of coercion to get the child to develop those talents. True, the same thing may happen—to a lesser degree—to a child now, but the cloning scenario is more problematic. A parent may force a naturally conceived child to practice piano hours on end, but will probably eventually give up if the child seems uninterested or tone deaf. More fervent attempts to develop the child's musical ability would likely occur if the parents chose (or even paid for) nuclear material from a talented pianist. And pity the poor child cloned from a famous basketball player. If he were to break his kneecap at age ten, would his parents consider him worthless? Would he consider himself a failure?

In attempting to cull out from the resulting child the favored traits of the loved one or celebrity who has been cloned, the social parents would probably limit the environmental stimuli that the child is exposed to. The pianist's clone might not be allowed to play baseball or just hang out with other children. The clone of a dead child might not be exposed to food or experiences that the first child had rejected. The resulting clone could be viewed as being in a type of "genetic bondage"[85] with improper constraints on his or her freedom.

Some scientists argue that this possibility will not come to pass because everyone knows that a clone will be different than the original. The NBAC report puts it this way: "Thus the idea that one could make through somatic cell nuclear transfer a team of Michael Jordans, a physics department of Albert Einsteins, or an opera chorus of Pavarottis, is simply false."[86] But this overlooks the fact that we are in an era of genetic determinism, in which newspapers daily report the gene for this or that, and top scientists tell us that we are a packet of genes unfolding.

James Watson, codiscover of deoxyribonucleic acid (DNA) and the first director of the Human Genome Project, has stated, "We used to think our fate was in the stars. Now we know, in large measure, our fate is in our genes."[87] Harvard zoologist Edward O. Wilson asserts that the human brain is not *tabula rasa* later filled in by experience, but "an exposed negative waiting to be slipped into developer fluid."[88] Genetics is alleged to be so important by some scientists that it prompted psychiatrist David Reiss at George Washington University to declare that "the Cold War is over in the nature and nurture debate."[89]

Whether or not this is true, parents may raise the resulting clone as if it were true. After all, the only reason people want to clone is to assure that the child has a certain genetic makeup. Thus, it seems absurd to think they will forget about that genetic makeup once the child comes into being. Elsewhere in our current social policies, though, we limit parents' genetic foreknowledge of their children because we believe it will improperly influence their rearing practices.

Cloning could undermine human dignity by threatening the replicant's sense of self and sense of autonomy. A vast body of developmental psychology research has signaled the need of children to have a sense of an independent self. This might be less likely to occur if they were the clone of a member of the couple raising them or of previous children who died.

The replicant individual may be made to feel that he is less of a free agent. Under such an analysis, it does not matter whether or not genetics actually determines a person's characteristics. Having a predetermined genetic makeup can be limiting if the person rearing the replicant or the replicant himself believes in genetic determinism.[90] In addition, there is much research on the impact of genetic information that demonstrates that a person's genetic foreknowledge about himself or herself (whether negative or positive) can threaten that individual's self-image, harm his or her relationships with family members, and motivate social institutions to discriminate against him or her.[91]

Even though parents have a constitutional right to make child-rearing decisions similar to their constitutional right to make childbearing decisions, parents do not have a right to receive genetic information about their children that is not of immediate medical benefit. The main concern is that a child about whom genetic information is known in advance will be limited in his horizons. A few years ago, a mother entered a Huntington's disease testing facility with her two young children. "I'd like you to test my children for the HD gene," she said. "Because I only have enough money to send one to Harvard."[92] That request and similar requests to test young girls for the breast cancer gene or other young children for carrier status for recessive genetic disorders raise concerns about whether parents' genetic knowledge about their child will cause them to treat that child differently. A variety of studies have suggested that there may be risks to giving parents such information.

" 'Planning for the future,' perhaps the most frequently given reason for testing, may become 'restricting the future' (and also the present) by shifting family resources away from a child with a positive diagnosis," wrote Dorothy Wertz, Joanna Fanos, and Philip Reilly in an article in the *Journal of the American Medical Association*.[93] Such a child "can grow up in a world of limited horizons and may be psychologically harmed even if treatment is subsequently found for the disorder."[94] A joint statement by the American Society of Human Genetics (ASHG) and the American College of Medical Genetics (ACMG) notes, "Presymptomatic diagnosis may preclude insurance coverage or may thwart long term goals such as advanced education or home ownership."[95]

The possibility that genetic testing of children can lead to a dangerous self-fulfilling prophecy led to the demise of one study involving testing children. Harvard researchers proposed to test children to see if they had the XYY chromosomal complement, which had been linked (by flimsy evidence) to criminality. They proposed to study the children for decades to see if those with that genetic makeup were more likely to engage in a crime than those without it. They intended to tell the mothers which children had XYY. Imagine the effect of that information—on the mother, and on the child. Each time the child took his little brother's toy, or lashed out in anger at a playmate, the mother might freeze in horror as the idea that her child's genetic predisposition was unfolding itself. She might intervene when other mothers would normally not, and thus distort the rearing of her child.

Because of the potential psychological and financial harm that genetic testing of children may cause, a growing number of commentators and advisory bodies have recommended that parents not be able to learn genetic information about their children. The Institute of Medicine Committee on Assessing Genetic Risks recommended that "in the clinical setting, children generally be tested only for disorders for which a curative or preventive treatment exists and should be instituted at that early stage. Childhood screening is not appropriate for carrier status, untreatable childhood diseases, and late-onset diseases that cannot be prevented or forestalled by early treatment."[96] The American Society of Human Genetics and American College of Medical genetics made similar recommendations.

A cloned child will be a child who is likely to be exposed to limited experiences and limited opportunities. Even if he or she is cloned from a person who has favored traits, he may not get the benefit of that heritage. His environment might not provide him with the drive that made the original succeed. Or so many clones may be created from the favored original that their value and opportunities are lessened. (If the entire NBA consisted of Michael Jordan clones, the game would be far less interesting and each individual less valuable.) In addition, even individuals with favored traits may have genes associated with diseases that could lead to insurance discrimination against the individuals cloned. If Jordan were to die young of an inheritable cardiac disorder, his clones would find their futures restricted. Banning cloning would be in keeping with philosopher Joel Feinberg's analysis that children have a right to an "open future."[97]

Some commentators argue that potential psychological and social harms from cloning are too speculative to provide the foundation for a governmental ban. Elsewhere, I have argued that speculative harms do not provide a sufficient reason to ban reproductive arrangements such as in vitro fertilization or surrogate motherhood.[98] But the risks of cloning go far beyond the potential psychological risks to the original whose expectations are not met in the cloning, or the risks to the child of having an unusual family arrangement if the original were not one of his or her rearing parents.

The risk here is of hubris, of abuse of power. Cloning represents the potential for "[a]buses of the power to control another person's destiny—

both psychological and physical—of an unprecedented order."[99] Francis Pizzulli points out that legal discussions of whether the replicant is the property of the cloned individual, the same person as the cloned individual, or a resource for organs all show how easily the replicant's own autonomy can be swept aside.[100]

In that sense, maybe the best analogy to cloning is incest. Arguably, reproductive privacy and liberty are threatened as much by a ban on incest as by a ban on cloning. Arguably the harms are equally speculative. Yes, incest creates certain potential physical risks to the offspring, due to the potential for lethal recessive disorders to occur. But no one seriously thinks that this physical risk is the reason we ban incest. A father and daughter could avoid that risk by contracepting or agreeing to have prenatal diagnosis and abort affected fetuses. There might even be instances in which, because of their personalities, there is no psychological harm to either party.

Despite the fact that risks are speculative—and could be counterbalanced in many cases by other measures—we ban incest because it is about improper parental power over children. We should ban the cloning of human beings through cloning—even if physical safety is established—for that same reason.

NOTES

1. See Michael Specter with Gina Kolata, "A New Creation: The Path to Cloning—A Special Report," *The New York Times*, March 3, 1997, at A1.

2. In 1993, embryologists at George Washington University split human embryos, making twins and triplets. See Kathy Sawyer, "Researchers Clone Human Embryo Cells; Work Is Small Step in Aiding Infertile," *The Washington Post*, Oct. 25, 1993, at A4. These embryos were not implanted into a woman for gestation. This procedure is distinguishable from cloning by nuclear transfer.

3. See Sharon Begley, "Little Lamb, Who Made Thee?," *Newsweek*, March 10, 1997, at 53–57. Seventeen months later, researchers at the University of Hawaii announced they cloned 50 mice from adult cells, representing three generations. Although their technique differed from that used to clone Dolly, the experiment proved once again that cloning can be achieved using adult differentiated cells. See Michael D. Lemonick, "Dolly, You're History," *Time*, Aug. 3, 1998, at 64. See also T. Wakayama, A. C. F. Perry, M. Zuccotti, K. R. Johnson, and R. Yanagimachi, "Full Term Development of Mice from Enucleated Oocytes Injected with Cumulus Cell Nuclei," 394 *Nature* 369 (July 1998). In July 1998, two Japanese researchers also announced that they had cloned eight calves using somatic cell nuclear transplantation cloning. See Yoko Kato, Tetsuya Tahi, Yusuke Sotomaru, Jun-ya Kato, Hiroshi Yasue, Yukio Tsunoda, "Eight Calves Cloned from Somatic Cells of Single Adult," 282 *Science* 2095, December 11, 1998. See Ronald Kotulak, "New Method May Advance Cloning Research," *Chicago Tribune*, July 24, 1998, at 4. See also I. Wilmut, A. E. Schnieke, J. McWhir,

A. J. Kind, and K. H. S. Campbell, "Viable Offspring Derived from Fetal and Adult Mammalian Cells," 385 *Nature* 810–813 (1997).

4. See T. Wakayama et al., *supra* note 3. See also Rick Weiss, "Scientists Clone Adult Lab Mice," *The Washington Post*, July 23, 1998, at A1.

5. *See Additional Protocol to the Convention for the Protection of Human Rights and Dignity of the Human Being*, Council of Europe, ETS No. 168, Paris, Jan. 12, 1998, *available at* <http://www.coe.fr/eng/legaltxt/168e.htm>.

6. *See* "Transcript of Clinton Remarks on Cloning," *U.S. Newswire*, Mar. 4, 1997, *available at* 1997 WL 571115.

7. *See* Letter from Dr. Harold T. Shapiro, Chairman, National Bioethics Advisory Commission, to President William J. Clinton (June 9, 1997), reprinted in National Bioethics Advisory Commission, *Cloning Human Beings: Report and Recommendations of the National Bioethics Advisory Commission*, at preface (1997), *available at*<http://bioethics.gov/pubs.html>.

8. Cloning Prohibition Act of 1997, H. R. Doc. No. 105–97 (1997). For the state laws banning cloning, *see* Cal. Health & Safety Code Ann. § 24185; LA R. S. 40:1299.36.2; Mich. Comp. Laws Ann. § 750.430a; R. I. Gen. Laws § 23–16.4.

9. § 24187(c) also allows the penalty to be measured by not more than two times the amount of any gross pecuniary gain if it is greater than the $1,000,000 cap set in § 24187(a). If the violator is an individual, he or she may be subject to a civil fine of up to $250,000. *See also*, R. I. Gen Laws 1956, § 23–16.4–3 modeled after the California statue.

10. Cal. Health & Safety Code Ann. § 16105.

11. Mich. Comp. Laws. Ann. § 750.430a(3).

12. Mich. Comp. Laws. Ann. § 333.16274(3)

13. The California statute is automatically repealed on January 1, 2003 unless another statute deletes or extends that date, Cal. Health & Safety Code § 24189. Similarly, the Rhode Island statute remains in effect until July 7, 2003, R. I. Gen Laws 1956, § 23–16.4–4 and Louisiana's statute expires after July 1, 2003, LA R.S. 40:1299.36.2.

14. *See* Cal. Bus. & Prof. Code § 2260.5; LA R.S. 40:1299.36.2(C); Mich. Comp. Laws. Ann. § 750.430a(2); R. I. Gen Laws § 23–16.4–1.

15. *See* LA R. S. 40:1299.36.2(C); Mich. Comp. Laws. Ann. § 750. 430a(2).

16. Cal. Bus. & Prof. Code § 2260.5. A twelve-member Human Cloning Advisory Committee was appointed by the California Department of Health Services to evaluate the implications of human cloning and report its findings to the legislature by December 31, 2001, press release *available at* <http://www.dhs.ca.gov; shopa/prssrels/1999/03–99.htm>.

17. § 24185 states in relevant part, . . . " 'clone' means the practice of creating or attempting to create a human being by transferring the nucleus from a human cell from whatever source into a human egg cell from which the nucleus has been removed. . . ."

18. *See* James A. Thomson et al, "Embryonic Stem Cell Lines Derived from Human Blastocysts," 282 *Science* 1145, Nov. 6, 1998. *See also*, Robert Lee Holtz, "Cow Eggs Used as Incubator in Cloning Boon," *L.A. Times*, Jan. 19, 1998 at A1.

19. S. B. 6538, 223rd Leg. (N.Y. 1999). See also A. B. 8341, 222nd Leg. (N.Y. 1999–2000) making a violation of the Cloning Prohibition and Research Protection Act punishable by a civil fine of $250,000.

20. See, e.g., *Griswold* v. *Connecticut*, 381 U.S. 379 (1965); *Eisenstadt* v. *Baird*, 405 U.S. 438 (1972).

21. *Planned Parenthood* v. *Casey*, 505 U.S. 833.

22. *Planned Parenthood* v. *Casey*, 505 U.S. 833, 857.

23. *Eisenstadt* v. *Baird*, 405 U.S. 438, 453 (1972).

24. 735 F.Supp. 1361 (N.D. Ill. 1990), *aff'd without opinion, sub nom.*, *Scholberg* v. *Lifchez*, 914 F.2d 260 (7th Cir. 1990), *cert. denied*, 498 U.S. 1068 (1991).

25. *Ibid.* at 1377 (citations omitted). The court also held that the statute was impermissibly vague because of its failure to define "experiment" or "therapeutic." *Id.* at 1376.

26. See Herbert Wray, Jeffrey L. Sheler, Traci Watson, "The World After Cloning," *U.S. News and World Report*, March 10, 1997, at 59.

27. Jay Katz, *Experimentation with Human Beings* 977 (1972).

28. Willard Gaylin, "We Have the Awful Knowledge to Make Exact Copies of Human Beings," *New York Times Magazine*, March 5, 1972, 56 at 48.

29. Rick Weiss, "Babies in Limbo," *Washington Post*, February 8, 1998, A2.

30. J. B. S. Haldane, "Biological Possibilities for the Human Species in the Next Thousand Years," in *Man and His Future* 337 (ed. G. Wolstenholme, 1963), cited in Francis C. Pizzulli, Note, "Asexual Reproduction and Genetic Engineering: A Constitutional Assessment of the Technology of Cloning," 47 *S. Cal. L. Rev.* 476, 490 n.66 (1974).

31. Fletcher, "Ethical Aspects of Genetic Controls," 285 *N. Eng. J. Med.* 776, 779 (1971).

32. J. B. S. Haldane, "Biological Possibilities for the Human Species in the Next Thousand Years," in *Man and His Future* 354–55, cited in Pizzulli, *supra* note 30 at 520.

33. Anita Manning, "Pressing a 'Right' to Clone Humans, Some Gays Forsee Reproduction Option," *USA Today*, March 6, 1997, at D1.

34. *Ibid.*; see also Liesl Schilinger, "Postcard From New York," *The Independent—London*, March 16, 1997 at 2.

35. Schilinger, *supra* note 34.

36. See Manning, *supra* note 33.

37. See Schilinger, *supra* note 34.

38. Pizzulli, *supra* note 30, at 550. Charles Strom, director of genetics and the DNA laboratory at Illinois Masonic Medical Center, argues that the high rate of embryo death that has occurred in animal cloning should not dissuade people from considering cloning as a legitimate reproductive technique. *See* Sheryl Stolberg, "Sheep Clone Researcher Calls for Caution Science," *Los Angeles Times*, March 1, 1997, at A18. Strom points out that all new reproductive technologies have been marred by high failure rates, and that it is just a matter of time before cloning is as economically efficient as any other form of artificial reproduction. See Schilinger, *supra* note 34.

39. John Robertson's Statement to the National Bioethics Advisory

Commission, March 14, 1997, at 83. This seems to be a reversal of Robertson's earlier position that cloning "may deviate too far from prevailing conception of what is valuable about reproduction to count as a protected reproductive experience. At some point attempts to control the entire genome of a new person pass beyond the central experiences of identity and meaning that make reproduction a valued experience." John Robertson, *Children of Choice: Freedom and the New Reproductive Technologies* 169 (Princeton, N.J.: Princeton University Press, 1994).

40. *Id.*

41. See June Coleman, "Playing God or Playing Scientist: A Constitutional Analysis of Laws Banning Embryological Procedures," 27 *Pac. L.J.* 1331, 1351 (1996).

42. George J. Annas, "Human Cloning: Yes: Individual Dignity Demands Nothing Less," 83 *ABA Journal* 80 (May 1997).

43. George Annas Testimony on Scientific Discoveries and Cloning: Challenges for Public Policy, before the Subcommittee on Public Health and Safety, Committee on Labor and Human Resources, United States Senate, March 12, 1977, at 4.

44. See, e.g., *Griswold* v. *Connecticut*, 381 U.S. 479 (1965); *Eisenstadt* v. *Baird*, 405 U.S. 438 (1972); *Roe* v. *Wade*, 410 U.S. 113 (1973); *Planned Parenthood of Southern Pennsylvania* v. *Casey*, 505 U.S. 833 (1992).

45. See *Lifchez* v. *Hartigan*, 735 F.Supp. 1361 (N.D. Ill), *aff'd without opinion, sub norm. Scholberg* v. *Lifchez*, 914 F.2d 260 (7th Cir. 1990), *cert. denied*, 498 U.S. 1068 (1991).

46. *Nature* press release, "Cloning: A Flock of Mice?," July 23, 1998 (*available at* http://www.students.washington.edu/radin/clonclon.htm).

47. See Weiss, *supra* note 4.

48. See J. Madeline Nash, "The Age of Cloning," *Time*, March 10, 1997, at 62–65; see also Peter N. Spotts and Robert Marquand, "A Lamb Ignites a Debate on the Ethics of Cloning," *The Christian Science Monitor*, Feb. 26, 1997, at 3.

49. See "The Law and Medicine," *The Economist*; March 1, 1997, U.S. edition, World Politics and Current Affairs section, 59; *see also* Pizzulli, *supra* note 30 at 484 (citing Briggs and King, "Transplantation of Living Nuclei from Blastula Cells into Enucleated Frogs' Eggs," 38 *Proc. Nat'l Acad. Sciences* 455 [1952]).

50. See Pizzulli, *supra* note 30 at 484, 487.

51. See "Horizon: Dawn of the Clone Age" (BBC television broadcast, Sept. 10, 1997), *transcript available at* <http://www.bbc.co.uk/horizon/cloneagetrans.shtml>.

52. See *id.*

53. See *id.*

54. See *id.*

55. See Angelika E. Schnieke et al., "Human Factor IX Transgenic Sheep Produced by Transfer of Nuclei from Transfected Fetal Fibroblasts," 278 *Science* 2130, 2130 (1997). "Transgenic" animals are those that carry and express a gene of another species.

56. See "Cloned Transgenic Lambs Produce Clotting Factor in Milk," *Biotechnology Newswatch*, Jan. 15, 1998, at 5.

57. See Schnieke et al., *supra* note 55, at 2132. This study determined that a "pregnancy" resulted when an embryo continued to develop 60 days after implantation, which is slightly less than half of this breed of lamb's normal gestational period. See *id.* at 2131–2132.

58. See *id.* at 2132. The fact that some of these lambs were delivered by cesarean section prompted an animal welfare organization, Compassion in World Farming, to criticize the Roslin team for using Scottish blackface sheep as surrogate ewes because these ewes are typically smaller than the Poll Dorset breed which are actually used to create the cloned embryos. See Nick Thorpe, "Scientists Baffled by Oversized Sheep Clones," *Scotsman*, July 28, 1997, at 1; see also Steve Connor, " 'Giant' Lambs Put Future of Cloning in Doubt," *Sunday Times (London)*, July 27, 1997, at 5.

59. See Thorpe, *supra* note 58; see also Connor, *supra* note 58.

60. See Pizzulli, *supra* note 30.

61. See Paul Recer, "Sheep Cloner Says Cloning People Would Be Inhumane," *Associated Press*, March 12, 1997 (reported testimony of Dr. Ian Wilmut and of Dr. Harold Varmus before the Senate on March 12, 1997, regarding the banning of human cloning research).

62. *Id.* (comments of Dr. Ian Wilmut, testifying that as of yet he does not know of "any reason why we would want to copy a person. I personally have still not heard of a potential use of this technique to produce a new person that I would find ethical or acceptable").

63. "Seed's Human Cloning Bid Draws Edgy World Reactions," *Medical Industry Today*, Jan. 8, 1998, *available in* LEXIS, News Library, CURNWS File (quoting Dr. Wilmut).

64. Shirley Tilghman Statement to National Bioethics Advisory Commission, March 13, 1997, at 173.

65. *Id.* at 147.

66. See Recer, *supra* note 61.

67. *Id.*; see also J. Madeleine Nash, "The Age of Cloning," *Time,* March 10, 1997, at 62–65.

68. See Recer, *supra* note 61; see also Jeremy Laurence and Michael Hornsby, "Warning on Human Clones," *Times*, Feb. 23, 1997; see also "Whatever Next?," *Economist*, March 1, 1997, at 79 (discussing the problems associated of having mitochondria of egg interact with donor cell).

69. See Terence Monmaney, "Prospect of Human Cloning Gives Birth to Volatile Issues," *L.A. Times*, March 2, 1997, at A1.

70. Chris Tenove, "Dolly May Grow Old Before Her Time, But a New Breed of Clone Are Taking Youth to Extremes," *New Scientist*, May 6, 2000.

71. Id. See also, Robert Lanza et al, "Extension of Cell Life-Span and Telomere Length in Animals Cloned From Senescent Somatic Cells," 288 *Science* 665–9, April 28, 2000.

72. Mara Bovsun, "Normal—Even Slow—Aging Seen in Cells From Cloned Calves," *Biotechnology Newswatch*, May 15, 2000.

73. Nash, *supra* note 48.

74. *Id.* See also Tilghman, *supra* note 64 at 145 (discussing the problem of mutations).

75. *Cloning Human Beings: Report and Recommendations of the Na-*

tional Bioethics Advisory Commission 9 (Rockville, Md.: National Bioethics Advisory Commission, June 1997).

76. See Tony Mauro, "Sheep Clone Prompts U.S. Panel Review," *USA Today*, Feb. 25, 1997, at A1.

77. See Pizzulli, *supra* note 30 at 502.

78. *Id.* at 512.

79. *Id.*

80. *Id.*; see also Mona S. Amer, Comment, "Breaking the Mold: Human Embryo Cloning and Its Implications for a Right to Individuality," 4 *UCLA L. Rev.* 1659, 1666 (1996) (citing Philip Elmer-Dewitt, "Cloning: Where Do We Draw the Line?" *Time*, Nov. 8, 1993, at 65).

81. See Catherine M. Valerio Barrad, Comment, "Genetic Information and Property Theory," 87 *Nw. U. L. Rev.* 1037, 1050 (1993).

82. *Id.*

83. *Id.*

84. See Amer, *supra* note 80.

85. Pizzulli, *supra* note 30.

86. NBAC Report, *supra* note 75 at 33.

87. Leon Jaroff, "The Gene Hunt," *Time*, March 20, 1989, at 63.

88. Tom Wolfe, "Sorry, But Your Soul Just Died," *Forbes ASAP*, Dec. 2, 1996, at 212.

89. Charles C. Mann, "Behavorial Genetics in Transition," 264 *Science* 1686 (June 1994).

90. There is much evidence of the widespread belief in genetic determinism. See, e.g., Dorothy Nelkin and Susan Lindee, *The DNA Mystique: The Gene as Cultural Icon* (W. H. Freeman & Company, 1995).

91. For a review of the studies, see Lori B. Andrews, "Prenatal Screening and the Culture of Motherhood," 47 *Hastings L.J.* 967 (1996).

92. Nancy Wexler, "Clairvoyance and Caution: Repercussions from the Human Genome Project," in Daniel J. Kevles and Leroy Hood, *The Code of Codes: Scientific and Social Issues in the Human Genome Project* 211–243, 233 (Cambridge, Mass.: Harvard University Press, 1992).

93. "Genetic Testing for Children and Adolescents: Who Decides?" 272 *JAMA* 875, 878 (1994).

94. *Id.* Similarly, the ASHG/ACMG statement notes, "Expectations of others for education, social relationships and/or employment may be significantly altered when a child is found to carry a gene associated with a late-onset disease or susceptibility. Such individuals may not be encouraged to reach their full potential, or they may have difficulty obtaining education or employment if their risk for early death or disability is revealed." American Society of Human Genetics and American College of Medical Genetics, "Points to Consider: Ethical, Legal, and Psychosocial Implications of Genetic Testing in Children and Adolescents," 57 *American Journal of Human Genetics* 1233–1241, 1236 (1995).

95. ASHG/ACMG statement, *supra* note 94.

96. Lori B. Andrews, Jane E. Fullarton, Neil A. Holtzman, and Arno G. Motulsky, eds., *Assessing Genetic Risks: Implications for Health and Social Policy* 276 (1994) [hereinafter, *Assessing Genetic Risks*].

97. NBAC Report, *supra* note 75 at 67, citing Joel Feinberg, "The

Child's Right to an Open Future," in *Whose Child? Children's Rights, Parental Authority, and State Power*, ed. W. Aiken and H. LaFollete (Towota, N.J.: Rowman and Littlefield, 1980).

98. Lori B. Andrews, "Surrogate Motherhood: The Challenge for Feminists," 16 *Law, Medicine and Health Care* 72 (1988).

99. Pizzulli, *supra* note 30 at 497.

100. *Id.* at 492.

24

First Dolly, Now Polly

Policy Implications of the Birth of a Transgenic Cloned Lamb

Andrea L. Bonnicksen

For all the attention given to Dolly, the first mammal cloned from a differentiated adult somatic cell, the real prize for scientists was Polly, a lamb cloned from sheep fetal cells who has a human gene in every cell of her body. Polly symbolizes potentially lucrative interventions in which cloning is a mechanism for efficiently modifying the genomes of animals for human medicine and commerce. She was what researchers wanted all along: a mammal for whom cloning facilitated genetic engineering.[1] To the scientists who enabled her conception, Polly was the "true goal" and Dolly was a mere "detour."[2]

Yet Polly, along with George, Megan, Morag, and other named and unnamed creatures who are the predecessors of cloning and transgenic combinations, is of interest not only for her economic promise but for the glimpse she gives of what might be on the horizon for human medicine. While Dolly provoked concerns about cloning and while the births of transgenic animals have raised concerns about interventions on the germ line, Polly suggests that procedures relating to cloning and germ-line gene therapy might intersect in heretofore unanticipated ways. This in turn suggests the inadequacies of current policy efforts in various countries in which cloning and germ-line interventions are addressed as discrete entities. These inadequacies are further underscored by another technique potentially on the horizon—egg cell nuclear transfer—that is neither fully a germ-line nor cloning procedure but that also brings cloning and germ-line issues together in unexpected ways.

With policy makers in Europe and North America weighing approaches to cloning regulation, it is timely to explore the merits of the technique-specific approach that has characterized cross-national policy discussions in procreative medicine to date. This chapter looks at the United States in particular, where legislators have considered bills to ban cloning and where policy makers will one day be in a situation to review funding policies on germ-line research,[3] to (1) contrast the dual, albeit unformed, policies emerging for germ-line interventions and cloning, and (2) illustrate how transgenic cloned animal studies and proposed egg cell nuclear transfer procedures point to the need for caution in framing separate policies for cloning and germ-line interventions.

DIVERGING POLICY, INTERSECTING TECHNIQUES

Germ-line interventions and cloning have long been on the minds of those watching developments in reproductive and genetic technologies. In recent years, numerous bioethics commissions have identified these techniques as especially troublesome,[4] and several nations have passed laws banning or limiting their development and use.[5] Given the ethical reservations expressed about both germ-line interventions and cloning, it is interesting that policy for each has diverged significantly, with the policy context for germ-line interventions more broadly supportive than that for cloning.

A prominent feature of germ-line policy in the United States has been its de facto permissive character. The federal government has responded to the prospect of germ-line gene therapy (GLGT) by neither actively promoting nor prohibiting it. Instead, following a 1990 policy, it declines to consider research proposals dealing with GLGT for humans and in so doing will not fund these projects. In that year, the Human Gene Therapy Subcommittee of the Recombinant DNA Advisory Committee (RAC) of the National Institutes of Health issued a "Points to Consider" document regarding somatic-cell gene research projects. One section related to germ-line interventions: "The RAC and its Subcommittee will not at present entertain proposals for germline alterations."[6] The government reiterated its hands-off approach in 1994 when it directed the Human Embryo Research Panel of the National Institutes of Health not to review the ethical acceptability of GLGT when it reviewed other interventions involving the human embryo.[7] This approach contrasts to that of several nations that have legislated against GLGT or sponsored commissions that recommended against it.[8]

The U.S. policy of withdrawal leaves the door open for future germ-line interventions. Nothing about federal administrative or statutory law would preclude germ-line interventions on human gametes or embryos, which are not funded, but neither are they illegal. The laws of the few states that restrict embryo research do not necessarily foreclose GLGT; they instead target nontherapeutic research on human embryos.[9] Moreover, there is room for the government to moderate its policy. The rule that removes germ-line research from federal funding review states that the RAC will not "at present" entertain such proposals,[10] which leaves matters open. In addition, the RAC periodically reviews its procedures for submitting somatic-cell gene-therapy proposals, and this might present an opportunity to revisit the germ-line language.[11] New interest in germ-line prospects suggests that a growing community of scientists may argue for relaxed funding rules.[12]

Conclusions about the ethics of GLGT vary cross-nationally, with some ethics commissions concluding that GLGT is ethically acceptable and others concluding it is not, largely on the basis of worries about genetic enhancement.[13] The Declaration on the Human Genome and Human Right, signed by 186 member nations of the United Nations Educational, Scientific, and Cultural Organization (UNESCO) in 1997, calls the human genome "in a symbolic sense . . . the heritage of humanity," but it does not specifically mention or proscribe GLGT.[14] On the other hand, the Council of Europe opened the Convention for the Protection of Human Rights and the Dignity of the Human Being with Regard to the Application of Biology and Medicine (Bioethics Convention) for the signature of its 40 member states. The convention would permit therapeutic intervention in the human genome only "if its aim is not to introduce any modification in the genome of descendants," which would bar GLGT.[15]

Cloning policy in the United States, which is newer and less formed, differs from GLGT policy in the United States in several ways. First, proposed and enacted cloning laws would be targeted and punitive rather than de facto permissive. The goal of critics of cloning is to forbid one or more forms of cloning. Following the announced birth of Dolly in 1997,[16] legislators in the United States introduced bills to Congress that would either bar federal funding of human cloning or ban the transfer of an embryo created by somatic cell nuclear transfer.[17] Even the no-funding proposals are more deliberate than for GLGT because they are part of isolated, cloning-specific bills rather than restrictions contained in lengthy appropriations measures. At the state level, nearly half of the states introduced approximately 40 bills designed to ban cloning. California, Michigan, Missouri, Louisiana, and Rhode Island passed anti-cloning statutes.[18]

Second, many proposed cloning laws would forbid with criminal penalties attached. A similar punitive tone appears at the international level, where the Council of Europe presented the Bioethics Convention to its member states. This Additional Protocol has been signed at the time of this writing by 23 member states of the Council of Europe that had signed the Bioethics Convention. It forbids cloning and imposes legally binding directives on signatory states to bring their laws into conformity with this provision unless doing so would conflict with preexisting legislation.[19]

The UNESCO document mentioned above (Declaration on the Human Genome and Human Right) is not legally binding. Still, the otherwise aspirational text of principles singles only one named technique, cloning, for approbation: "Practices which are contrary to human dignity, such as reproductive cloning of human beings, shall not be permitted."[20]

Third, in contrast to GLGT, there are few signs that a moderation of views about cloning is on the horizon.[21] Cloning is distinguished from GLGT by its muted medical rationale. Its primary purpose would be to allow conception by infertile couples, but critics dismiss this rationale as not compelling in light of numerous alternative methods available for conception. Germ-line gene therapy, on the other hand, would have medical purpose of preventing the birth of a child with a serious genetic disease. Moreover, if its safety were established, it might appeal to those who want to correct rather than discard embryos found to have a deleterious gene.

In summary, emerging policies for cloning and GLGT in the United States differ in their deliberateness, flexibility, degree of optimism/pessimism, and level of government involvement. At the same time that policy is diverging, however, scientific developments are bringing cloning and GLGT closer together. In the case of transgenic cloned animals, this means cloning and GLGT are intersecting. The same may one day happen in human medicine if the techniques underlying reproductive cloning and GLGT intersect for therapeutic rather than reproductive goals. It might also happen in the case of egg cell nuclear transfer, where the technique in question is not quite cloning and not quite GLGT.

Transgenic Cloned Animals

The splicing of genes into animal genomes to produce transgenic animals for pharmacologic and medical use has been a growing business since the early 1990s, yet the transgenic industry has been limited by inefficient traditional genetic engineering methods in which genes are injected into embryo cells individually and only a small number of embryos assimilate and express the injected gene. In addition, many of those animals that take up the gene do so only in some cells, so an additional generation must be bred using eggs or spermatozoa that contain the gene to produce an animal with the gene in every cell.[22]

The scientists' vision has been to refine cloning so that many cells can be available for gene splicing and transgenic interventions can be completed in one generation. To produce Polly, researchers added a human gene and a marker gene to the skin cells of a sheep fetus. They then used only cells that had taken up the human gene for somatic cell nuclear transfer.[23] As a member of the research team stated, "the main interest . . . was to find a method for the production of offspring by nuclear transfer from cells that could be maintained in culture and used as a route to precise genetic modification in animals for transmission through the germ line."[24] Four other lambs were born at the same time as Polly, but she was the only lamb with expressed transgenes.

Nearly a year after Polly's birth, scientists in the United States reported the birth of calves that had been cloned from fetal cells and transgenically manipulated. After inserting human genes into the fetal cells, the scientists transferred the nuclei of 276 of the cells to 276 enucleated cow eggs. Thirty-three embryos survived a week's growth in the laboratory and 28 were transferred to the uteri of surrogate cows. Four calves were born, but one died shortly after birth.[25] The three remaining calves, ACT3, ACT4, and ACT5, shared the same genome, which had been artificially manipulated. Different research groups have experimented with cell source (embryo, fetal, adult) and species of animal, which points to the prospect of rapid developments in the field of cloning and genetic technologies.[26]

The fact that Polly, Morag, Megan, and other creatures with personal monikers are fading in the wake of the impersonally named ACT calves foretells a normalcy of genetic and cloning combinations in biotechnology. This provokes many queries, from the ethics of generating large numbers of transgenic animals to the safety of using the tissues and organs of genetically modified pigs for transplantation to humans. In the area of human medicine, cloning could be a vehicle for germ-line interventions by providing great numbers of cells for a more efficient and faster expression of transgene. The potential intersection of cloning and germ-line interventions in animals suggests that if GLGT in humans were to gain credence, therapeutic cloning to facilitate germ-line therapy (and not to facilitate procreation) may be one step in the process.

Egg Cell Nuclear Transfer

Imagine a young woman who has learned that a disease recurring on her mother's side of the family has been diagnosed in her and has a genetic origin. The disease, mitochondrial encephalomyopathy with lactic acidosis and strokelike episodes (MELAS), results in neuromuscular degeneration at midlife, seizures, heart failure, and death-generally before age 40.[27] The disorder is linked to mutations in the mitochondrial DNA (mtDNA), which resides in the cytoplasm of all body cells and eggs, but not in spermatozoa.

If this woman reproduces, the mutation will be passed to her child because the mtDNA in the cytoplasm of the woman's egg will be replicated with the cell divisions that follow fertilization. The woman wants a genetically related child, but she does not want to pass a lethal disease to her child. Might she have a child genetically related to her but without the disease?

In 1995 a group of researchers in the United States published a clinical protocol to address such a situation, calling the proposed technique in vitro ovum nuclear transplantation (IVONT).[28] In this still untested technique, the genome-containing nucleus would be extracted from a woman's unfertilized egg and fused to a donor cell from which the nucleus has been removed and discarded. This cut-and-pasted egg would then be fertilized through in vitro fertilization (IVF), and the resulting embryo transferred to the woman for pregnancy. This technique would bypass MELAS by discarding the woman's disease-linked mtDNA and using a donor's presumably healthy mtDNA in its place. The woman's child would therefore be related to her genetically but he or she would not suffer from the disease that afflicted the woman's family.

This protocol—here referred to as egg cell nuclear transfer—would be complex but not excessively difficult. Egg cell nuclear transfer is a fertile topic for ethical and policy review due to its relation to germ-line gene therapy and cloning. Because it will affect all dividing cells, including germ cells, and will therefore affect future generations, it is a germ-line intervention. Nevertheless, it is not the type of germ-line intervention commentators and policy makers have had in mind when warning of the dangers of genetic engineering. It is also not cloning—no genome is replicated—but it shares the nuclear transfer procedure that is necessary for embryo or somatic cell cloning. Its use would therefore create a relatively uncontentious precedent that might moderate attitudes toward GLGT. Such use would also provide an opportunity to refine techniques that are essential to cloning.

Researchers have also proposed using egg cell nuclear transfer to improve fertilization rates in assisted conception.[29] If low fertilization rates for older women are due to programming errors in the cytoplasm of the older women's eggs, then a proposed solution is to fuse the patient's nucleus with enucleated eggs donated by younger women.[30] In addition, egg cell nuclear transfer might be used for egg freezing in infertility treatment. Success in freezing eggs has largely eluded practitioners in infertility programs, but if the usual inability of eggs to survive a freeze/thaw is due to cytoplasmic damage, the egg's nucleus could be thawed, removed, and transferred to an enucleated donor egg with fresh cytoplasm.[31]

The proposal to use egg cell nuclear transfer in the face of mitochondrial disease places the procedure in the genetics arena because it is intended to circumvent a genetic disease and because the authors labeled it as the first proposed human germ-line protocol. When called a first step to germ-line genetics, it directs attention to issues raised for germ-line interventions in general, such as whether egg cell nuclear transfer should be used to circumvent non–life-threatening diseases or to enhance capabilities.

When egg cell nuclear transfer is placed in the context of infertility treatment, however, it directs attention to somewhat different issues. It seems intuitively reasonable to suggest that techniques introduced to increase the odds of pregnancy are less "suspect" than techniques designed to avoid genetic disease. The prospect of egg cell nuclear transfer as an adjunct of infertility treatment makes it possible that the procedure, when designed for egg freezing or conception by older women, will be the quiet way station for GLGT and cloning. While egg cell nuclear transfer should not necessarily be held hostage to concerns about GLGT and cloning, its presence if developed clinically should be acknowledged in policy discussions.

In the United States a California law enacted in 1997 demonstrates the difficulty of rule making by technique when a full scientific overview is absent. Designed to restrict human cloning, the law is written so broadly that it precludes egg cell nuclear transfer. In relevant part it defines cloning as "the practice of creating or attempting to create a human being by transferring the nucleus from a human cell from whatever source into a human egg cell from which the nucleus has been removed for the purpose of, or to implant, the resulting product to initiate a pregnancy that could result in the birth of a human being."[32] The insertion of "from whatever source" could be interpreted to mean nuclei cannot be taken from egg cells, even though this would not be cloning.

Germ-line Gene Therapy

Scientists are looking with new interest at the prospect of GLGT in humans, as seen, for example, by a conference held in the United States at the University of California at Los Angeles in 1998. At this conference, which was held to provide a realistic assessment of the potential of GLGT over the next 20 years, some of the country's top geneticists outlined ways GLGT might unfold.[33] Their comments indicated ways that the concept of GLGT may need to be reconceptualized. For example, if GLGT can be conducted with artificial chromosomes that are not transmissible unless "turned on" by the consent of the patient in the new generation, this would eliminate an important reservation about GLGT; namely, that it would affect the genomes of future generations of individuals without their consent.[34] In the more immediate future, egg cell nuclear transfer may blur the lines, too, because it would affect inheritance; if successful, the intervention would transform a genetic disease from one certain to appear in a family to one certain not to appear,[35] thereby making it a germ-line intervention. Still, the inheritability of egg cell nuclear transfer differs from traditional notions of GLGT because it would (1) target mtDNA rather than nuclear DNA and (2) move a nucleus from one place to another rather than penetrating and correcting nuclear substructures. The intriguing question, then, is whether concerns expressed about GLGT apply with equal potency to egg cell nuclear transfer.

One line of thinking holds that it does. According to this view, the type of DNA in question is beside the point; the critical matter is inheritability. As Baccheta and Richter note in a response to the proposed protocol of Rubenstein et al., germ-line interventions have an unbounded duration in contrast to somatic cell interventions, which—because they affect the treated individual only—have a bounded duration.[36] According to this argument, the cautions raised against germ-line interventions apply equally to substituting mtDNA and correcting nuclear DNA; both can introduce mistakes with deleterious effects on future generations.

An alternate line of thinking holds that egg cell nuclear transfer is less complex and less risky than nuclear DNA splicing, and is thus a matter of lesser concern. According to Rubenstein et al., GLGT can be divided into three escalating levels or degrees of cellular penetration. As the degree of penetration into the cell grows, so does the level of necessary experimentation, cost, and risk, and thereby the number of ethical concerns.[37]

In this categorization, egg cell nuclear transfer is a "level 1" germ-line intervention, with minimum penetration into the cell only to remove intact nuclei from the patient's and donor's eggs. A theoretical "level 2" germ-line intervention, chromosomal transplantation, would involve two levels of penetration. Technicians would penetrate the cell and then the nucleus to remove a chromosome with a DNA mutation and substitute for it another chromosome from a "chromosome bank." Germ-line gene therapy is a "level 3" germ-line intervention. In this case, a chromosome itself is penetrated to splice into it a corrected DNA sequence.

Disaggregating GLGT by degree of penetration opens the door for more refined thinking about the concept of gene therapy that to this point has been more or less monolithic, aside from distinctions between the type of cell targeted (germ line versus somatic) and the goal of the intervention (therapeutic versus enhancement).[38] Yet while a disaggregation may be useful, one may question whether degree of penetration is the best criterion. Rubenstein et al. presume that the greater the cell penetration, the greater the ethical concern (because of safety, cost, and complexity). However, DNA splicing is not necessarily riskier than cytoplasmic substitution; moreover, transplanting nuclei between egg cells may actually be more intrusive than DNA splicing to correct a specific sequence of base pairs. In essence, one ends up with a simple distinction between genome types—mtDNA or nuclear DNA—which is conceptually richer rather than the degree of penetration as a variable.

A case can be made for drawing a line for ethical and policy analysis on the basis of genome type. Nuclear DNA is more central to genomic studies than mtDNA. Nuclear DNA, which contains from 50,000 to 100,000 genes, is enormously powerful; even minute splices of DNA inserted into animal embryos can create animals with unique medical conditions, while the deletion of just three base pairs of nitrogen bonds can produce the life-threatening illness of cystic fibrosis; and other infinitesimally small DNA mutations are associated with serious disease. As Cherfas has put it, "DNA is the primary determining force of every living thing. . . . It contains . . . all the information needed to construct and maintain the complex machinery of the living cell."[39]

Mitochondria, in contrast, are subcellular structures with a separate physiology and evolutionary source from nuclear DNA. Evolved from bacteria, mitochondria have their own genetic structure, and they replicate separately from nuclear DNA.[40] In contrast to the tens of thousands of genes of nuclear DNA, mtDNA has only 37 genes. In addition, mtDNA mutations can profoundly affect a person's health, but mtDNA is not thought to code directly for behavior and traits, which means that nuclear DNA is more intimately associated with the hallmarks of distinctively human traits than is mtDNA.

On the other hand, the line between genome types may be an artifact. First, the physiology of nuclear DNA is closely linked to that of mtDNA in that the former plays a not yet well understood programming role. Second, the mitochondrial role in aging and mental acuity is under investigation, so it may be premature to conclude that mtDNA are mere bacterial descendants with little ethical significance.

If one accepts that interventions on nuclear DNA are more significant than those on mtDNA, then national laws and commission recommendations may be overbroad. Policy recommendations to this point have been made and implemented with nuclear DNA splicing in mind. For example, when recommending against GLGT, the Council for Responsible Genetics defined "germ line manipulation" as follows: "Techniques are now available to change chromosomes of animal cells by *inserting new segments of DNA into them.* . . . If [this insertion] is performed on sperm or eggs before fertilization, or on the undifferentiated cells of an early embryo, it is called germ cell or germ line gene modification."[41]

Similarly, Canada's Royal Commission on New Reproductive Technologies recommended against national funding of "research involving alteration of the DNA or human zygotes," and it defined gene therapy as that "aimed at curing a disease due to a defective gene, either by *insertion of a normal gene or by correction* of the abnormal one."[42]

Using such definitions, commissions and governments have proposed or enacted policy with varying levels of permissiveness. The German Embryo Protection Act of 1990 is a restrictive law making it a criminal act to "manipulate the genetic information of a human germ cell" or to "use genetically manipulated germ cells for the purpose of fertilization."[43] In contrast, the British Human Fertilisation Act of 1990 is a permissive law that nevertheless excludes licenses for research that would "[alter] the genetic structure of any cell" while it becomes part of an embryo except if it conforms with regulations.[44]

The U.S. policy, which removes research on "germ line alterations" from federal funding, defines a germ-line alteration as one in which a "specific attempt is made to introduce genetic changes into the germ (reproductive) cells of an individual, with the aim of changing the set of genes passed on to the individual's offspring."[45]

These policies are likely to encompass egg cell nuclear transfer because they broadly define germ-line alterations, but the restriction is not explicit inasmuch as the guidelines were crafted without egg cell nuclear transfer in mind. Examples of other national laws are Sweden's, which forbids embryo research "to develop methods aimed at causing heritable genetic effects,"[46] and Austria's, which does not permit "interventions involving the germline."[47] While the implicit regulation of egg cell nuclear transfer might conform with the intentions of policy makers, it presents a situation in which a technique is limited by default rather than by deliberate design. With advances in GLGT certain to proceed and to take unexpected forms, the need arises to examine existing policy approaches in light of a likely variety of proposed germ-line interventions.

CONCLUSIONS

Policy on assisted conception has traditionally been developed in a technique-by-technique manner. In the United States, this has taken the form of a hands-off policy for GLGT and proposed punitive laws for reproductive cloning. In Germany and other European nations and in some international documents, this has taken the form of anticipatory laws that forbid categories of techniques, including GLGT and cloning.

Knoppers has identified two errors that result from technique-specific laws, which contrast to aspirational policies that appeal to principles. With the error of overinclusion, legitimate techniques are inadvertently forbidden within the law's broad sweep. With the error of underinclusion unlisted techniques are assumed to be permissible when, in fact, no consensus exists about their ethical acceptability.[48] As this chapter demonstrated, another problem with technique-specific laws is that the categories used to define techniques become dated as advances in biotechnology foretell potential applications in human medicine.

Since Hotchkiss coined the term "genetic engineering" in 1965 to acknowledge advances in molecular biology,[49] the science, ethics, and policy of genetic interventions has undergone a more refined analysis. Beginning in the early 1990s, commentators began to question the supposed onerousness of germ-line interventions and to call for a more neutral assessment of germ-line prospects.[50] Around that time also, germ-line therapy gained proponents who argued the merits of using germ-line interventions to prevent disease,[51] and several years later others began to examine the merits of germ-line enhancement, long presumed to be the most worrisome use of genetic knowledge.[52]

While some commentators envisioned the forms germ-line interventions would take, Rubenstein et al. have proposed a different germ-line protocol—the substitution of mitochondrial DNA through nuclear transfer between human eggs. Whether or not egg cell nuclear transfer is imminent, its hypothetical presence has theoretical, clinical, and policy implications.

In the theoretical sense, egg cell nuclear transfer encourages more layers to be added to the concept of germ-line gene therapy because it demonstrates that germ-line proposals may take forms different from those predicted. In the clinical sense, it presents a backdoor approach to germ-line therapy by presenting a relatively nonthreatening technique designed to address a specific medical purpose. In the policy arena, it provokes queries about the wisdom and longevity of sunset clauses, funding restrictions, working definitions, and the guiding principles around which policy is framed.

Realistically, one may expect the following in the next several years: a not-quite-germ-line technique such as egg cell nuclear transfer will arise that confers specific benefits on a distinct group of patients, will be privately funded, and will produce a successful pregnancy or birth. Its promised beneficence will be difficult to resist, and the first step toward the DNA splicing associated with GLGT will have been taken, despite long-standing apprehension about it. Government action is unlikely unless the procedure results in a child with birth defects.

The birth of Polly foretells a future in which cloning procedures are used to facilitate germ-line interventions. This would present a backdoor approach to cloning by presenting cloning as a means to a therapy-related end, not to the more contentious, end of procreation. This in turn has implications for proposed laws that would either ban somatic cell nuclear transfer altogether or just ban it for procreation. The California statute, for example, leaves it unclear whether cloning for GLGT would be permissible. In order to prevent genome replication, it forbids cloning to "create a human being" through nuclear transfer done to "initiate a pregnancy." Would it also forbid nuclear transfer to produce a germ-line change in order to initiate a single pregnancy and not the birth of cloned individuals?

One course of action for policy making is to continue targeting techniques such as cloning. Another is to examine sub-categories of techniques in order to identify issues that transcend individual techniques. For example, rather than crafting new policy each time a new development attracts public attention, a paradigm of preconception therapy could be generated. Next, principles could be developed about how ethically to manage innovations, such as rules determining when if ever it is safe to move from animal to human research or how developments in animal biotechnology can be flagged early in development for public scrutiny.

This is not to say rules governing cloning and GLGT are unnecessary; on the contrary, rules appear to be more important than ever in light of escalating advances in cloning research.[53] It is to say that the legislative and parliamentary chambers, with their air of finality and inflexibility, are inappropriate places for first-generation rules to be crafted. Voluntary moratoria and professional sector oversight are among the alternative avenues in which first-generation rules and limits can be developed either as preparation for or substitution of statutory requirements.

This chapter considered the ramifications of the interview reported in the first chapter of this volume, in which the scientists' reaction to the birth of a cloned lamb was more muted than that of the public and the reaction of scientists to the birth of a transgenic cloned lamb was more expressive. In the rush to impose moral judgment on techniques such as reproductive cloning, we must be careful not to overlook more modest and seemingly less newsworthy developments that may indicate more accurately what is on the horizon for human procreative medicine.

NOTES

Parts of this chapter appeared in earlier form as Andrea L. Bonnicksen, "Transplanting Nuclei between Human Eggs: Implications for Germ-Line Genetics," *Politics and the Life Sciences* 17 (March 1998):3–10.

1. Elizabeth Pennisi, "Transgenic Lambs From Cloning Lab," *Science* 277 (August 1, 1997):631.

2. See chapter 1 in this volume.

3. Andrea L. Bonnicksen, "The Politics of Germline Therapy," *Nature Genetics* 19 (May 1998):10–12.

4. See, e.g., Mary Warnock, *A Question of Life: The Warnock Report on Human Fertilisation and Embryology* (Oxford: Basil Blackwell, 1985); Royal Commission on New Reproductive Technologies, *Proceed With Care: Final Report of the Royal Commission on New Reproductive Technologies* (Ottawa: Minister of Government Services Canada, 1993).

5. See, e.g., Andrea L. Bonnicksen, "National and International Approaches to Human Germ-Line Gene Therapy," *Politics and the Life Sciences* 13 (February 1994):39–49; Andrea L. Bonnicksen, "Procreation by Cloning: Crafting Anticipatory Guidelines," *Journal of Law, Medicine & Ethics* 25 (1997):273–282; Bartha Maria Knoppers, "Cloning: An International Comparative Overview." In National Bioethics Advisory Commission, *Cloning Human Beings: Report and Recommendations of the National Bioethics Advisory Commission. Volume II: Commissioned Papers*. Rockville MD, 1997, pp. G1–G13.

6. NIH Guidelines for Recombinant DNA Research *Federal Register* 55 (40):7438–7448 (March 1, 1990).

7. National Institutes of Health, *Final Report of the Human Embryo Research Panel* (Bethesda, Md. National Institutes of Health, September 1994).

8. See, for example, "German Embryo Protection Act (October 24th, 1990): Gesetz zum Schutz von Embryonen," reprinted in *Human Reproduction* 6(4)(1991):605–606; "Denmark Law no. 460 of June 10 1997 on artificial fertilization in connection with medical treatment, diagnosis, and research, etc. (Lovtidende, 1997, Part A, June 11 1997, No. 84, pp. 2195–2198). Den. 97.25." Reprinted in *International Digest of Health Legislation* 483/4)(1997):321–333.

9. Lori Andrews and Nanette Elster, "International Regulation of Human Embryo Research: Embryo Research in the US," *Human Reproduction* 13(1)(1998):1–3.

10. NIH Guidelines, p. 7444.

11. Bonnicksen, "Politics of Germline Therapy."

12. Program on Science, Technology and Society at the Center for the Study of Evolution and the Origin of Life, "Engineering the Human Germline." University of California Los Angeles, March 20, 1998.

13. For reviews of reports, see LeRoy Walters, "Ethics and New Reproductive Technologies: An International Review of Committee Statements," *Hastings Center Report* 17 (June 1987): 3S–9S; John C. Fletcher and W.

French Anderson, "Germ-Line Gene Therapy: A New Stage of Debate," *Journal of Law, Medicine & Health Care*, 20(1–2)(spring-summer 1992):26–39.

14. Declaration on the Human Genome and Human Right (http://www.unesco.org/ibc/uk/genome/projet/index.htm).

15. Convention for the Protection of Human Rights and the Dignity of the Human Being with Regard to the Application of Biology and Medicine: Convention on Human Rights and Biomedicine. European Treaties, ETS No. 164 (1997).

16. I. M. Wilmut, A. E. Schnieke, J. McWhir, A. J. Kind, and K. H .S. Campbell, "Viable Offspring Derived from Fetal and Adult Mammalian Cells," *Nature* 385 (February 27, 1997): 810–813.

17. Bills variously would bar (1) funding to create an embryo through somatic cell nuclear transfer (e.g., H.R. 922 as amended) (2) funding to create a human (i.e., transfer an embryo created through somatic cell nuclear transfer to a woman's uterus for possible pregnancy) (e.g., S. 368), (3) the act of creating an embryo through somatic cell nuclear transfer (e.g., S. 1601), or (4) the act of transferring an embryo so created to a woman's uterus (e.g., H.R. 923).

18. California (Cal. Stats. 1997. Ch. 688 [S.B. 1344]; Michigan (H.B. 5475, S. B. 67, H.B. 4962; http://www.statenews.com/editions/o61098/plcloning.html). For citations of bills introduced in spring 1997, see National Bioethics Advisory Commission, *Cloning Human Beings: Report and Recommendations of the National Bioethics Advisory Commission* (Rockville: National Bioethics Advisory Commission, June 1997), p. 104.

19. Additional Protocol to the Convention for the Protection of Human Rights and Dignity of the Human Being with Regard to the Application of Biology and Medicine, on the Prohibition of Cloning Human Beings. European Treaties ETS No. 168 (1998). Http://www.coe.fr/eng/legaltext/. For a list of the signatories, see http://www.coe.fr/tableconv/168t.htm. Nations can sign the original Bioethics Convention with a "reservation" that they will not comply with a particular article if they have preexisting legislation. F. William Dommel, Jr., and Duane Alexander, "The Convention of Human Rights and Biomedicine of the Council of Europe," *Kennedy Institute of Ethics Journal* 7 (September 1997):259–276, 274.

20. Declaration on the Human Genome and Human Right.

21. But see John A. Robertson, "Human Cloning and the Challenge of Regulation," *New England Journal of Medicine* 339(2)(July 9, 1998):119–122; Carson Strong, "Cloning and Infertility," *Cambridge Quarterly of Healthcare Ethics* 7(3)(summer 1998):279–293.

22. Gina Kolata, "Holstein Calves Cloned From Cells, Paper Says," *New York Times* (April 23, 1998):A9.

23. Pennisi, "Transgenic Lambs."

24. See chapter 1 in this volume.

25. Jose B. Cibelli, Steve L. Stice, Paul J. Golueke, Jeff J. Kane, Joseph Jerry, Cathy Blackwell, F. Abel Ponce de Leon, James M. Robi, "Cloned Transgenic Calves Produced from Nonquiescent Fetal Fibroblasts," *Science* 280 (May 22, 1998):1256–1258; Kolata, "Holstein Calves."

26. Gary B. Anderson and George E. Seidel, "Cloning for Profit," *Science* 280 (May 29, 1998):1400–1401.

27. D. S. Rubenstein, D. C. Thomasma, E. A. Schon, and M. J. Zinaman, "Germ-Line Therapy to Cure Mitochondrial Disease: Protocol and Ethics of In Vitro Ovum Nuclear Transplantation," *Cambridge Quarterly of Healthcare Ethics* 4(3)(1995):316–339; see also A. L. Bonnicksen, "Transplanting Nuclei Between Human Eggs: Implications for Germ-Line Genetics," *Politics and the Life Sciences* 17(1)(1998):3–10.

28. Rubenstein et al., "Germ-Line Therapy."

29. Gina Kolata, "Scientists Face New Ethical Quandaries in Baby-Making," *New York Times* (August 19, 1997):Bill.

30. Ibid.; J. Zhang, J. Grifo, A. Blaszczk, L. Meng, A. Adler, A. Chin, L. Krey, NYU Medical Center Program for IVF, New York, N.Y. and Oregon Regional Primate Research Center, Beaverton, Ore. (1997). "In Vitro Maturation (IVM)of Human Preovulatory Oocytes Reconstructed by Germinal Vesicle (GV) Transfer." Paper presented at the 53rd annual meeting of the American Society for Reproductive Medicine, Cincinnati (Program Supplement):S1.

31. Carol M. Warner, "Genetic Engineering of Human Eggs and Embryos: Prelude to Cloning," *Politics and the Life Sciences* 17(1)(1998):33–34.

32. S.B.1344. Cal.Ch.1.4, Sec.24185(c).

33. The proceedings were published in Gregory Stock and John Campbell, eds. *Engineering the Human Germline: An Exploration of the Science and Ethics of Altering the Genes we Pass to Our Children.* New York:Oxford University Press, 2000.

34. Ibid.

35. Rubenstein et al., "Germ-Line Therapy," p. 329; National Institutes of Health, p. 92.

36. M. D. Bacchetta and G. Richter, "Response to 'Germ-Line Therapy to Cure Mitochondrial Disease: Protocol and Ethics of In Vitro Ovum Nuclear Transplantation' by Donald S. Rubenstein, David C. Thomasma, Eric A. Schon, and Michael J. Zinaman (CQ Vol4, No3)," *Cambridge Quarterly of Healthcare Ethics* 5(3)(1996):450–457, p. 456.

37. Rubenstein et al., "Germ-Line Therapy," p. 327.

38. See, e.g., W. F. Anderson, "Human Gene Therapy: Scientific and Ethical Considerations," In R. F. Chadwick, ed., *Ethics, Reproduction and Genetic Control* (London: Croon Helm, 1987).

39. J. Cherfas, *Man-Made Life* (New York: Pantheon Books, 1982), p. 3.

40. Ruth Hubbard and Elijah Wald, *Exploding the Gene Myth* (Boston: Beacon Press, 1993), p. 164.

41. "Position Paper on Human Germ Line Manipulation Presented by Council for Responsible Genetics, Human Genetics Committee, Fall, 1992," *Human Gene Therapy* 4(1)(1993):35–37, p. 35 (emphasis added).

42. Royal Commission on New Reproductive Technologies, p. 942, 1157 (emphasis added).

43. "German Embryo Protection Act (October 24th, 1990):Gesetz zum Schutz von Embryonen," *Human Reproduction* 6(4)(1991): 605–606, p. 605.

44. "Human Fertilisation and Embryology Act 1990," *International Digest of Health Legislation* 42(1)(1991): 69–85.

45. NIH Guidelines, p. 7444.

46. "Sweden. Law No. 115 of 14 March 1991 concerning measures for the purposes of research or treatment in connection with fertilized human oocytes. (Svensk forfattningssamling, 1991, 26 March 1991, 2 pp.) Swed. 93.2," *International Digest of Health Legislation* 44(1)(1993):58.

47. "Austria. Federal Law of 1992 (Serial No. 275) regulating medically assisted procreation (the Reproductive Medicine Law), and amending the General Civil Code, the Marriage law, and the Rules of Jurisdiction. Date of entry into force: 1 July 1992. *(Bundesgesetzblatt für die Republik Osterreich,* 4 June 1992, No. 105, pp. 1299–1304. Austr. 93.)," *International Digest of Health Legislation* 44(2)(1993):247–248.

48. Knoppers (see note 5).

49. R. D. Hotchkiss, "Portents for a Genetic Engineering," *Journal of Heredity* 56(5)(1965):197–202.

50. See e.g., LeRoy Walters, "Human Gene Therapy: Ethics and Public Policy," *Human Gene Therapy* 2(1991):115–122; R. M. Cook-Deegan, "Human Gene Therapy and Congress," *Human Gene Therapy* 1(1990):163–170; J. C. Fletcher and W. F. Anderson, "Germ-Line Gene Therapy: A New Stage of Debate," *Law, Medicine and Health Care* 20(1–2)(1992):26–39.

51. See, e.g., Burke Zimmerman, "Human Germ-Line Therapy: The Case for Its Development and Use," *Journal of Medicine and Philosophy* 6(6)(1991):593–612; Ronald Munson and Lawrence H. Davis, "Germ-Line Gene Therapy and the Medical Imperative," *Kennedy Institute of Ethics Journal* 2(2)(1993):137–138.

52. See, e.g., David Resnick, "Debunking the Slippery Slope Argument Against Human Germ-Line Gene Therapy," *Journal of Medicine and Philosophy* 19(1)(1994):23–40; William Gardner, "Can Human Genetic Enhancement Be Prohibited?" *Journal of Medicine and Philosophy* 20(1)(1995):65–84; Ted Peters, " 'Playing God' and Germline Intervention," *Journal of Medicine and Philosophy* 20(4)(1995):365–386.

53. See, e.g., T. Wakayama, A. C. F. Perry, M. Zuccotti, K. R. Johnson, and R. Yanagimachi, "Full-term Development of Mice From Enucleated Oocytes Injected with Cumulus Cell Nuceli," *Nature* 394 (July 23, 1998): 369–374.

25

Ethical Aspects of Genetic Modification of Animals

Opinion of the Group of Advisers on the Ethical Implications of Biotechnology of the European Commission

Group of Advisers on the Ethical Implications of Biotechnology

1.1. The opinion addresses the genetic modification of animals brought about by DNA technology, including transfer of genes (transgenesis) and the deletion of genes ("knock-out").

1.2. Genetic modification of animals is a developing technology which will add to rather than replace existing techniques. It can be used:

- in fundamental biomedical research to improve our genetic and physiological knowledge;
- to make models of human diseases;
- for the production of proteins or other substances for therapeutic aims;
- as an alternative source of tissues and organs for "xeno-transplantation";

- to obtain or to improve desired features of farmed animals including fish, such as disease resistance and food production.

1.3. In this context, international competition is strong because practical applications are to be anticipated, both:

- medical (e.g., models of cystic fibrosis, production of alpha-1-antitrypsin for the treatment of acute respiratory distress syndrome); and
- agricultural (e.g., improved meat quality in pigs, better wool production in sheep, fish with increased growth rate).

1.4. Man has always been trying to modify nature, for instance by animal breeding. However, modern animal biotechnology differs from traditional breeding, more specifically by:

- the methods used: direct intervention on a micro level (cells, DNA, genes) by highly sophisticated technologies;
- the range of the new methods: breaking through species and even kingdom barriers;
- the speed and precision of such modern technology;
- the introduction of new objectives: for instance, xenotransplantation.

1.5. In our societies, besides the potential benefit we can expect from these techniques, there are concerns, for example:

- harm to animal health and welfare;
- impact on human health;
- animals are being used as mere instruments for human benefit and interests;
- it is seen as an infringement of animal "integrity" or of the "intrinsic" or "inherent" value of animals;
- it is seen as "unnatural," for example, because it transgresses species boundaries;
- it is seen as taking environmental risks the consequences of which are difficult to calculate;
- it is seen as a slippery slope toward eugenic applications on human beings.

1.6. Some of these public concerns about genetic modification of animals may be based on lack of information. They also reflect fundamental philosophical questions about our society and the place of science and technology therein, the place of humans in nature, the relation between humans and animals, and in particular the extent to which animals may be subordinated to human needs.

1.7. The religious and philosophical approach of Western countries traditionally gives a specific place to humans in their relationship with nature and animals, as humans are taken to have a "superior" position. This approach is shared by most people in Europe even though society has experienced, in recent decades, movements which advocate a more bio-oriented way of thinking.

1.8. In the pluralistic societies of the European Union, a complete consensus on moral and philosophical issues is not likely. However, we all may agree that vertebrate animals, as sentient beings, deserve ethical awareness and that genetic modification of animals means an increase of scientific power and thus of responsibility.

On the map of this new technology, the ethical pathways are not yet clearly marked, which implies the adoption of a policy of prudence.

1.9. The use of genetically modified animals or products issuing from them may have potential implications for humans, which should be taken into account, namely concerning:

- our security as a consumer,
- our right to be informed,
- the protection of our environment (including biodiversity).

1.10. Since it involves additional ethical issues, the controversial question of legal protection of genetically modified animals requires a separate opinion.

THE OPINION

2.1. Genetic modification of animals may contribute to human well-being and welfare, but is acceptable only when the aims are ethically justified and when it is carried out under ethical conditions.

2.2. These conditions would include the following:

- the public's right to protection against risks as well as their right to adequate information;
- human responsibility for animals, nature, and environment, including biodiversity;
- the duty to avoid or minimize animal suffering since unjustified or disproportionate suffering is unacceptable;
- the duty of reducing, replacing, and when possible refining the experimentation adopted for the use of animals in research.

2.3. In view of the consequences this technology may have for the health of humans and animals, for the environment and society, a policy of great prudence is required.

2.4. The scope of this policy should apply to:

- the making of genetically modified animals;
- the use and care of these animals;
- the release of these animals;
- putting genetically modified animals and their products onto the market (including import/export).

2.5. Concerning these points, licensing bodies in all member states should have the task of assessing research projects and applications in the light of the above mentioned principles.

2.6. The task of these licensing bodies would include at least the assessment of:

- the objectives: transparency and ethical acceptability;
- the risks: human health, environmental impact;
- animal health, welfare, and care;
- the proportionality of means and ends concerning genetic modifications of animals;
- the quality of the procedures;
- the possibility of alternatives.

2.7. Great care should be taken to prevent the release into the environment of genetically modified animals capable of surviving and breeding in the wild.

2.8. Research into the ethical, philosophical, and social aspects of animal biotechnology should be stimulated. Research into and monitoring of the possible consequences of animal biotechnology for the public and the environment should be supported at the European level.

2.9. There should be appropriate and understandable information for the public about genetic modification of animals and their products. The European institutions should strive to:

- systematically bridge the information gap;
- stimulate the dialogue between research, industry, and the public;
- create platforms for participation of the public in decision-making.

COMMENTARY
by Egbert Schroten

At an international workshop on Transgenic Animals and Food Production in Stockholm in May 1997 Peter Sandoe, a Danish philosopher, character-

ized the difference between Europe and the United States in attitudes toward biotechnology as a difference between "why?" and "why not?" To do
so, of course, sins against the eleventh commandment, "Thou shalt not gen-
eralize," but the distinction he draws is a concise way to highlight the differences in policy making in matters of biotechnology.

To analyze these differences in more detail would be interesting, but
that is not my intention here. One main reason might be the way of looking
at the role of government. In the United States public policy is more or less
a "policy of the market." If safety is guaranteed, let the market decide! In
Europe, government(s) is (are) expected to take a more active steering role
in the introduction of technology in general and of biotechnology in particular. Although the European Commission, in its so-called "White Paper"
(1994), identified biotechnology as one of the key technologies of the coming decades and economically very important, the market is not the only
place for policy making. A European Parliament Resolution of 1996 argued,
in response to the White Paper, that in public policy concerning biotechnology, sustainability should be emphasized and public concern should be
taken into account.

A recent article in *Nature* (Vol. 387, June 26, 1997, pp. 845–847) confirms that the latter part of this resolution is realistic. The article was written
by an international team of researchers working as part of a Concerted Action of the European Commission (B104-CT95-0043), administered on behalf of Directorate General XII by Andreas Klepsch, based on a survey called
"Eurobarometer on Biotechnology" (46.1). This article argues that public
concern in Europe cannot be reduced to a lack of information about developments in biotechnology. On the contrary, public acceptance in that area
is not necessarily increased by more knowledge (p. 845). What becomes
clear in the survey is "an increasing lack of confidence in national political
institutions" (p. 846). Moreover, the crucial role of moral doubt is emphasized. Irrespective of people's views on the usefulness and riskiness of biotechnology they "act as a veto" (p. 845). And moral doubts are shown to
appear particularly concerning the production of genetically modified food,
transgenic (research) animals, and xenotransplants.

This latest Eurobarometer survey confirms results of earlier surveys (in
1991 and 1993), and may serve to explain why, in European public policy,
the acceptance of biotechnology is not left to the marketplace. Ethical, social, and cultural aspects are taken into account as well. In 1991, the European Commission set up the Group of Advisers on the Ethical Implications of Biotechnology (GAEIB). This independent advisory body of the
commission is charged to identify and define the ethical issues raised by
biotechnology; to assess, from an ethical point of view, the impact of the
activities of the European Community in that area; to advise the commission
on the ethical aspects of biotechnology; and to ensure that the general public is kept properly informed.

In short, E.U. public policy concerning biotechnology takes place
within the triangle of economic interests: the market, public acceptance,

and the ethical values in a plural society. The work of the GAEIB should be seen against this background. Since 1992 ten "Opinions" have been published, one on "Ethical Aspects of Genetic Modification of Animals" (1996). In the request by the E.C. a number of ethical issues are raised: the transfer of human genes into animals, animal heath and welfare, genetic variation, and the extent to which animals may be subordinated to the needs of people. Also raised is the policy question, "Should community-wide guidelines consisting of ethical criteria be developed to assess research projects in the field of transgenic animals? Guidelines could have the advantage of signaling ethical problems at an early stage. Compliance with such guidelines could be a condition for participation in community-sponsored research in this field." In other words, the suggestion is that (at least in animal biotechnology) research and development should be assessed from the beginning, not only at the end when products are entering the marketplace.

The opinion could speak for itself. It could be characterized as a plea for a "policy of great prudence" in animal biotechnology, based on ethical conditions such as human safety, well-being, and welfare; human responsibility for animals, nature, and environment; the duty to avoid or minimize animal suffering; and the duty to reduce, replace, and refine the use of animals in research: "licensing bodies in all member states should have the task of assessing research projects and applications in the light of the above mentioned principles" (2.5). The recent Opinion on Ethical Aspects of Cloning Techniques (May 1997) refers to this position explicitly (2.1–2.3).

Let me finish by returning to public concern and public acceptance. Time and again the GAEIB stresses the importance of informing public opinion and creating platforms for public participation in decision making (e.g., paragraph 2.9). Science and technology take place not in a void but in the context of society. And they have an enormous impact on society. Therefore, society has a right to know what is going on in science and technology and to ask critical questions. People have a right to ask for openness and quality of research and for transparency of public policy. A democratic society is a participatory society.

26

Ethical Aspects of Cloning Techniques

Opinion of the Group of Advisers on the Ethical Implications of Biotechnology of the European Commission

The following points aim to shed light on the cloning debate by giving information on the scientific aspects of cloning and the ethical problems relating to them.

> 1.1. Cloning is the process of producing "genetically identical" organisms. It may involve division of a single embryo, in which case both the nuclear genes and the small number of mitochondrial genes would be "identical," or it may involve nuclear transfer, in which case only the nuclear genes would be "identical." But genes may be mutated or lost during the development of the individual: the gene set may be identical, but it is unlikely that the genes themselves would ever be totally identical. In the present context, we use the term "ge-

netically identical" to mean "sharing the same nuclear gene set."

1.2. It is inherent in the process of sexual reproduction that the progeny differ genetically from one another. In contrast, asexual reproduction (cloning) produces genetically identical progeny. This is a common form of reproduction in plants, both in nature and in the hands of plant breeders and horticulturists. Once a desired combination of characteristics has been achieved, asexual reproduction is the best way of preserving it. Asexual reproduction is also common among some invertebrate animals (worms, insects). Asexual reproduction in plants and invertebrates usually takes place by budding or splitting.

1.3. The first successful cloning in vertebrate animals was reported in 1952, in frogs. Nuclei from early frog embryos were transferred to unfertilized frog eggs from which the original nuclei had been removed. The resulting clones were not reared beyond the tadpole stage. In the 1960s, clones of adult frogs were produced by transfer not only of nuclei from early embryos but also of nuclei from differentiated larval intestinal cells. Later, clones of feeding tadpoles were obtained by nuclear transfer from differentiated adult cells, establishing that differentiation of cells involving selective gene expression does not require the loss or irreversible inactivation of genes. Nuclear transfer in frogs has not yet generated an adult animal from cells of an adult animal.

1.4. Nuclear transfer can be used for different objectives. Nuclear transfer in mice has been used to show that both a female and a male set of genes are required for development to birth. If the two pronuclei, taken from fertilized eggs and transferred into an enucleated egg, are only maternal or only paternal, normal development does not occur. This is not cloning, since the single embryo formed is not identical to any other embryo and the objective is not to multiply individuals.

1.5. Nuclear transfer has also been used for cloning in various mammalian species (mice, rabbits, sheep, cattle), but until recently only nuclei taken from very early embryos were effective, and development was often abnormal, for reasons that are not fully understood.

1.6. In contrast, cloning by embryo splitting, from the 2-cell up to the blastocyst stage, has been extensively used in sheep and cattle to increase the yield of progeny from genetically high-grade parents. Because of the different pattern of early development, embryo splitting is much less successful in mice. From a scientific point of view, it would probably not be very effective in the human, although monozygotic (one-egg) twins and higher multiples occur naturally at a low incidence.

1.7. In 1996, a new method of cloning sheep embryos was reported, which involved first establishing cell cultures from single embryos. Nuclei from the cultured cells were transferred to enucleated unfertilized sheep eggs, particular attention being paid to the cell cycle stage of both donor and host cells, and the eggs were then artificially stimulated to develop. Genetically identical normal lambs were born.

1.8. Cell cultures were then established not only from embryonic and fetal stages, but also from mammary tissue taken from a 6-year-old sheep. Nuclear transfer was carried out as before, and in 1997 it was reported that several lambs had been born from the embryonic and fetal transfers and one lamb named Dolly (out of 277 attempts) from the adult nuclear transfer. It is not known whether the transferred nucleus was from a differentiated mammary gland cell or from a stem cell.

1.9. From the point of view of basic research, this result is important. If repeatable it may allow greater insight into the aging process, how much is due to cell aging, and whether or not it is reversible. Such work may also increase our understanding of cell commitment, the origin of the cancer process, and whether it can be reversed, but at the present time the research is at a very early stage. Dolly may have a shortened life span or a greater susceptibility to cancer: if she is fertile, her progeny may show an increased abnormality rate, owing to the accumulation of somatic mutations and chromosomal damage.

CONCERNING THE APPLICATIONS OF ANIMAL CLONING

1.10. Potential uses of cloning animals are reported to include:

- in the field of medicine and medical research, to improve genetic and physiological knowledge, to make models for human diseases, to produce at lower cost proteins like milk proteins to be used for therapeutic aims, to provide a source of organs or tissues for xenotransplantation;
- in agriculture and agronomical research, to improve the selection of animals or to reproduce animals having specific qualities (longevity, resistance, . . .) either innate, or acquired by transgenesis.

1.11. From the point of view of animal breeding, the technology could be useful, in particular if it increases the medical and agricultural benefits expected from transgenesis (genetic modification of animals). By using genetic modification and selection in cultured cell lines, rather than in adult animals, it could become possible to remove genes, such as those provoking allergic reactions, as well as adding genes, for the benefits of human health.

1.12. Furthermore, transgenesis is an uncertain process: different transgenic animals express the introduced gene in a different manner and to a different extent, and do not "breed true." Cloning of adult animals of high performance, if it is possible, would reduce the number of transgenic animals needed and would allow human pharmaceuticals, for example, to be produced at a lower cost than would be possible otherwise.

1.13. If the use of cloning became more widespread in the animal breeding industry, for example, to bring the level of the general herd up to the level of the elite breeding populations, there is a danger that the level of genetic diversity could fall to an unacceptable degree. The introduction of artificial insemination in cattle raised similar problems.

CONCERNING HUMAN IMPLICATIONS

1.14. A clear distinction must be drawn between reproductive cloning aimed at the birth of identical individuals, which in humans has never been performed, and nonreproductive cloning, limited to the in vitro phase.

1.15. In considering human implications, we must again distinguish between cloning by embryo splitting and cloning by nuclear replacement (see 1.1). We must also distinguish between nuclear replacement as a means of cloning and nuclear replacement as a therapeutic measure, for example, to avoid the very serious consequences of mitochondrial disease. The latter situation, which would require an enucleated donor egg containing normal mitochondria, as it need not involve the production of genetically identical individuals, will not be considered further here although we appreciate that it will raise ethical problems of its own.

1.16. Embryo splitting in the human is the event that gives rise to monozygotic (one-egg) twins and higher multiples. It has been discussed in the context of assisted reproduction, as a means of increasing the success rate of IVF, but there is no evidence that it has ever been used for this purpose, nor that it would be effective if it were so used, because of the pattern of early development of the human embryo.

1.17. Monozygotic twins show us that genetically identical individuals are far from identical: they may differ from one another not only physically but also psychologically, and in terms of personality. Individuals cloned by nuclear transfer from an adult cell would of course be even more different from their donor, since they would have different mitochondrial populations, they would be different in age, and they would have had a different environment both before and after birth and a different upbringing. We are not just our genes.

1.18. There is no ethical objection to genetically identical human beings per se existing, since monozygotic twins are not discriminated against. However, the use of embryo splitting, or the use of human embryo cells as nuclear donors, deliberately to produce genetically identical human beings raises serious ethical issues, concerned with human responsibility and instrumentalization of human beings.

1.19. However, research involving human nuclear transfer could have important therapeutic implications, for example, the development of appropriate stem cell cultures for repairing human organs. It could also provide insights into how to induce regeneration of damaged human tissues. If such research resulted in embryonic development, the serious and controversial ethical issues concerning human embryo research would of course arise. Any attempt to develop methods of human reproductive cloning would require a large amount of human experimentation.

1.20. If adult cells were to be used as nuclear donors, we are still ignorant of the possible risks: whether the cloned individuals would have a shorter life span, a greater susceptibility to cancer, whether they would be fertile, and if so, whether they or their offspring would suffer from an abnormal rate of genetic abnormalities. Furthermore, the procedure would be immensely costly: each attempt would require several eggs and an available uterus, and many attempts would be unsuccessful. The issues of human responsibility and instrumentalization of human beings are even more ethically acute in this context.

THE OPINION CONCERNING CLONING OF ANIMALS

2.1. Research on cloning in laboratory and farm animals is likely to add to our understanding of biological processes, in particular aging and cell commitment, and hence may contribute to human well-being. It is ethically only acceptable if carried out with strict regard to animal welfare, under the supervision of licensing bodies.

2.2. Cloning of farm animals may prove to be of medical and agricultural as well as economic benefit. It is acceptable only when the aims and methods are ethically justified and when it is carried out under ethical conditions, as outlined in the GAEIB's Opinion No. 7 on the Genetic Modification of Animals.

2.3. These ethical conditions include:

- the duty to avoid or minimize animal suffering since unjustified or disproportionate suffering is unacceptable;
- the duty of reducing, replacing, and, when possible, refining the experimentation adopted for the use of animals in research;
- the lack of better alternatives;
- human responsibility for animals, nature, and the environment, including biodiversity.

2.4. Particular attention should be paid to the need to preserve genetic diversity in farm animal stocks. Strategies to incorporate cloning into breeding schemes while maintaining diversity should be developed by European institutions.

2.5. Insofar as cloning contributes to health, special attention should be paid to the public's right to protection against risks as well as their right to adequate information. Furthermore, if the costs of production are reduced, consumers should also benefit.

CONCERNING HUMAN IMPLICATIONS

2.6. As far as reproductive cloning is concerned, many motives have been proposed, from the frankly selfish (the elderly millionaire vainly seeking immortality) to the apparently acceptable (the couple seeking a replacement for a dead child, or a fully compatible donor for a dying child, or the attempt to perpetuate some extraordinary artistic or intellectual talent). Considerations of instrumentalization and eugenics render any such acts ethically unacceptable. In addition, since these techniques entail increased potential risks, safety considerations constitute another ethnical objection. In the light of these considerations, any attempt to produce a genetically identical human individual by nuclear substitution from a human adult or child cell ("reproductive cloning") should be prohibited.

2.7. The ethical objections against cloning also rule out any attempt to make genetically identical embryos for clinical use in assisted reproduction, either by embryo splitting or by nuclear transfer from an existing embryo, however understandable.

2.8. Multiple cloning is a *fortiori* unacceptable. In any case, its demands on egg donors and surrogate mothers would be out of the realms of practicality at the present time.

2.9. Taking into account the serious ethical controversies surrounding human embryo research: for those countries in which nontherapeutic research on human embryos is allowed under strict licence, a research project involving nuclear substitution should have the objective either to throw light on the cause of human disease or to contribute to the alleviation of suffering, and should not include replacement of the manipulated embryo in a uterus.

2.10. The European Community should clearly express its condemnation of human reproductive cloning and should take this into account in the relevant texts and regulations in preparation as the Decision adopting the Vth Framework Programme for Research and Development (1998–2002) and the proposed directive on legal protection of biotechnological inventions.

GENERAL REMARKS

2.11. Further effort must be made to inform the public, to improve public awareness of potential risks and benefits of such technologies, and to foster informed opinion. The European Commission is invited to stimulate the debate involving public, consumers, patients, environment and animal protection associations, and a well structured public debate should be set up at European level. Universities and high schools should also be involved in the debate at European level.

2.12. These new technologies increase the power of people over nature and thus increase their responsibilities and duties. Along the line of the promotion by the European Commission of research on the ethical, legal, and social aspects of life sciences, the commission should continue to foster ethical research on cloning-related areas, at a European level.

COMMENTARY
by Anne McLaren

In Lewis Carroll's *Alice Through the Looking-Glass*, Humpty Dumpty said: "When *I* use a word, it means just what I choose it to mean—neither more nor less." One may not agree with this approach, but it may well be preferable to using a word: (a) with only a vague idea of its meaning, or (b) with many possible meanings.

Consider cloning. DNA can be cloned, cells can be cloned, computers can be cloned. The Group of Advisers on the Ethical Implications of Biotechnology (GAEIB) appreciates the importance of defining words to avoid confusion. They define cloning in the present context as the process of producing "genetically identical" organisms. They then define what they mean by "genetically identical."

In outlining the scientific background to cloning, the opinion emphasizes that not all cloning involves nuclear transfer, nor does all nuclear transfer imply cloning. The sheep Dolly was produced by transfer of an adult nucleus into an enucleated egg, and hence was a clone of the nucleus donor; but human monozygotic (so-called "identical") twins are also clones, produced by embryo splitting, a procedure quite widely used in cattle breeding. If the nucleus of an egg from a woman with abnormal mitochondria were transferred for therapeutic reasons to an enucleated egg containing normal mitochondria, the procedure would not constitute cloning, since no genetically identical organisms would be produced. Careless wording by legislators can cause unintended harm.

In nonhuman animals, GAEIB recognizes that cloning by nuclear transfer could be of value both for research on such fundamental questions as cell commitment and differentiation, and the cellular basis of aging (since Dolly's nuclear donor was already six years old, is Dolly now one or seven years old?), and also for possible applications of benefit to medicine and agriculture. Compared with the present approach through transgenic technology, nuclear transfer from cell lines could substitute selection between cells for selection between animals, and thus reduce the amount of animal experimentation required for a particular beneficial outcome.

The conditions under which the group regards animal experimentation in general as ethically acceptable have been outlined in an earlier opinion. Clearly, strict regard must be paid to animal welfare, under the supervision of licensing bodies, as well as to the objectives of the research. There is a duty to avoid or minimize animal suffering: unjustified or disproportionate suffering is ethically unacceptable.

The human implications of cloning are more remote but ethically more serious. The group distinguished between reproductive cloning, which it defined as cloning (either by nuclear transfer or embryo splitting) aimed at the birth of genetically identical individuals, and nonreproductive cloning, involving only in vitro studies, concerned, for example, with tissue repair and regeneration. Any human embryo research raises profound ethical problems that GAEIB has not as yet addressed; but for those countries in which nontherapeutic human embryo research is permitted under strict license, the opinion insists that nuclear transfer should be included only if the aim of the project is to alleviate suffering or to throw light on the causes of human disease, and in no case should a manipulated embryo be replaced in the uterus.

Much discussion in the group was devoted to the ethical status of human reproductive cloning. Most of the arguments against cloning by deliberate embryo splitting apply *a fortiori* to cloning by nuclear transfer. The most immediate ethical argument against cloning concerns risk, the possibility of doing harm. The risks of producing fetal mortality and congenital abnormalities by embryo splitting may be great; for nuclear transfer, they would be far greater, and could include increased risks of cancer and possible inherited defects passed on to future generations.

But even if a time ever came when such procedures were considered safe enough to attempt clinically, would it be ethically acceptable to clone humans? Although the group was unanimously of the opinion that it would *not* be acceptable, conclusive ethical arguments proved hard to marshal. It has been argued that the unique identity of human beings must be protected; but monozygotic twins are at least as identical genetically as any deliberately cloned human beings would be, while all the important influences of upbringing and environment that make monozygotic twins not identical would ensure that individuals produced by nuclear transfer were still more different from one another and from their donor. If we do not wish to impugn the unique identity of each monozygotic twin, it is hard to base a convincing argument against cloning on this concept. It is often stated that cloning would offend against human dignity; but dignity is hard to define in this context. Some applications of cloning would be unacceptable on eugenic grounds, but not all. Certainly the degree of instrumentalization, using human beings as means to an end, that would be involved in any reproductive cloning procedure was regarded by the group as ethically unacceptable; but this is a relative argument, since in other contexts we are prepared to accept, to some degree, using people as a means to an end. Where should one draw the line?

The view that cloning people would be ethically unacceptable is not confined to GAEIB. It is shared by very many people, all over the world. For that reason alone, it would not be appropriate for regulatory bodies to allow human reproductive cloning. But discussions about cloning should be encouraged to continue. In the last two paragraphs of its opinion, GAEIB calls on the European Commission both to foster debate among universities, high schools, and the general public, and to promote research into the ethics of cloning-related areas.

27

Cloning Issues in Reproduction, Science, and Medicine

A Report by a Joint Working Group of the
UK Human Genetics Advisory Commission and the
UK Human Fertilisation and Embryology Authority

BACKGROUND AND COMMENTARY
Onora O'Neil & Ruth Deech

News of the UK Human Genetics Advisory Commission's first meeting was overshadowed by the Roslin Institute's announcement of Dolly the sheep, the first vertebrate cloned from another adult. Since that event, cloning has been a subject of intense debate. While the major motivation for the work that resulted in Dolly was to develop methods for the genetic improvement of livestock, subsequent commentary has understandably been marked by fears that this technique might be open to misuse.

The Human Genetics Advisory Commission (HGAC) was established in December 1996 to provide independent advice to UK Health and Industry ministers on issues arising from developments in human genetics that have wider social, ethical or economic consequences. It also advises on ways to build understanding of the new genetics.

The Human Fertilization and Embryology Authority (HFEA) was created by the Human Fertilisation and Embryology Act 1990 and has been operating since 1991. Its major function is to license all clinics offering treatments involving the use of embryos created outside the body (in vitro fertilization) or donated eggs or sperm (e.g., donor insemination). The HFEA also licenses the storage of eggs, sperm, and embryos and all permitted research on human embryos. The HFEA has a statutory duty to keep the whole field of fertility treatment and research under review and make recommendations to the government, if asked to do so. It maintains a confidential database recording all treatments.

Although the creation of human offspring by cloning (or "reproductive" cloning) is prevented in the United Kingdom either explicitly through the 1990 act (nuclear substitution of an embryo) or by the HFEA through its licensing system, both the HGAC and the HFEA recognized that this was a subject on which reasoned and thoughtful discussion was needed. A joint public consultation exercise was held to explain the issues and stimulate wide and informed debate. More than 1,000 copies of the document were issued, and it was also accessed from the HGAC's website. Nearly 200 responses were received—about 40% from individual members of the public and the rest from a wide range of constituencies, including professional bodies, religious organizations, and lay groups, many of whom organized their own discussion groups or otherwise canvassed views.

The HGAC/HFEA joint report was submitted to ministers and published in December 1998. It set out the scientific developments, described the current legal and administrative arrangements covering the parameters within which the HFEA may issue treatment, storage and research licenses, and presented the findings of the consultation exercise.

The consultation had resulted in concerns being expressed about human reproductive cloning, including safety and other ethical issues, and supported a total ban on its use. The report concluded that, as the Human Fertilisation and Embryology Act 1990 had proved effective in dealing with new developments relating to human cloning, then these safeguards should be recognized as fully adequate to forbid human reproductive cloning in the United Kingdom. However, it was further suggested that the government might wish to consider the possibility of introducing legislation that would explicitly ban human reproductive cloning regardless of the techniques used, so that the full ban would depend not upon the decision of a statutory body (the HFEA) but would itself be enshrined in statute.

Many respondents to the consultation saw benefit in new therapeutic uses of cell nucleus replacement (CNR) techniques that might be developed to treat serious medical conditions. Some of the therapeutic advances now being developed were never envisaged when the 1990 act was drafted. Therefore, the report also recommended that the secretary of state for health should consider specifying in regulations two further purposes for which the HFEA might issue licenses for research so that potential benefits can clearly be explored. First, for developing methods of therapy for mitochondrial disease (which would not involve cloning) and second for developing in vitro cell-based methods of therapy for diseased or damaged tissues or organs.

Some respondents expressed the view that there is a need for international legislation to prohibit human reproductive cloning. The report reviewed international developments in this area and highlighted the difficulties that exist in finding mutually acceptable and internationally agreed definitions for even quite simple concepts.

Views expressed about genetic identity indicated that there were fewer concerns about entitlement to a unique genetic identity. These nevertheless highlighted collateral issues such as confidentiality and consent—issues that are not unique to human genetics.

The consultation process itself drew attention to the need for greater public understanding of human genetics, and the report mentions some initiatives arising out of the consultation which promoted wider public debate.

Finally, in recognition of the pace of scientific advances in the area of human genetics, the report recommended that the issues be examined again in five years' time in the light of developments and public attitudes toward them.

It is hoped that the report will contribute to improved understanding of the issues around human cloning and CNR techniques, and how they might best be addressed in the future. The potential benefits of this technology should be maximized while at the same time concerns are recognized and adequate safeguards are implemented. It is now up to government to decide what action, legislative or otherwise, should be taken.

I. INTRODUCTION

1.1. In February 1997 Dolly the sheep, the first vertebrate cloned from a somatic cell of an adult animal, generated considerable public interest and much media comment. A major motivation for the work was to improve methods for the genetic improvement of livestock. The technology could also be used to improve the efficiency of production of transgenic livestock, with potential benefits in, for example, increasing production of human proteins in the milk of transgenic animals (e.g., proteins used to treat blood clotting disorders such as hemophilia). Although hailed as a remarkable scientific development the birth of Dolly raised concerns both nationally and internationally about the implications and use of this technology, particularly the possibility of cloning human beings. Reactions were worldwide and resulted in several international initiatives.

1.2. Dolly was the result of a collaborative project between the Roslin Institute and PPL Therapeutics PLC to test the suitability of different sources of cells for nuclear replacement. She was derived from cells taken from the udder of a 6-year-old Finn Dorset ewe which were then cultured in the laboratory. Two hundred, seventy-seven of these cells were then fused with 277 unfertilized eggs from which the nucleus had been removed to create "reconstructed eggs." This process resulted in 29 viable reconstructed eggs, each with a nucleus from the adult animal, which were then implanted into surrogate Blackface ewes. One gave birth to Dolly.[1]

1.3. However, Dolly was not the first sheep to be created using nuclear replacement technology. In 1996, it was reported that sheep embryos had been cloned using nuclear replacement technology and had resulted in the birth of two genetically identical sheep, Megan and Morag.[2] The difference between Dolly and Megan and Morag was the nuclear donor source: Dolly was derived from an adult sheep, Megan and Morag from a sheep embryo.

1.4. At the end of the year following the birth of Dolly, the Roslin Institute and PPL Therapeutics PLC announced the birth of Polly.[3] She was a transgenic sheep produced by transfer of the nucleus of a cultured fetal fibroblast. She carried a human gene for blood-clotting factor IX, which is used for treatment of hemophilia.

1.5. The announcement of Dolly the sheep prompted an inquiry by the House of Commons Science and Technology Committee, which published its findings in March 1997 in a report "The Cloning of Animals from Adult Cells."[4] The committee believed that concerns over the cloning of Dolly may have overshadowed potential benefits, and trusted that the Human Genetics Advisory Commission would advise on the implications of the work for human genetics. It also recommended that work which would create "experimental human beings" should not be carried out, and suggested that Parliament should reaffirm a ban on human reproductive cloning.

1.6. This last point was addressed in a reply to a parliamentary question given by the minister of state for public health on June 26, 1997,[5] who said, "We regard the deliberate cloning of human beings as ethically unacceptable. Under United Kingdom law, cloning of individual humans cannot take place whatever the origin of the material and whatever technique is used."

1.7. The Human Genetics Advisory Commission (HGAC) was established in December 1996 in response to a report by the House of Commons Science and Technology Committee, as a nonstatutory advisory body. It provides independent advice to UK health and industry ministers on issues arising from developments in human genetics that have social, ethical, and/or economic consequences. The commission was also asked to advise on ways to build public understanding of the new genetics. The commission is charged with setting its own priorities, although from time to time it may be requested to provide urgent advice to ministers.

1.8. The Human Fertilisation and Embryology Authority (HFEA) took up its powers in August 1991 as a result of the passage of the Human Fertilisation and Embryology Act 1990 (HFE Act). The first statutory body of its type in the world, the HFEA's creation reflected public and professional unease about the potential future of human embryo research and infertility treatments, and a widespread desire for statutory regulation of this ethically sensitive area. The HFEA's principal tasks are to license and monitor those clinics that carry out in vitro fertilization, donor insemination and human embryo research. The HFEA regulates the storage of gametes and embryos and keeps a register of all licensed treatments carried out in the UK. It also keeps under review information about the development of human embryos and the provision of treatment services and activities governed by the HFE Act and advises the secretary of state for health if asked about those matters.

1.9. The HGAC and the HFEA decided to explore ways of holding a public consultation exercise into the implications of cloning developments. A joint working group, consisting of members of both bodies, was established to consider the planning, drafting, distribution, and analysis of a joint consultation paper on the issues for human genetics arising from advances in cloning technology.

1.10. The government response to the Science and Technology Committee's report was published in December 1997.[6] It reiterated the minister of state for public health's statement, explained that the HGAC and the HFEA were exploring ways of holding a public consultation exercise on cloning, and said that the government would consider carefully, in the light of developments, whether the legislation needed to be strengthened in any more specific way, taking into account the views of members of Parliament, the HGAC, the HFEA, and the responses received to a more general consultation on the broader issues.

1.11. The HGAC/HFEA consultation paper[7] was published on January 29, 1998, and comments were invited by April 30, 1998. The paper differentiated between the different concepts which are all broadly termed as "cloning" and sought views on these different meanings, the ethical implications raised by the possibility of human cloning, including the safety of the technique if it were applied to humans, and the ethical concerns raised by cloning in specific circumstances. It also raised questions about more abstract concepts such as individuality and human dignity.

1.12. Since the consultation paper was circulated, there has been independent confirmation that Dolly was cloned from the cells of an adult sheep[8] and cloning by nuclear replacement using nuclei from adult cells has been successfully extended to mice[9] and also to cattle[10] (two cloned calves born recently in Japan, using a procedure similar to that which produced Dolly). It is therefore clear that the birth of Dolly was not a "one off" event, due to some unique concatenation of circumstance. The feasibility of nuclear replacement cloning is therefore not confined to one species. Although the efficiency of the procedure is very low in mice, as it is in sheep, it is likely to increase more rapidly with research on mice. The molecular genetics of mice is better understood and, moreover, the gestation time is shorter.

2. CONSULTATION

2.1. The consultation document "Cloning Issues in Reproduction, Science and Medicine," attracted a great deal of interest both nationally and internationally. It was launched at a press conference attended by representatives of the national and international press and was widely reported, with coverage including details of where copies of the document could be obtained. More than 1,000 copies were distributed. We believe the document also reached a larger audience through the HGAC website and by others copying and circulating the paper.

2.2. The consultation sought general comments about how the technology that led to Dolly might develop and the opportunities and problems that might be raised by human reproductive cloning and other applications of nuclear replacement technology. It also invited views on priorities for the future and the ethical settings in which these scientific developments are taking place. It sought comments on any other ethical issues raised by human cloning that respondents identified. It was requested that responses be structured around replies to six questions. Respondents were also invited to make suggestions about what advice might be offered to ministers on ways to build public confidence and understanding of the new developments in genetic techniques.

2.3. Nearly 200 responses were received—about 40% from individuals, and the rest from a wide range of constituencies— scientists, clinicians, academics, religious groups, ethicists, lawyers, industry, and lay groups. The latter included groups of individuals in local communities that had come together to discuss the issues raised and to share views and concerns. Responses from a number of constituencies involving groups of individuals sometimes did not reach a consensus, but reflected a range of opinions.

2.4. Responses varied enormously, some expressing a horror at the very idea of any form of cloning without addressing the specific issues raised in the document, while others were very detailed in their consideration of the issues. Some respondents provided additional questions and arguments to those raised in the consultation document. This wide range or views was welcome, reflecting common reactions, misunderstandings, and hopes and fears about this rapidly developing technology. A number of respondents congratulated the authors of the document for setting out the issues clearly, but others were concerned about the language used to express some issues. For example, it was suggested that the differentiation between reproductive and therapeutic cloning and the positive and negative phrasing of the questions may have had some influence on the responses received.

2.5. The widest spread of opinions occurred in responses from individuals and religious groups, but the overall balance of opinion was generally reflected throughout the various constituencies where sufficient numbers of responses from each grouping made comparisons possible. There were a small number of "don't know" answers to most questions, where respondents did not feel that they had sufficient knowledge to express a view.

3. THE LEGISLATIVE AND ADMINISTRATIVE CONTEXT

Background

3.1. The history of the Human Fertilisation and Embryology Act 1990 is usually traced back to 1978 with the birth of Louise Brown, the world's first IVF baby. Although the 1990 act also covers treatments such as donor insemination, the development of IVF treatment and human embryo research was the main impetus for the development of the current legislation and the HFEA. An important milestone was the publication in 1984 of the Warnock Report.[11] It was the Warnock Report that first suggested the setting up of a statutory body to oversee the practice of certain fertility treatments and human embryo research. [Editor's note: See chapter 21 for Baroness Warnock's views on the regulation of new technology specifically in the context of cloning.]

3.2. Without a license from the HFEA the 1990 act makes it a criminal offense to carry out any treatment using human embryos outside the body, or to use donated gametes; also to store any eggs, sperm, or embryos and to undertake any research into human embryos. The act also sets out the parameters within which the HFEA may issue treatment, storage and research licenses. For example, the HFEA may not license any research project involving human embryos after the primitive streak has appeared or after the 14th day of development (whichever is the sooner).

3.3. Around 23% of respondents indicated that in their view any form of embryo research/manipulation was simply wrong because they believe that the embryo possesses the full moral status of a human being. There were 24% who thought that the 14-day limit was arbitrary, some of them considering that it could be open to the possibility of extension. Both these points of view are questioning decisions enshrined in the 1990 act. Both the HGAC and HFEA have respect for these opinions. However, the relevant issues were fully debated, both in Parliament and by the wider public, at the time of the passing of the HFE Act. While the decisions then reached did not command universal ethical assent, they are the basis of the present policy and they necessarily form the framework within which the HGAC/HFEA must make their recommendations to ministers. It would not be appropriate to use this limited inquiry into cloning to reopen the wider questions relating to work with human embryos.

Cloning

3.4. The 1990 act (section 3 (3)(d) expressly prohibits one type of cloning technique, namely the nuclear substitution of any cell that forms part of an embryo. Further, the act (section 3(1)) requires a license from the authority for any creation, use or storage of a human embryo outside the body. The technique used to create Dolly involves nuclear substitution into an egg and not into an embryo. Thus it is not specifically covered by section 3(3)(d). Some have argued that, as fertilization is not involved, section 3(1) also does not apply. The Department of Health and the HFEA have taken counsel's advice on this issue. As a result, both ministers and the authority reject this position and are content that the act does allow the HFEA to regulate nuclear replacement into an unfertilized egg through its licensing system.

Existing HFEA Requirements

Licensing

3.5. Each center in the UK that offers clinical treatments involving in vitro fertilization or donor insemination, storage of gametes or embryos, or that carries out research involving the use of human embryos must be licensed by the HFEA. All licensed centers are subject to an annual inspection. All licensing decisions are taken by Authority License Committees, which may refuse to grant a license or to renew a license. They also have the power to suspend or revoke a license, which has already been granted. Where there is the possibility that a criminal offense has been committed a License Committee will decide what action should be taken, including, whether the police should be involved or the matter referred to the director of Public Prosecutions.

3.6. The HFEA is required by law to produce an annual report. This provides general information about the number of licensed clinics and the range of licensed research, and discusses the clinical, scientific, and ethical issues around current and anticipated developments in reproductive technology. The report also gives information about any investigations the HFEA has carried out in respect of alleged or apparent breaches of the act.

3.7. The approval of a properly constituted independent ethics committee is a necessary prerequisite to the authority considering an application for a research license. In addition, all applications for research licenses are submitted for peer review. Peer reviewers comment on a number of issues including the importance of the work's originality and justification for it. Their recommendations are submitted to a License Committee, which has the responsibility of deciding whether a license should be granted and on what conditions.

Policy

3.8. Before the consultation document was issued the HFEA had stated its policy of not permitting human reproductive cloning, a stance with which the HGAC was fully in agreement.

3.9. The authority's policy on embryo splitting was developed in 1994 following reports in 1993 that a team of researchers in America had used embryo splitting in a research project. The authority agreed that it would not permit embryo splitting to be used in treatment cycles and would only permit embryo splitting in research projects where the aim of the project did not include increasing the number of embryos for transfer. Members have subsequently agreed on a similar line for the nuclear replacement of eggs, that is, that no reproductive cloning involving the technique would be licensed, but that research with a nonreproductive aim would be considered.

3.10. Research applications involving the creation or use of human embryos created through the nuclear replacement of eggs are likely to be some way off for a variety of reasons. In line with general HFEA policy, further research using animal embryos is needed before the use of human embryos in research would be appropriate.

---

4. HUMAN REPRODUCTIVE CLONING

4.1. Section 1 of this report quotes the minister of state for public health's statement in June 1997 (see paragraph 1.6). Section 3 explains the legal position in detail. Nevertheless, the Working Group felt it essential as part of the consultation exercise to gauge public opinion on this issue and to ascertain the reasons for that opinion. In consequence, a specific question about whether the creation of a clone of a human person would be an ethically unacceptable act was included in the consultation document.

4.2. The consultation document set out a number of scenarios where cloning technology could be applied to make a "copy" of another human being, envisaging single or multiple "copies" of a living or a dead fetus, baby, child, or adult. These included parents who might wish to "replace" an aborted fetus, dead baby, or child killed in an accident: produce a sibling to be a compatible tissue or organ donor for a child dying from, say, leukemia or kidney failure; or an individual attempting to "cheat death" by using cloning technology. Mention was also made of the possibility of selecting characteristics in offspring or to assist human reproduction in the case of infertile couples or lesbian couples. Views were sought on the acceptability of cloning in all, or any, of these circumstances.

4.3. The response to the consultation was conclusive. There was very little support for reproductive cloning, though there were a few who saw benefit in certain circumstances, mainly in connection with infertility treatment. Of the respondents, 80% thought it was an ethically unacceptable procedure, an opinion that was endorsed within each of the different groupings of those responding.

4.4. Safety is itself an ethical issue. Nuclear replacement in animals is at present very inefficient. Few of the reconstructed embryos develop, some develop abnormally, some die at or soon after birth. In humans, the wastage of human eggs and the high risks of miscarriage and congenital malformation alone would exclude any realistic prospect of reproductive cloning. However, since issues of efficiency and safety may eventually be resolved, it is necessary to analyze further the reasons why human reproductive cloning is so widely judged to be ethically unacceptable.

4.5. A central ethical issue is the widely accepted moral principle that human beings may never be treated merely as means to an end, but only as an end. Many of the suggested reasons for which reproductive cloning might be employed have a strongly instrumental character to them, for they contemplate bringing human beings into existence for reasons outside the persons themselves. Examples would be the "replacement" of a lost relative or the making available of compatible tissue for transplantation into another. It would be morally demeaning and psychologically damaging for someone to learn that the primary reason for their existence lay not in their own value, but in their utility for another purpose, as the substitute for someone else, or for the benefit of someone else. Moreover, in the case of attempted "replacement," the action would be based on the fallacious equation of a person with his or her genome (see Section 6).

4.6. These particular ethical arguments would not apply to the possible use of cloning as an extreme measure to relieve infertility in a case where nuclear replacement seems the only way to produce an embryo for implantation which incorporated genetic material derived from one of the intending parents. In these latter circumstances there would be good reason to suppose that the person brought into being would be highly valued for his or her own sake. However, other ethical considerations would also be relevant. The relief of the pain of infertility is, in general, a good end, but it is not an absolute end to be achieved without regard to the ethical acceptability of the means employed for that achievement. The wish for genetic offspring is a natural human aspiration, but this has to be held in balance with other desirable aspects of human well-being, and it cannot be given an overriding priority above all other considerations. While the desire for children, and feelings of solidarity with kin, are the source of much human good, too exclusive an emphasis on genetic connection can lead to distortions.

4.7. The use of reproductive cloning to relieve infertility would involve risks likely to be ethically unacceptable for human use in the foreseeable future (see paragraph 4.4 above). There are further ethical difficulties about the source of the genetic material that could be used in nuclear replacement to relieve infertility. If the nucleus was derived from one of the parents, this would generate an unbalanced genetic relationship of an entirely unprecedented kind within the family. A child cloned in this way would have a unique set of family relations, as he or she would inherit a *complete* genetic makeup from one of the "parents" and have no genetic connection with the other. This complete genetic identity between the child and one parent would constitute a novel situation of which there is no previous experience and there must be uncertainties and doubts about the effects this would have on the family and the child. For these psychological and social reasons, there must be serious ethical doubts about the propriety of bringing about such a set of relationships. All these considerations give rise to serious ethical concerns about reproductive cloning as a means to relieve infertility. There is a difference from donor insemination, in which an entirely new individual is conceived as a result of the fertilization of a gamete produced from one parent with a donor gamete, so there is a partial connection with one parent and none with the other.

4.8. In relation to using reproductive cloning as a means to relieve infertility it is also necessary to consider the wider question of public policy. Decisions about what may be done involve not only the couple themselves and their medical advisors but also society as a whole. For any type of infertility treatment to function satisfactorily there has to be a degree of social acceptance of the measures being taken. It is quite clear that human reproductive cloning is unacceptable to a substantial majority of the population. A total ban on its use for any purpose is the obvious and straightforward way of recognizing this. The results of the consultation fully support government policy in this respect.

5. THERAPEUTIC USE OF CELL NUCLEUS REPLACEMENT (CNR)

5.1. The consultation document attempted to distinguish between reproductive cloning or the production of an entire animal from a single cell by asexual reproduction, and what was termed "therapeutic cloning"—applications of nuclear replacement technology that do not involve the creation of genetically identical individuals. This definition led to some criticism. Some thought that the distinction between reproductive and therapeutic cloning was arbitrary, others just responded negatively to anything described as "cloning," and some were upset at the description of identical twins as "a natural form of cloning." It is clear that the term "cloning" carries an automatic stigma for many because of its association with imagery such as that portrayed in *Brave New World*.¹² To avoid this confusion this section of the report has been headed "Therapeutic Use of Cell Nucleus Replacement (CNR)" and concentrates on new techniques which might be developed to treat serious medical conditions. This could include (see paragraph 5.10) a use of CNR that would not involve any form of cloning.

5.2. There was a significant response, primarily from individuals, rejecting all research using human embryos. However, the Human Fertilisation and Embryology Act permits licensed research on human embryos up to 14 days of development (this stage of development immediately precedes the primitive streak stage at which development of individual embryos is established and cell determination for the future fetus sets in). An amendment to prohibit the creation of embryos for research was defeated by a large majority in the House of Commons and by a very large majority in the House of Lords. Thus the production of a human embryo by CNR for research purposes could be permitted, provided that the research project was licensed by the HFEA according to the strict criteria that such a license demands.

5.3. The most likely objective of a research project involving the use of CNR would be to create a cultured cell line for the purposes of cell or tissue therapy. People who have tissues or organs damaged by injury or disease (e.g., skin, heart muscle, nervous tissue) could provide their own somatic nuclei and by using these to replace nuclei in their own or donated eggs, individual stem cells (not embryos) could be produced in culture. These cells could then be induced (by exposure to appropriate growth factors) to form whichever type of cell or tissue was required for therapeutic purposes with no risk of tissue rejection and no need for treatment of the patient with immunosuppressive drugs.

5.4. For some processes somatic noncloned cultured cells can be used for some kinds of tissue repair. They have a disadvantage, however, in that, depending on the age of the individual, they may have a limited lifespan. Other approaches to the treatment of degenerative disease and the repair of tissue damage, avoiding the risk of tissue rejection, may have been perfected before any success has been achieved with the types of embryo research outlined in paragraph 5.3 above. It may prove possible to treat tissues in such a way as to abolish their antigenicity, or custom-made "humanized" transgenic animals may provide tissues that can be successfully transplanted to any recipient, without the need for permanent treatment with immunosuppressive drugs. However, these possibilities are again speculative, and unlikely to be available for clinical testing for a decade or two. It would therefore seem unwise to rule out absolutely any lines of research not involving reproductive cloning that might prove of therapeutic value.

5.5. Some research has already been licensed by the HFEA into the possible generation in vitro of stem cell lines from human embryos for the purposes of analyzing the factors that affect the development of embryos fertilized and grown in vitro and assessing their development potential. There is a recent report from the United States of the successful derivation from human embryos (fertilized in vitro and donated for research) of cell lines resembling stem cells in many respects.[13] It therefore seems likely that applications for research projects involving CNR in human oocytes may be received by the HFEA within the next few years. Any such project would require a source of oocytes donated for research which at present are not widely available. However in vitro techniques for maturing very immature oocytes from human ovaries are being devised, and these could possibly provide a source of enucleated oocytes adequate for research use.[14]

5.6. The eventual clinical use of such procedures would be to provide immunologically compatible tissues for the treatment of degenerative diseases of, for example, the heart, liver, kidneys, and cerebral tissue, or to repair damage to skin or bone. The potential value of such techniques to human medicine is enormous. However, restrictions have been placed on the authority in schedule 2 of the HFE Act on the circumstances in which a license may be issued.

5.7. Schedule 2 of the 1990 act states that the HFEA cannot authorize a research project "unless it appears to the Authority to be necessary or desirable for one the following purposes:

(a) promoting advances in the treatment of infertility;
(b) increasing knowledge about the causes of congenital disease;
(c) increasing knowledge about the causes of miscarriage;
(d) developing more effective techniques of contraception;
(e) developing methods for detecting the presence of gene or chromosome abnormalities in embryos before implantation.

or for such other purposes as may be specified in regulations."

5.8. The 1990 act further requires that such licenses can only be granted if the HFEA is satisfied that any proposed use of embryos is necessary for the purposes of the research.

5.9. Since therapeutic approaches to disease or tissue damage are not at present included in the purposes for which research can be licensed under the HFE Act (see paragraph 5.7), the making of new regulations would be required to extend the scope of the act to include these purposes.

5.10. One potential future application of CNR would be for the avoidance of mitochondrial diseases. These life-threatening and debilitating diseases are caused by defects in the mitochondria, which are small organelles located in the cytoplasm of each cell. Defective mitochondria are transmitted from the mother, in her egg cells, to all her offspring. It has been suggested that a woman suffering from such a disease could have a healthy child if the nuclear material from one of her eggs was transferred before fertilization into a donor egg from which the nuclear material had been removed. Nuclear replacement between eggs has been successful in animals. It is not followed by the high incidence of embryonic mortality and abnormal development that characterizes nuclear replacement procedures using somatic nuclei that require genetic reprogramming (as with Dolly). Nuclear replacement from one egg to another is not cloning, since after fertilization the embryo is not identical to the mother, nor to any other embryo. Similarly, because the use of CNR for the avoidance of mitochondrial disease is not at present included in the purposes for which research can be licensed under the HFE Act (see paragraph 5.7), the making of new regulations would be required to extend the scope of the act to include this purpose.

5.11. A significant number of respondents expressed fears and reservations about the possible commercialization of therapeutic uses of CNR techniques. Similar anxieties arise in connection with any major advance in medical intervention. There is an understandable desire on the part of the public that curative procedures should not simply be exploited as sources of financial gain for their developers, but that there should be respect for the public good and corresponding access to these techniques for those who would benefit from them. A balance has to be struck between affording a reasonable recompense to those who have exercised initiative (and undertaken the risk involved in major and costly development programs) and ensuring that the needs of the sick are properly met. The system of patenting is intended to provide a degree of such safeguard, for it requires that knowledge relevant to the new invention is available in the public domain. While granting the discoverer a limited period of protected benefit. There does not seem to be any reason why developments in the field of nuclear replacement therapy should differ significantly from other kinds of medical advances in this respect.

5.12. It has been questioned whether the 14-day limit for human embryo research could be breached by serial nuclear transfer. The HFEA and the HGAC take the view that this is not the case. Whether the nucleus to be replaced in an enucleated oocyte is taken from an adult or from another embryo, the clock is put back to the beginning, embryonic development starts over again, and the primitive streak stage specified in the act would still not be reached within the 14-day time limit.

---

6. GENETIC IDENTITY

6.1. Question 3 in the consultation document addressed the issue of genetic identity. In our view, persons are more than their genes, their nature and character being substantially influenced by their nurture and life experiences. Personhood derives from a humanity that is expressed through relationships with others. This is made clear by the unquestionable individuality enjoyed by identical twins, despite their having exactly the same genome, together with the common properties that flow from that. It is clear from this example that the existence of a clone (in this case naturally occurring) is not in itself a threat to an individual's identity. Therefore, the claim that each person is entitled to a unique genetic makeup is a correspondingly questionable assertion. Of itself, it could not prove an adequate ethical objection to human reproductive cloning, though one should note that the latter could produce genetically identical persons in different generations, which is impossible naturally and which could raise novel problems of which there is no prior experience to enable evaluation.

6.2 This was acknowledged by half the respondents to question 3 in the consultation document. Nevertheless, the responses indicated that there was uneasiness in relation to issues collaterally related to this question. These seemed to center on three points:

(i) The right to confidentiality with respect to knowledge of one's genetic makeup. This is a part of the general medical ethical requirement of confidentiality about a patient's health information. We fully support this requirement.

(ii) The right to exercise a veto on the manipulative use that another might make of an individual's genome. Whether in relation to participation in a research programme or in relation to other use of genetic material selected from a person, there is a clear obligation to obtain full and informed consent for any such usage.

(iii) A new kind of right is being asserted where it is also claimed, as some respondents suggested, that a person's genetic makeup should not be directly determined by the deliberate choice of another. This proposition requires careful evaluation. It would be one thing to require a person to be genetically identical to another (the reproductive cloning that we have rejected), or for someone to be given genes that made him seven feet tall because his parents wanted a basketball champion ("designer babies" an issue lying outside the scope of this report). It would be another thing to ensure, when and where possible, that a disease gene be eliminated.

6.3. As far as the issues relevant to this report are concerned, it does not seem that new issues arise beyond those covered already by the requirements of ethical medical practice.

7. INTERNATIONAL

7.1. The consultation document gave brief details of the laws affecting cloning in several European countries. These invariably post-date the Human Fertilisation and Embryology Act

1990, which was the first to address the issues and formed the basis of much of the legislation in other countries that followed it.

7.2. Some who responded to the consultation document saw benefit in the international harmonization of legislation relating to cloning. There have been a number of significant international agreements to prohibit human reproductive cloning. These include:

(i) a UNESCO Declaration on the Human Genome and Human Rights, unanimously adopted on November 11, 1997,[15] of which Article 11 states, "Practices which are contrary to human dignity, such as reproductive cloning of human beings, shall not be permitted":

(ii) a protocol forbidding the cloning of human beings developed under the Council of Europe Convention on Human Rights and Biomedicine.[16]

(iii) a European Commission Directive on the Legal Protection of Biotechnological inventions prohibits the issue of any patent on work leading to intentional cloning of human beings.[17]

(iv) at the advisory level, the European Commission's former "Group of Advisers on the Ethical Implications of Biotechnology" called for the commission to express clear condemnation of human reproductive cloning. The group has been expanded to form the "European Group of Ethics in Science and New Technologies" which met for the first time in February 1998. The new group is composed of 12 experts, and its remit has been enlarged to cover all new technologies as well as scientific research: and

(v) at the fifty-first session of the World Health Assembly, meeting in Geneva (May 11–16, 1998) a resolution was adopted on the social implications of cloning on human health and circulated to member states.[18]

7.3. In the United States, following the announcement of the birth of Dolly the sheep in February 1997, President Clinton asked the National Bioethics Advisory Commission for advice within 90 days on the ethical, legal, and scientific issues surrounding human cloning. The publication of its report, "Cloning Human Beings,"[19] contributed to the public debate in the United States but as yet no legislation had resulted. In addition, in early November 1998 President Clinton instructed the same commission to review the implications of the potential genetic reprogramming of human nuclei transferred into enucleated cow eggs.

7.4. It will be for the government to determine the extent to which any further specific international initiatives are supported by the United Kingdom, taking account of the strength of feeling about human reproductive cloning demonstrated by this consultation. It will be necessary to consider how carefully international initiatives have been drafted lest they should preclude the therapeutic use of cell nucleus replacement as well as human reproductive cloning. There are often difficulties over finding mutually acceptable and agreed definitions for even quite simple concepts.

8. ENCOURAGING PUBLIC UNDERSTANDING

8.1. The HGAC has consistently sought to bring issues of public policy in relation to developments in human genetics before the public in a clear and accurate way to facilitate discussion. The consultation document, in a section headed "Building Public Confidence," specifically asked for any suggestions respondents might have on what advice ministers might be given in respect of the implications of human cloning. A small but significant number of those who responded to that particular question (13%) interpreted "building public confidence" as attempting to manipulate public opinion.

8.2. While a number of respondents specifically commented that the document had set out quite complex matters in a clear and lucid way. There was some concern that question 2 had been phrased in a way that would elicit a positive response, that the term "therapeutic cloning" had been chosen because it implied a benefit, and that questions 4 and 5 had used "ethically unacceptable" and "ethically acceptable" in a way that confused the issues. In fact, the intention of the consultation document was to raise the issues in a clear and accurate way that would elicit considered public response.

8.3. There is no evidence that the way in which the questions were phrased caused actual misunderstandings. Most respondents who addressed the individual questions took great care in giving detailed answers to make their views extremely clear—very few responses were limited to a "yes" or "no." The distinction between reproductive cloning and other work is valid, but at section 5 above, account has been taken of a wide range of comments about the terminology used in drawing this distinction.

8.4. Many respondents recognized that there was a need for greater public understanding of the issues. A number thought that more education was needed. Others considered that this was unlikely to produce immediate benefits and that more informed debate was required rather than scientific education, with a responsible role for the media and with leaflets and imaginative presentations as catalysts.

8.5. This view is not confined to the United Kingdom in the United States of America, the National Bioethics Advisory Commission report "Cloning Human Beings," recommended that "federal departments and agencies concerned with science should co-operate in seeking out and supporting opportunities to provide information and education to the public in the area of genetics, and on other developments in the biomedical sciences, especially where these affect important cultural practices, values and beliefs."

8.6. Publication of the HGAC/HFEA consultation document itself has promoted wider debate about the issues, including, for example:

(i) the Wellcome Trust commissioned research into public attitudes to cloning, based on 7 focus groups representing various interests within society;[20]

(ii) the Workers Education Association ran a course on modern genetics which included discussion of the consultation document by all 29 students;

(iii) a parish downloaded the consultation document from the website, gathered 25 members for discussion, summarized their views and submitted them, posting them, on its own website too; and

(iv) chairman and members of the HGAC and HFEA have addressed audiences, given interviews, and written articles explaining the background to the consultation and the issues raised.

8.7. The HGAC held its first national conference, "Human Genetics: Learning for the Millennium and Beyond," at the Royal Society in London on October 16, 1998. This was aimed at those who work in education, to discuss issues that they had identified as raised by human genetics, how these matters are taught, and what practical steps could be taken to improve knowledge and understanding. The report of the conference will be published in the near future. The HFEA discussed cloning at its 1997 conference and is often the media's first point of contact on this issue.

8.8. Both HGAC and HFEA regularly review communications needs and consider all opportunities to explain often complex issues in simple terms to the widest possible audience.

9. SUMMARY OF CONCLUSIONS AND RECOMMENDATIONS

9.1. The Human Fertilisation and Embryology Act 1990 is on the statute book, and despite the concerns of some respondents, the purpose of the consultation was not to reopen old debates about it. What needed to be considered was the effectiveness of the act in dealing with new developments concerning cloning. The difficulty is in considering the appropriateness of controls in a rapidly changing area where it is difficult to envisage just what direction developments will take and what problems might be encountered. The safeguards provided by existing legislation are dealt with in some detail in section 3 of this report.

9.2. The legal position is clear (section 3). The government has explicitly ruled out reproductive cloning, and the HFEA has stated its policy that it will not license the use of nuclear replacement for this purpose. HGAC and HFEA *recommend* that these safeguards be recognized as being wholly adequate to forbid human reproductive cloning in the United Kingdom. The government may, nevertheless, wish to consider the possibility of introducing primary or secondary legislation *explicitly* banning reproductive cloning regardless of the technique used, when there is an opportunity to do so in the legislative program, so that the full ban would not depend upon the decision of a statutory body (the HFEA) but would itself be enshrined in statue.

9.3. When the 1990 HFE Act was passed, the beneficial therapeutic consequences that could potentially result from human embryo research were not envisaged. We therefore *recommend* that the secretary of state should consider specifying in regulations two further purposes to be added to the list in paragraph 3(2) of schedule 2 (as described in paragraph 5.7 of this report), being:

- *developing methods of therapy for mitochondrial diseases*
- *developing methods of therapy for diseased or damaged tissues or organs.*

9.4. We have concluded that, as far as the issues relevant to this report are concerned, it does not seem that new issues arise regarding the protection of genetic identity beyond those covered already by the requirements of ethical medical practice, such as confidentiality and consent. We therefore make no additional recommendations.

9.5. The international situation and initiatives undertaken in several international fora are outlined in section 7. It will be for the government to determine the extent to which any further specific international initiatives are supported by the United Kingdom. It is not for the HGAC or the HFEA to make specific recommendations. Nevertheless, account should be taken of the strength of feeling about human reproductive cloning demonstrated by this consultation. We attach importance to careful consideration about the difficulties we have mentioned (see paragraph 7.4) over finding mutually acceptable and agreed definitions for even quite simple concepts.

9.6. The responses received to the consultation document indicated the need for more education and informed debate about the new genetics. Section 8 refers to work already undertaken in response to this need. (See section 8.7)

9.7. Finally, because of the pace of scientific advances in the area of human genetics, the HGAC and the HFEA believe that the issues need to be kept under regular review to monitor scientific progress. We therefore *recommend* that the issues be reexamined again in, say, five years' time, in the light of developments and public attitudes toward them in the interim.

[Editor's postscript: The report comprising this chapter was published in December 1998. In response, the UK government asked its Chief Medical Officer, Prof. Liam Donaldson, to form another expert committee with the specific remit of examining the question of therapeutic cloning, including stem cell studies. In August 2000, the committee issued a report.

Under the existing legislative scheme—the Human Fertilisation and Embryology Act and its implementing regulations—embryos up to 14 days old may be used for specific research purposes, including improving infertility treatment and preimplantation diagnosis of genetic and chromosomal disorders. The Donaldson Report recommended that the Human Fertilisation and Embryology Authority be allowed to license a wider range of research projects in order to explore the therapeutic potential of stem cells.

Among the report's specific recommendations were the following: Research using embryos, created by in vitro fertilization and by cell nuclear replacement, should be permitted, subject to the controls of the Human Fertilisation and Embryology Act of 1990, in order to increase understanding about human disease and disorders and their cell-based treatments. When licensing research using embryos created by cell nuclear replacement, the Human Fertilisation and Embryology Authority should ensure that there are no other means for meeting the objectives of the research.

Gamete donors must give specific consent for the use of their gametes in a research project to derive stem cells. Research using cell nuclear replacement in human eggs, which are subsequently fertilized by human sperm, should be permitted in order to increase the understanding of, and develop treatments for, mitochondrial diseases. The transfer of an embryo created by cell nuclear replacement into the uterus of a woman (so-called reproductive cloning) should remain a criminal offense.

The government's response, published on the same day, included the following points: It would introduce regulations for debate in both Houses of Parliament to extend the purposes for which embryos can be used in research governed by the Human Fertilisation and Embryology Act. It proposed that members of Parliament decide on these regulations in a so-called free vote (not bound by party positions) because the issue raised questions of conscience. Finally, the government stated that it would introduce legislation in Parliament that would reinforce the ban on reproductive cloning.]

On January 22, 2001—despite unexpectedly serious and spirited opposition—the House of Lords voted to approve the regulations, and to set up a Select Committe that will further examine the ethical issues raised by therapeutic cloning. At the urging of Baroness Warnock, the committee will also attempt to address the deepening distrust of science on the part of the general public.

NOTES

1. "Viable Offspring Derived from Foetal and Adult Mammalian Cells," *Nature* **385**, 881: 1997.

2. "Sheep Cloned by Nuclear Transfer from a Cultured Cell Line," *Nature* **380**, 64–66: 1996.

3. "Transgenic Sheep Expressing Human Factor IX," *Science* **278**, 2130–2133: 1997.

4. The Science and Technology Committee, Fifth report of Session 1996–97 (HC373-L).

5. Official Report, June 26, 1997, column 615.

6. Governmental Response to the Fifth Report of the House of Commons Select Committee on Science and Technology, 1996–97 Session (Cm 3815).

7. "Cloning Issues in Reproduction, Science and Medicine—A Consultation Document," January 1998.

8. "DNA Microsatellite Analysis of Dolly," *Nature* **394**, 329: 1998.

9. "Full-term Development of Mice from Enucleated Oocytes Injected with Cumulus Cell Nuceli," *Nature* **394**, 369–374: 1988.

10. ". . . . As Japanese Announce Cloned Calf Twins," *Nature* **394**, 114: 1998.

11. Report of the Committee of Inquiry into Human Fertilisation and Embryology, HMSO. July 1984 (Cm. 9314).

12. *Brave New World*, Aldous Huxley, 1932 (New York: Harper and Row).

13. "Embryonic Stem Cell Lines Derived from Human Blastocysts," Science 282, 1145–1147: 1998.

14. Within the confines of its remit HFEA will be looking at stem cell technology.

15. "Universal Declaration on the Human Genome and Human Rights" published by UNESCO, November 1997.

16. Council of Europe (1997). Additional Protocol to the Convention for the Protection of Human Rights and Dignity of the Human Being with regard to the Application of Biology and Medicine, on the Prohibition of Cloning Human Beings. Strasbourg: Council of Europe 1997.

17. European Parliament and Council Directive on the Legal Protection of Biotechnological Inventions COM(97)446 final.

18. WHO 51st World Health Assembly, implementation of resolution WHA50.37-A51/6 Add.1-8 April 1998.

19. "Cloning Human Beings" report and recommendations of the National Bioethics Advisory Commission: Rockville, Maryland, June 1997.

20. "Public Perceptions on Human Cloning," Wellcome Trust, December 1998.

28

Stem Cells

From the House of Lords Select Committee Report on Stem Cell Research

---

INTRODUCTION
Richard Harries

The Human Fertilisation and Embryology Act of 1990 allowed research on human embryos, under strict regulation, for five purposes connected with reproduction. In 2001 Parliament passed regulations adding three new purposes to the original five in the act:

a. increasing knowledge about the development of embryos
b. increasing knowledge about serious disease
c. enabling such knowledge to be applied in developing treatments for serious disease.

Because of unease in the House of Lords about these additional purposes a Select Committee was set up to look into the issues more thoroughly than is possible in an ordinary debate. The House of Lords rejected an amendment which would not have passed the regulations until a Select Committee had reported. The committee consisted of members of the House of Lords with a broad range of experience and expertise. It had access to excellent scientific and legal advice throughout its meetings. Because members of the committee were not themselves practicing scientists they had to work hard to get on top of the science and it is this perhaps which gives chapter 2 of the report, on the basic science of stem cells, a helpful clarity.

The first crucial question which the Select Committee had to consider was whether stem cells derived from adults could do all that was necessary. Was it in fact necessary to use stem cells derived from embryos at all? One of the problems is that advances in this field are being reported almost every week, sometimes showing that adult stem cells can do a great deal more than was previously thought and sometimes suggesting that this is not quite the case. In any case we took with the utmost seriousness the possibilities of adult stem cells and urged that research money be put into this field. Nevertheless, we were not convinced by the arguments that adult stem cells alone would suffice. First of all, we do not know at this stage what the real therapeutic potential of either embryonic stem cells or adult stem cells really is. That will only be know in the next five or ten years. It seemed wrong to us to close down, at this stage, what could possibly turn out to be of enormous therapeutic benefit. Even more weighty, in our view, was the need to do basic research on embryonic stem cells in order that the processes of differentiation and dedifferentiation might be better understood. If adult stem cells are to realize their full therapeutic benefit, it can only be on the basis of what they do being fully understood and it is embryonic stem cells that, as it were, provide the benchmark by which the various processes can be measured. They are pluripotent (that is, capable of differentiating into all the cells that make up the human body), whereas adult stem cells are much more sharply focused and therefore limited.

It was also primarily for reasons of basic research that we supported cell nuclear replacement (CNR). When the committee was set up high hopes were held out by some campaigning bodies for the therapeutic potential of CNR. In the course of the committee's work it began to appear that this potential may not be quite what was originally envisaged. Nevertheless, CNR involves a process of dedifferentiation and understanding this process could prove crucial in realizing the potential of whatever stem cells are finally used for therapeutic purposes. So again we thought it important to support basic research on CNR with the important proviso that this research was absolutely essential and that the required information could not be reached by any other route. Chapter 5 of the Select Committee's report on Cell Nuclear Replacement and Cloning is not reproduced here but chapter 3, which seeks to assess from a scientific point of view the potential advantages and limitations of ES cells and adult stem cells, is.

In addition to assessing the advantages and limitations of embryonic stem (ES) cells and adult stem cells the other crucial question concerns the moral status of the early embryo. After much debate we fixed on the term "early embryo" as the one that seemed least to prejudge the issue by use of inappropriate terminology.

The committee had the deepest respect for those who take the view that the early embryo is a human being in the fullest sense from the moment of fertilization but put forward some arguments that seem to count against that view. Against the argument that as the early embryo has the potential to become a person, it enjoys the full rights of a human being and should be accorded the respect owed to a human being the committee suggested that

> Those who deny the force of the potentiality argument argue that the fact that a person has the potential to qualify as a member of some class in the future, if certain conditions are met, does not confer the rights that belong to members of that class of being until those conditions are met. A medical student is a potential physician, and if he or she qualifies may practise as such; but the potentiality alone does not confer a right to practise. A child is a potential voter but has no claim to be treated as a voter until reaching the age of 18.

In the end however it was not the ethical arguments in themselves which were conclusive, but the fact that Parliament, after widespread debate, had already passed acts not only allowing research on embryos in 1990 but allowing IVF (in vitro fertilization) and abortion under certain circumstances. The committee therefore concluded

> Whilst respecting the deeply held views of those who regard any research involving the destruction of a human embryo as wrong and having weighed the ethical arguments carefully, the Committee is not persuaded, especially in the context of the current law on social attitudes, that all research on early human embryos should be prohibited.

The Warnock Committee, on the basis of whose recommendations the 1990 HFEA Act was passed, recommended that "The embryo of the human species should be afforded some protection in law." The effect of this is that research on embryos should only be undertaken if there is no other way of obtaining the desired research results. It is therefore an obligation of the Human Fertilisation and Embryology Authority to assure itself that this is the case before granting a licence for research. So the law, combined with the regulatory authority, ensures that the early embryo is respected, even if it is not accorded the absolute respect and right to life that we recognize in a baby or adult.

The Select Committee reported in February 2002. The government made its response in July and the House of Lords debated both the report and the response in November 2002.

SUMMARY OF CONCLUSIONS AND RECOMMENDATIONS

Background

1. The Select Committee was appointed in March 2001 to review issues arising out of the Human Fertilisation (Research Purposes) Regulations

2001. The regulations extended the purposes for which research on human embryos could be undertaken under licence from the Human Fertilisation and Embryology Authority (HFEA) from purposes concerned with reproduction in the Human Fertilisation and Embryology Act 1990 to three additional purposes:

(a) increasing knowledge about the development of embryos,

(b) increasing knowledge about serious disease, or

(c) enabling any such knowledge to be applied in developing treatments for serious disease.

It is important to keep in mind that the regulations are concerned only with research, not with treatment.

2. The Committee considered a considerable body of oral and written evidence and examined both the scientific and ethical aspects of its terms of reference in depth.

3. Concerns had been expressed about the regulations on three main grounds:

(d) that they were unnecessary, because developments in adult stem cell research made research on early human embryos unnecessary;

(e) that they were unethical as they permitted the use of early human embryos for wide-ranging research purposes; and

(f) that they represented a significant step on the path to human reproductive cloning.

We examined each of these issues.

Possible Alternatives to Research on Early Human Embryos

4. Stem cells are cells found in the embryo (ES cells), but also in many parts of the human body, which have the capacity to develop ("differentiate") into different cell types. As such, they have great potential for use in therapies to regenerate tissues in a wide range of serious, but common, diseases. Recent developments in research on adult stem cells have led some to claim that work on ES cells is no longer necessary. We examined in detail the potential of stem cells for developing new therapies and the relative advantages and disadvantages of adult stem cells and ES cells. Adult stem cells have great therapeutic potential and research on them should be strongly encouraged. Nevertheless there is a clear scientific case for continued research on ES cells, in order that the full potential of adult stem cells for therapy can be realized and because it is likely that some therapies will need to use ES cells.

The Status of the Early Embryo

5. The public debate reflects strongly differing views on whether or not the early embryo should be given the full protection due to a person. We

set out the arguments in the body of the report. The Committee believes that the issue cannot be looked at in isolation but must take account of the law as it has developed over the last 30 years. That is the background to the Committee's conclusion that whilst respecting the deeply held views of those who regard any research involving the destruction of the early embryo as wrong, and having weighed the ethical arguments carefully, it is not persuaded, especially in the context of the current law and social attitudes, that all research on human embryos should be prohibited.

6. Most research on early embryos uses "surplus" embryos left over from IVF treatment. But the 1990 Act allows embryos to be created for research. The number created has been much smaller than the number of surplus embryos donated for research. In the Committee's view embryos should not be created specifically for research purposes unless there is a demonstrable and exceptional need that cannot be met by the use of surplus embryos.

7. The 14-day limit on research on early human embryos should remain.

Cell Nuclear Replacement and Cloning

8. Cell nuclear replacement (CNR) involves the replacement of the nucleus of an egg with the nucleus of a cell from another individual (to produce an embryo that is the "clone" of the donor). The implantation of such an embryo in a woman (commonly called "reproductive cloning") was made a specific criminal offense by the Human Reproductive Cloning Act 2001. There have been calls to prohibit the use of CNR for research purposes as well. The majority scientific view seems to be that CNR is more likely to be used as a research tool, which would assist the understanding of the behavior of adult stem cells and how they might be manipulated, than as the basis for general therapies in its own right. In the committee's view that is a sufficiently serious and important objective, particularly if the potential of adult stem cells is to be realized, to justify the use of CNR, if licensed by the HFEA, provided that (as with embryos created by IVF for research) embryos are not created by CNR unless there is a demonstrable and exceptional need that cannot be met by the use of surplus embryos.

9. We have examined the issues surrounding reproductive cloning, mainly because of the fear that allowing CNR for research purposes would increase the likelihood of its being used to try to produce a cloned baby. There are very strong scientific and ethical objections to reproductive cloning. The committee unreservedly endorses the legislative prohibition on it and calls on the government to support any moves to negotiate an international ban. However, we do not believe that the risk of reproductive cloning is such as to justify prohibiting the use of CNR for research. The HFEA has an excellent record in ensuring that IVF clinics comply with the law, and the committee is satisfied that its regulatory powers, now reinforced by

a specific statutory prohibition, provide sufficient protection against the development of CNR leading to reproductive cloning in the United Kingdom.

Future Legislation and Regulation

10. The committee also considered a number of other issues arising out of the Regulations.

Regulation

11. The HFEA's role is crucial to the effective regulation of research on human embryos and the maintenance of public confidence in the regulatory regime. The government should keep the funding of the HFEA under review and ensure that it is commensurate with its increased responsibilities. It is also important that there should be closer monitoring of the outcomes of research licensed by the HFEA, and the committee invites the HFEA and the Department of Health to consider how such a review might be undertaken and updated on a regular basis.

The Wording of the Regulations

12. There is no definition of "serious disease" in the regulations, which could cause uncertainty. We invite the Department of Health to draw up guidance on the matter.

13. There is not a perfect match between the basic research on stem cells that currently needs to be undertaken and the wording of the purposes in the regulations. While the regulations can be construed without strain so as to encompass basic research, the government should consider putting the matter beyond doubt when a legislative opportunity arises.

A Stem Cell Bank

14. Stem cell "lines" derived from a single early human embryo can be maintained in culture, in principle indefinitely. As more of these lines are developed it is important that a stem cell bank should be set up for research purposes as a matter of urgency to ensure that there is a single body responsible for the custody of stem cell lines, ensuring their provenance and purity and monitoring their use. In that way stem cell lines can be made widely available to reputable researchers and an overview maintained of their use. Over time this will reduce the need for research on early human embryos.

Informed Consent

15. To ensure that informed consent to the donation of embryos for research is freely given—and seen to be freely given—it should be standard

practice that the prospective researcher is not the same as the person giving the IVF treatment. The "immortality" of stem cell lines makes the operation of procedures for giving informed consent particularly important where the research is intended to lead to their generation. The committee recommends that the authorities concerned ensure that the implications arising from the "immortality" of stem cell lines are fully covered in obtaining informed consent from donors.

Conclusions and Recommendations

The committee's detailed conclusions and recommendations are as follows:

Stem Cell Research

(i)   Stem cells appear to have great therapeutic potential for the treatment of many disorders that are both common and serious and for the repair of damaged tissue.

(ii) Until recently most research on stem cells has focused on ES cells from animals and the derivation of ES cell lines from them; cell lines from human ES cells have the potential to provide a basis for a wide range of therapies.

(iii) Recent research on adult stem cells, including stem cells from the placenta and umbilical cord, also holds promise of therapies; and research on them should be strongly encouraged by funding bodies and the government.

(iv) To ensure maximum medical benefit it is necessary to keep both routes to therapy open at present since neither alone is likely to meet all therapeutic needs.

(v)   For the full therapeutic potential of stem cells, both adult and ES, to be realized, fundamental research on ES cells is necessary, particularly to understand the processes of cell differentiation and dedifferentiation.

(vi) Future developments might eventually make further research on ES cells unnecessary. This is unlikely in the foreseeable future; in the meantime there is a strong scientific and medical case for continued research on human ES cells.

Status of the Early Embryo

(vii) Whilst respecting the deeply held views of those who regard any research involving the destruction of a human embryo as wrong and having weighed the ethical arguments carefully, the committee is not persuaded, especially in the context of the current law and social attitudes, that all research on early human embryos should be prohibited.

(viii) Fourteen days should remain the limit for research on early embryos.

(ix) Embryos should not be created specifically for research purposes unless there is a demonstrable and exceptional need which cannot be met by the use of surplus embryos.

Cell Nuclear Replacement and Cloning

(x) Basic research is a necessary step to developing treatments and facilitating the potential use of adult stem cells and should be permitted under the regulations in the same way as more directly applied research to which it is designed to lead, provided that it is subject to strict regulation.

(xi) Although there is a clear distinction between an IVF embryo and an embryo produced by CNR (or other methods) in their method of production, the committee does not see any ethical difference in their use for research purposes up to the 14 day limit.

(xii) Even if CNR is not itself used directly for many stem cell–based therapies, there is still a powerful case for its use, subject to strict regulation by the HFEA, as a research tool to enable other cell-based therapies to be developed. However, as with embryos created by IVF for research, CNR embryos should not be created for research purposes unless there is a demonstrable and exceptional need which cannot be met by the use of surplus embryos.

(xiii) If CNR is permitted in certain limited circumstances, oocyte nucleus transfer should also be allowed for research purposes.

(xiv) Given the high risk of abnormalities the scientific objections to human reproductive cloning are currently overwhelming.

(xv) There are further strong ethical objections in addition to those based on the risk of abnormalities, although not all the arguments deployed against reproductive cloning are equally valid. The most powerful are the unacceptability of experimenting on a human being and the familial and child welfare considerations arising from the ambiguity of the cloned child's relationships.

(xvi) The committee unreservedly endorses the legislative prohibition on reproductive cloning now contained in the Human Reproductive Cloning Act 2001.

(xvii) The HFEA has an excellent record in ensuring that IVF clinics comply with the law, and we are satisfied that its regulatory powers, now reinforced by a specific statutory prohibition, provide sufficient protection against the development of CNR leading to reproductive cloning in the United Kingdom.

(xviii) The government should take an active part in any move to negotiate an international ban on human reproductive cloning.

Legislation and Regulation

(xix) At an appropriate time, perhaps toward the end of the decade, the government should undertake a further review of scientific developments, particularly of the progress of adult stem cell research and therapies, and of the development of stem cell banks, with a view to determining whether research on human embryos is still necessary.

(xx) The government should keep the funding of the HFEA under review and ensure that its resources are commensurate with its increased responsibilities.

(xxi) The HFEA and the Department of Health should consider how a review of the outcomes of research licensed under the Act might be undertaken and updated on a regular basis.

(xxii) The Department of Health should examine with the HFEA the possibility of drawing up indicative guidance as to what constitutes serious disease.

(xxiii) When the government bring forward legislation they should consider making express provision for such basic research as is necessary as a precursor for the development of cell-based therapies.

(xxiv) The separation of clinical and research roles should be standard practice for donation of eggs or embryos. The prohibition in the United Kingdom of payment to donors for gametes has been an important element in preventing undesirable commercialization of this aspect of assisted reproduction and should be strictly maintained.

(xxv) The Department of Health should consider either establishing a body similar to the Gene Therapy Advisory Committee with oversight of clinical studies involving stem cells, or extending the membership and remit of GTAC to achieve the same ends. The committee sees no other special need at present for additional regulation of the use of stem cells in the treatment of patients.

(xxvi) The Department of Health's proposals to establish a stem cell bank overseen by a steering committee, responsible for the custody of stem cell lines, ensuring their purity and provenance and monitoring their use, are endorsed. As a condition of granting a research license, the HFEA should require that any ES cell line generated in the United Kingdom in the course of that research is deposited in the bank. Before granting any new license to establish human ES cell lines, the HFEA should satisfy itself that there are no existing ES cell lines in the bank suitable for the proposed research.

(xxvii) The HFEA should ensure that the implications arising from the "immortality" of stem cell lines are fully covered in obtaining informed consent from donors giving embryos for the potential establishment of ES cell lines for research. To prevent future restrictions in using ES cell lines (and therefore minimize the need to generate new ES cell lines) the HFEA should not permit ES cell lines be generated from donated embryos unless informed consent places no specific constraint on their future use. Where parents wish to restrict the type of research which can be undertaken, for example specifically for reproductive purposes, the embryos donated should be used for purposes other than the generation of ES cell lines.

STEM CELLS

1. In this chapter we seek to explain what stem cells are, and assess the potential of stem cell research to generate new therapies. We then examine in chapter 3 the relative scientific advantages and disadvantages of research on ES and adult stem cells.

What Are Stem Cells?

2. In the human body new cells are generated by cell division. Most specialized cells do not themselves divide but are replenished, often via intermediate less-specialized cell types, from populations of stem cells. Stem cells have the capacity to undergo an asymmetric division such that one of the two "daughter" cells retains the properties of the stem cell while the other begins to "differentiate" into a more specialized cell type (see box 1). Stem cells are thus central to normal human growth and development, and are also a potential source of new cells for the regeneration of diseased or damaged tissue. Stem cells are found in the early embryo, in the fetus, in the placenta and umbilical cord, and in many (possibly most) tissues of the body. As stem cells from different tissues have different physical properties, they are difficult to identify by their physical characteristics alone. Stem cells from different tissues, and from different stages of human development, vary in the number and types of cells to which they normally give rise.

Box 1   Differentiation

In the course of human development a single cell, the fertilized egg, ultimately gives rise to more than 200 cell types (blood cells, neural (brain) cells, liver cells, etc.) which make up the human body. This process, whereby less-specialized cells turn into more-specialized cell types, is called "differentiation." As all cells in the body have (with few exceptions) the same genes, differentiation occurs, in large part, by switching on ("expressing") or switching off ("repressing") different subsets of these genes. Thus, differentiated cell types express different subsets of genes. For example, red blood cells express the gene making hemoglobin (the protein which carries oxygen around the body) but neural cells do not. In general as cells become more specialized (differentiated) the subset of genes that they can express becomes more restricted.

3. We describe in chapter 4 the process of human embryonic development. At the earliest stages after fertilization (up to about the eight-cell stage) all the cells are "totipotent" (i.e., they have the capacity to develop into every type of cell needed for full human development, including the extra-embryonic tissues such as the placenta and umbilical cord). After about five days the blastocyst stage is reached. Within this ball of 50 to 100 cells there is the inner cell mass comprising about a quarter of the cells, from which a unique class of stem cells, embryonic stem cells (ES cells), can be derived. Unlike any other type of stem cell yet identified, ES cells have the innate capacity ("potential") to differentiate into each of the 200 or so cell types which make up the human body, and are described as "pluripotent." The potential of different types of stem cells is described in more detail in box 2.

4. As development proceeds beyond the blastocyst, stem cells comprise a progressively decreasing proportion of cells in the embryo, fetus, and human body. However, many, if not most, tissues in the fetus and human body contain stem cells which, in their normal location, have the potential to differentiate into a limited number of specific cell types in order to regenerate the tissue in which they normally reside. These stem cells, described as "multipotent," have a more restricted potential than ES cells, in that they normally give rise to some but not all the cell types present in the human body. Extra-embryonic tissues such as the placenta and umbilical cord also contain multipotent stem cells with the same genetic makeup as the cells of the embryo.

5. ES cells are a very specific class of stem cell which can be derived from the blastocyst. Other stem cells from later in the development of the early embryo or fetus, sometimes also (confusingly) referred to as embryonic stem cells, are not known to be pluripotent. Indeed, other embryonic, fetal and extra-embryonic stem cells are more akin to adult stem cells than to

---

Box 2   The Potential of Stem Cells

Stem cells from different sources differ in their potential for differentiation, i.e., in the number of cell types to which they can normally give rise. Stem cells which can give rise to all the cells required for human development, including extra-embryonic tissues, are described as "totipotent." Stem cells which can give rise to multiple, but not all, cell types are generally referred to as "multipotent." For example, hematopoietic (blood) stem cells from the bone marrow are multipotent as they give rise to the several different cell types present in blood but do not normally develop into, e.g., neural cells. Sometimes the term "pluripotent" is used interchangeably with "multipotent" and this can cause confusion. We use "pluripotent" to refer to a stem cell which can give rise to every cell type in the human body, in contrast to "multipotent," which refers to stem cells which give rise to many, but not all, cell types in the body. As pluripotent cells cannot give to the extra-embryonic tissues they are not totipotent.

ES cells. (It has been suggested that it is more appropriate to refer to all stem cells in the body, whether embryonic, fetal, or adult as "somatic" to distinguish them from ES cells.[1]) The use of different definitions, both in the scientific literature and in the evidence we received, can be confusing but is perhaps inevitable in a rapidly moving scientific field where hard and fast boundaries cannot always be drawn.[2]

The Potential of Stem Cells for Developing New Therapies

6. Because of their ability to reproduce themselves, and to differentiate into other cell types, stem cells offer the prospect of developing cell-based treatments, both to repair or replace tissues damaged by fractures, burns, and other injuries and to treat a wide range of very common degenerative diseases, such as Alzheimer's disease, cardiac failure, diabetes, and Parkinson's disease. These are some of the most common serious disorders, which affect millions of people in the United Kingdom alone, and for which there is at present no effective cure. Stem cell treatments, unlike most conventional drug treatments, have the potential to become a lifelong cure.

---

1. See memorandum by Professor Angelo Vescovi (pp. 473–475).
2. This is exemplified in recent debate over the efficacy or otherwise of stem cell transplants for Parkinson's disease. A 2001 study (reported in the *New England Journal of Medicine* (344:710–719)) suggesting that such a treatment had unwelcome side-effects has been cited by some as grounds for concern about the safety of embryonic stem cells. However, although these experiments were carried out with stem cells from an embryo, they were in fact from 7- to 8-week embryos and were therefore fetal and not ES cells.

7. This potential has given stem cell research a high profile and is leading to significant interest and investment in academic, medical and commercial research throughout the world. The main funding bodies gave evidence on the level of their investment in stem cell research (much of it in work on animals):

The Biotechnology and Biological Research Council has invested about £17 million in stem cell research over the last ten years.

The Chief Executive of the Medical Research Council (MRC), Professor Sir George Radda, told us that the MRC gives stem cell research a very high strategic priority and supports it to the tune of about £4.5million a year.

Since 1995 the Wellcome Trust has awarded some 15 project and programme grants specifically for stem cell research, totalling about £4.5 million. Although this is only just over half of one per cent of the total Trust spend, the Director of the Trust, Dr Mike Dexter, envisages many more applications in the future.

Although the amounts so far invested are relatively modest, all the funding bodies saw this as a major growth area.

8. The simplest way of using stem cells for therapy is by implanting a tissue which contains appropriate stem cells into an individual in whom that tissue is diseased or damaged, so that the transplanted stem cells regenerate the various cell types of that tissue. This type of therapy is in routine clinical use for treating patients with leukemia and other blood disorders by introducing hematopoietic stem cells, for example by bone marrow transplants. Despite the fact that such treatments have been successfully applied for about 20 years, few other examples of this type of approach have been developed. This is because the haematopoietic system is unusual in its accessibility and in the fact that it has evolved specifically to continuously replenish cells in the blood at high rates.

9. Recent scientific advances have opened up the possibility of treating a much wider range of disorders by isolating and growing stem cells in the laboratory. In some cases it may be possible to administer stem cells directly to an individual, in such a way that they would migrate to the correct site in the body and differentiate into the desired cell type in response to normal body signals. However, currently it seems more likely that stem cells will be grown and induced to differentiate into a defined cell type in the laboratory prior to implantation. In the longer term it may also be possible to induce stem cells to differentiate into several cell types, generating whole tissues, prior to implantation. For these approaches a much greater understanding of differentiation and developmental "signals" will be required.

10. None of our witnesses seriously questioned the therapeutic potential of stem cells for a wide range of disorders. There were differences of view as to when such therapies might be realized. Most witnesses believed that the introduction of effective stem-cell–based therapies would be a gradual process over the next five to twenty years, requiring much basic and

clinical research prior to clinical application. This is a normal timespan for the development of any new treatment. Even "conventional" drugs therapies take five to fifteen years and several hundred million pounds of investment to reach the patient.

The Research Path to Therapeutic Application

11. Any potential new treatment for disease requires a great deal of scientific and clinical research before it can be made available to patients. Three necessary steps can be distinguished. First, basic scientific research is required to establish what may or may not be possible, and to identify the best approaches to take and any pitfalls to be avoided. (The types of research questions which must be answered if stem cell therapies are to be developed effectively are set out in paragraph 13 below.) Second, preclinical studies in animals (normally mice) and small-scale clinical studies in human volunteers must be carried out to gain "proof of principle" for each new therapy and to ascertain whether it is safe and whether or not there may be significant side-effects. Third, large-scale clinical trials are required to determine whether the therapy is of real clinical benefit and to further assess and assure safety. In the development of most therapies there is an iterative process between the first and second stages, during which blind alleys are eliminated and the best approaches refined. The great majority of potential stem cell–based therapies are still at the first stage of this process, basic scientific research.

12. Stem cell research is currently subject to very rapid change and our report can reflect only the current state of knowledge. From the evidence we have received we are clear that over the next few years most studies on stem cells, whether adult, fetal, or embryonic in origin, will be basic research. This research will not in itself be therapeutic, but will be undertaken with the aim of gaining the understanding necessary if stem cells are to be used widely for therapeutic benefit. The potential for stem cell therapies to last the life of the individual patient makes it particularly important to ensure that any safety issues are identified and resolved satisfactorily. Only after considerable advances in understanding processes such as the control of differentiation will it be possible fully and safely to exploit stem cells to treat or cure individuals.

13. There is unlikely to be a single approach to the use of stem cells in therapy: different disorders are likely to require different types of stem cell and different therapeutic approaches. For example, for some treatments it may be possible to transplant whole tissues without isolating stem cells (as with bone marrow transplants), whereas for others it may be more effective to purify and grow stem cells in the laboratory prior to differentiation and reimplantation. In order to exploit stem cells to the full it is likely to be necessary to

identify and characterize the specific stem cells to be used. Currently stem cells are primarily defined only by their biological function; if stem cells are to be isolated and purified for therapeutic purposes, scientists must be able to identify unique characteristics which will allow them to be isolated routinely, efficiently, and reliably from amongst the millions of cells in a tissue;

isolate and purify the required stem cells in sufficient numbers to be useful. Stem cells often form a very small proportion of cells in a tissue;

grow stem cells in the laboratory under "clean" conditions so that they (or cells derived from them) can be transplanted back into patients; doctors must be certain that the properties of the cell have not changed in the laboratory, and that there are no contaminants that might cause harm if used to treat patients;

show that stem cells, once isolated from their normal location and grown in the laboratory, do not undergo unwanted changes in their properties. For example, all stem cells have the potential to divide, and it is therefore important to ensure that any manipulation does not alter the control of this division process and create a risk of generating cancerous cells;

"direct" stem cells to differentiate efficiently into specific cell type(s) required for therapeutic purposes, and ensure that this process does not give rise to any inappropriate cell type. Scientists still know little about the signals which direct differentiation;

understand the differentiation process so that when a stem cell has been induced to differentiate into a specific cell type, scientists can be sure that that cell is indistinguishable from normal cells of the same type in the body and will integrate properly with them;

understand the dedifferentiation process so that, if an adult stem cell is dedifferentiated to enhance its normal potential, scientists can be sure that this has been achieved accurately and that the signals it originally acquired during differentiation have been erased;

understand how stem cells get to and remain in their proper location in the body, so that when they (or cells derived from them) are transplanted into the body they do not migrate to inappropriate locations;

avoid immunological rejection of any implanted cells.

14. Until recently it has generally been considered that in mammalian cells the process of differentiation is irreversible. However, it has been demonstrated in animals that it is possible to reprogram ("dedifferentiate") the

genetic material of a differentiated adult cell by CNR (see chapter 5). Following this seminal finding, many studies have also suggested that adult stem cells may have greater "plasticity"[3] than previously suspected: they may be reprogrammed to give rise to cell types to which they do not normally give rise in the body. The potential of specialized cells to differentiate into cell types other than those to which they normally give rise in the body is little short of a revolutionary concept in cell biology. It has significantly increased the possibilities for developing effective stem cell-based therapies.

Immunological Rejection of Stem Cell-Based Therapies

15. Immunological rejection is a particularly important consideration for stem cell–based therapies. The human body possesses an immune system which recognizes cells that are not its own and rejects them. The immune system has evolved primarily as a protection against microorganisms that cause disease. However, the body also rejects human cells or tissues that do not belong to it. Immune rejection is one of the major causes of organ transplant failure and is one of the problems which will need to be overcome for any stem cell–based therapy to be effective. There are three main ways of avoiding or repressing immune rejection of transplanted cells or tissues:

> *The use of immuno-suppressant drugs.* These drugs have been refined over many years, as part of organ transplantation research. However, they are not always effective; they must normally be taken over the lifetime of the patient; and they leave the patient open to infection.
>
> *Using "matching" tissues.* Sometimes during transplantation it is possible to get a matched tissue type, usually from a near relative. This is often sought for bone marrow transplants. Finding a matching donor is unlikely to be a useful approach for most cell-based therapies. However, because stem cells can, in principle, be cultured indefinitely, it might be possible to establish stem cell banks of sufficient size to comprise stem cells with a reasonable (though never perfect) match to the majority of individuals in the population. If this proved possible, the appropriate matching stem cell from the bank could be selected and differentiated into the cell type required for therapy. Several thousand stem cell lines would be needed to obtain matches to the majority of the British population comparable with those achieved with matched bone marrow transplants.
>
> *Using the individual's own cells or tissues.* This would be the surest means of avoiding immune rejection. Adult stem cells isolated

---

3. The capacity of a cell to develop into different cell types.

from an individual, and then used to treat him or her, offer one possible way of achieving this, although not in all circumstances. Alternatively CNR could be used to generate cells or tissues that match those of the patient, although it is generally thought that this approach is unlikely to provide the major therapeutic route (see chapter 5).

POTENTIAL ADVANTAGES AND LIMITATIONS OF ES CELLS AND ADULT STEM CELLS

1. We have received much evidence on the relative advantages and disadvantages of ES cells compared with adult stem cells for the development of stem cell–based therapies. The main scientific considerations are summarized in the following paragraphs.

ES Cells

2. A great deal of research has been undertaken on ES cells from animals, particularly mice, over many years. In the last three or four years researchers around the world, including in Australia, India, Singapore, Sweden, and the United States, have used similar methods to establish human ES cell lines from blastocysts. Three research licenses have been granted by the HFEA (for purposes permitted by the 1990 Act) which could result in human ES cell lines being derived in the United Kingdom, but we were told that at the time of writing none had yet been derived.

Potential Advantages

3. Research on mice has shown that it is possible to isolate pluripotent ES cells from the blastocyst, culture and multiply them in the laboratory, in principle indefinitely, and induce them to differentiate into a wide range of different cell types. Cultured ES cells from some mouse strains are routinely re-implanted into a blastocyst, and then into a mother, to give rise to normal offspring. This demonstrates that ES cells, at least those from mice, can be grown and manipulated safely in culture, and that they can generate all cell types in the body.

4. This research has shown that ES cells have significant potential for developing new therapies. First, they are at present the only stem cells that can be readily isolated and grown in culture in sufficient abundance to be useful. Second—at least for mice—they can be used to generate a normal animal, which indicates that they are unaltered and potentially safe for therapeutic use. Third, ES cells have the potential to regenerate all normal cell types in the body—the only cell type currently known to have this potential. Finally, because ES cells are undifferentiated it is not necessary to de-differentiate ES cells prior to differentiation into a new cell type.

Possible Limitations

5. The most significant potential scientific limitation on the therapeutic use of ES cells is the problem of immune rejection. Because ES cells will not normally have been derived from the patient to be treated, they run the risk of rejection by the patient's immune system. Three main approaches to overcoming this problem were described elsewhere.

6. It has been argued that, because ES cells have the potential to differentiate into all cell types, it might be difficult to ensure that, when used therapeutically, they did not differentiate into unwanted cell types; or undergo chromosomal alterations which generated tumors. It is clearly essential to guard against these risks, but there is no reason to believe that this is a significantly greater risk for ES cells than for other stem cells.

7. Current methods for growing human ES cell lines in culture are adequate for research purposes, but the requirement for co-culture of human ES cells with animal materials necessary for growth and differentiation would preclude their use in therapy. This problem, which applies to all cells grown in culture, is unlikely to be insoluble.

Adult Stem Cells

8. The potential of adult stem cells for therapeutic application is illustrated by the use of hematopoietic stem cells to treat leukemias and other blood disorders. As discussed, this type of whole tissue transplant probably has limited general applicability. However, recent studies suggesting that various adult stem cells have much greater potential for differentiation than previously suspected (see box 3) have opened up the possibility that other routes to adult stem cell therapy might be available.

Potential Advantages

9. The developments referred to above suggest that adult stem cells may have greater therapeutic potential than had previously been thought. Their most significant potential scientific advantage is that, at least for some disorders, they might be isolated from the individual to be treated and therefore avoid rejection by the immune system when transplanted back into that same individual for therapeutic purposes.

Possible Limitations

10. Even if much of the potential of adult stem cells is realized, there are circumstances where they are unlikely to be useful. The isolation of some types of adult stem cells for therapy, for example the isolation of neural cells from a patient's brain, would be impractical. Similarly, where a person suffers from a genetic disorder or some types of cancers, adult

Box 3   Increased Plasticity of Adult Stem Cells

Recently it has been observed that some relatively specialized stem cells can be induced (at least under some conditions) to give rise to a wider range of cell types than had been expected. For example, it has been reported that stem cells from blood, which in the body normally give rise only to blood cells, can be induced to differentiate into neural cells. This process might occur in one of the following ways:

the original stem cell might dedifferentiate to pluripotency and then be reprogrammed to generate the second cell type; or

the original cell might change into the second cell type without going through a dedifferentiated intermediate stage, a process sometimes called "transdifferentiation."

Little is known about such increased "plasticity," which is based on observations from which plasticity is inferred rather than on an understanding of the processes involved.

stem cells isolated from that individual will retain the damaging genetic alterations underlying the disease and so be of little therapeutic value.

11. If adult stem cells are to be of general utility, it will be necessary to learn how to isolate them, grow them in culture and differentiate them into new cell types. The isolation and growth of adult stem cells have to date proved very difficult. Stem cells generally represent a very small proportion of cells in adult tissues. Unambiguous identification is difficult as their presence in a tissue or mixture of cells is generally inferred from a research observation rather than indicated by any specific biochemical marker which might aid their purification.[4] Although there are several reports of "enrichment"[5] of adult stem cells, there are few, if any, reports of adult stem cells being purified to homogeneity (i.e., where no other cell types are present). It has been suggested that some adult stem cells retain many of their characteristics only as a result of the presence of signals from other surrounding cells, and that maintenance in culture may therefore be difficult.

12. Current understanding of the potential of adult stem cells for redifferentiation is still very limited. Although many studies suggest that such

4. For example, when an adult tissue is transplanted from one individual to another (normally carried out using mice), the transplanted tissue is observed to give rise to cells of a type not present in the original tissue. From this observation it is inferred that the transplanted tissue contains adult stem cells with the potential to differentiate into new cell types, but the stem cells themselves have not necessarily been identified or isolated.

5. Increasing the proportion of stem cells in a sample by removing some of the non–stem cell material.

processes occur, there is often a degree of ambiguity, for example whether or not the multiple new cell types arise directly from a single adult stem cell with increased potential for differentiation, or from several different stem cells each with a limited but different potential for differentiation. Moreover, it is not yet known whether adult stem cells give rise to cells of different tissue types by transdifferentiation, or by dedifferentiation to a pluripotent cell, which then differentiates into the new cell types (see box 2).[6] The control and safety of dedifferentiation is a major challenge and one about which little is yet known.

13. The efficiency of differentiation of transplanted adult stem cells is, to date, very poor. For example, although transplantation of bone marrow into mice suffering from muscular dystrophy can lead to new, repaired muscle fibers, the efficiency is several orders of magnitude below that which would be therapeutically useful. Much research is still required to determine whether the efficiency can be enhanced.

14. In their natural location in the body adult stem cells do not exhibit great potential for differentiation into new cell types but have evolved to give rise only to specific cell lineages. Indeed, if they exhibited increased potential or plasticity in their natural position in the body this would have disastrous consequences: the "wrong" cell types might develop into the "wrong" tissues. The feasibility of manipulating adult stem cells to undergo dedifferentiation and redifferentiation along pathways which they do not normally exhibit, and the consequences of doing so, are as yet uncertain.

Do Developments on Adult Stem Cells Make Research on ES Cells Unnecessary?

15. Research on adult stem cells is at a very early stage. Without a great deal of further research it will not be clear to what extent their therapeutic potential will be realized, or for what type and proportion of potential applications adult stem cells will be applicable. Although almost all the scientists who gave evidence to us were excited by recent studies on adult stem cells, most sounded a note of caution: many of the published studies

---

6. Research by Dr. I.S. Abuljadayel has been cited on several occasions, including in the debate in the House of Lords on the Regulations (22 January 2001, Col 37) as supporting this proposition. As Dr. Abuljadayel's work has not been published, we invited her to submit it to the Committee as evidence, which she kindly did (pp 296–306) along with other supporting material. Briefly Dr. Abuljadayel claims that blood cells can be induced to dedifferentiate to a pluripotent stem cell (a process she calls "retrodifferentiation") which can then be directed to redifferentiate into different cell lineages. If this claim were borne out, it would be a major breakthrough. We have taken advice on Dr. Abuljadayel's work and we are satisfied that it does not lead us to modify our conclusions. We note that her manuscript was submitted for publication to four of the leading scientific journals, but was not accepted for publication.

are still open to multiple interpretations or require replication; and there are many crucial scientific issues to be resolved.

16. We received evidence from a number of individuals arguing that recent developments in research on adult stem cells demonstrated their therapeutic potential and made research on ES cells unnecessary.[7] However, the evidence from the great majority of scientific and medical research organizations, and the experts on adult stem cells whom we consulted, did not support that view. They did not see adult stem cells and ES cells as alternatives but as complementary pathways to therapy. They argued that relatively little is yet known, and that substantially more research on both adult and ES cells is needed before the best routes for therapies can be ascertained; that, despite increasing optimism, it is still not known to what extent it will be possible to exploit adult stem cells therapeutically and in the meantime other avenues should not be closed off; and that, even if much of the potential of adult stem cell–based therapies is realized, it is unlikely that adult stem cells will fulfil all therapeutic needs.

17. Although adult stem cells may ultimately fulfil many therapeutic needs, the strong weight of evidence is that the full potential of adult stem cell research and its therapeutic application is unlikely to be realized without research on ES cells. This is because, apart from CNR, ES cells provide the only realistic means at present of studying the mechanisms and control of the processes of differentiation and dedifferentiation. If stem cell therapies (whether using ES or adult stem cells) are to be of clinical benefit and of demonstrated safety, a much clearer understanding of these processes is required. The utility of ES cells for studying them is clearly demonstrated by advances made from animal studies. Most future studies probably can and will be undertaken using mouse (or other animal) ES cells rather than human ES cells. Nevertheless, if safe and reliable therapies are to be developed, a comparison with human ES cells must eventually be made.

18. ES cells are needed for this purpose, partly because of the relative ease with which they can be isolated, maintained in culture, and differentiated into other cell types; and partly because they are the only fully undifferentiated pluripotent cell type available for study. If scientists are to dedifferentiate adult stem cells to pluripotency, prior to redifferentiation into a new cell type for therapeutic purposes, they must know whether they have done this correctly and whether the process is safe. Differentiation involves "marking" the genetic material in a number of ways. These "markings" (including chemical changes to the DNA and the interaction of specific proteins with it) are "remembered" during cell division. If an adult stem cell is to be dedifferentiated prior to redifferentiation for therapeutic purposes, these markings must be correctly erased. In fact it is not fully

---

7. Notably from Dr. Elizabeth Allan, who submitted a memorandum comprehensively documenting research studies involving adult stem cells (pp 306–359).

established that CNR produces complete differentiation and erasing of these markings, as recent discussion of Dolly the sheep illustrates.[8]

19. It may be that, in time, scientific understanding of the processes involved and developments in the manipulation of adult stem cells will make research on ES cells redundant. The committee is not convinced that this point has yet been reached or will be reached in the near future. These issues were exhaustively considered in the United States last year by the National Institutes of Health. Its comprehensive report reviewed the state of the science as of 17 June 2001.[9] Emphasizing that ES and adult stem cells are different, its concluding paragraph reads:

> Predicting the future of stem cell applications is impossible, particularly given the very early stage of the science of stem cell biology. To date, it is impossible to predict which stem cells—those derived from the embryo, the foetus, or the adult—or which methods for manipulating the cells, will best meet the needs of basic research and clinical applications. The answers clearly lie in conducting more research.

20. Because of the importance of this issue we also asked a number of internationally renowned adult stem cell experts for their views. We received replies from Professor Helen Blau, of the Stanford University School of Medicine; Dr Jonas Frisen, of the Karolinska Institute, Stockholm; Professor Nadia Rosenthal, of the European Molecular Biology Laboratory, Monterotondo-Scalo; and Professor Angelo Vescovi, Director of Research at the Stem Cell Research Institute, Milan. They are published in the volume of evidence (pp 472–478). They were unanimous that there was a need for research on both adult and ES cells. Dr Frisen expressed his view as follows:

> My opinion is that adult stem cells are clearly different from ES cells, and that there are no scientific data suggesting the opposite. Although I believe everyone would agree that it would be very good if adult stem cells had the same potential as embryonic, this is unfortunately today only wishful thinking. I find it very important today to work on both embryonic and adult stem cells. This will ensure that potential therapies are not delayed.

21. Of all the scientific issues relevant to our inquiry we have given more attention to recent developments in adult stem cell research than to any other. Scientific developments in this field are so rapid that it is difficult

---

8. Dolly was created by dedifferentiating an adult cell nucleus by inserting it into an egg from which the original nucleus had been removed. It is remarkable that this can be achieved at all, although it is still not known whether this dedifferentiation was "perfect." It has been suggested that some of the properties of the differentiated cell have not been fully erased and that the biological age of Dolly might therefore be greater than her birth age. The same caveats could be applied to dedifferentiation of adult stem cells.

9. *Report on Stem Cell Research*, National Institutes of Health, 2001.

to make any firm predictions with confidence. This in itself suggests that avenues of research should not be closed off prematurely.

Conclusions

22. Based on the evidence we have heard, our conclusions on the research and therapeutic potential of ES cells and adult stem cells are as follows:

Stem cells appear to have great therapeutic potential for the treatment of many disorders that are both common and serious and for the repair of damaged tissue.

Until recently most research on stem cells has focused on ES cells from animals and the derivation of ES cell lines from them; cell lines from human ES cells have the potential to provide a basis for a wide range of therapies.

Recent research on adult stem cells, including stem cells from the placenta and umbilical cord, also holds promise of therapies; and research on them should be strongly encouraged by funding bodies and the government.

To ensure maximum medical benefit it is necessary to keep both routes to therapy open at present since neither alone is likely to meet all therapeutic needs.

For the full therapeutic potential of stem cells, both adult and ES, to be realized, fundamental research on ES cells is necessary, particularly to understand the processes of cell differentiation and dedifferentiation.

Future developments might eventually make further research on ES cells unnecessary. This is unlikely in the foreseeable future; in the meantime there is a strong scientific and medical case for continued research on human ES cells.

INDEX